MEANING, LOGIC AND LUDICS

MEANING, LOGIC AND LUDICS

Alain Lecomte

Université de Paris VIII, France

ICP

Imperial College Press

Published by

Imperial College Press
57 Shelton Street
Covent Garden
London WC2H 9HE

Distributed by

World Scientific Publishing Co. Pte. Ltd.
5 Toh Tuck Link, Singapore 596224
USA office: 27 Warren Street, Suite 401-402, Hackensack, NJ 07601
UK office: 57 Shelton Street, Covent Garden, London WC2H 9HE

British Library Cataloguing-in-Publication Data
A catalogue record for this book is available from the British Library.

MEANING, LOGIC AND LUDICS

ISBN-13 978-1-84816-456-7
ISBN-10 1-84816-456-4

Typeset by Stallion Press
Email: enquiries@stallionpress.com

Printed in Singapore by B & Jo Enterprise Pte Ltd

To my three first grandchildren Shanti, Balbino and Sacha

Preface

This book originates from the desire to summarize some advances in computational linguistics which have been made during the last two decades. Computational linguistics, in the sense developed here, is not simply "natural language processing", but a way of reflecting the fact that language is, in many respects, *computing*. To understand language amounts to applying algorithms which are not only implemented "in the mind", but also given with our whole background as humans living in society, sharing interaction-based knowledge and endowed with a high-level competence in handling symbols. This book therefore mixes several approaches coming from at least three disciplines: linguistics, logic and computer science. If we adopt the idea that logic is a part of mathematics, that also puts this book in this field. In many respects, language is concerned with mathematical structures. This has for instance been emphasized by the mathematician Jim Lambek who wrote, as early as 1958, a seminal paper entitled "The Mathematics of Sentence Structure". Generative grammar also emerged in the 1950s from mathematical reflections on "What is a language?" This aspect has been lost in its more recent evolutions where mathematics is suspected of spurious formalization which brings nothing valuable to the comprehension of language seen as a mental organ. I think it is a pity, since even nowadays biologists examine mathematical theories.

To sum up this book, I could say that it starts from the present state of the art in formal semantics, strongly based on the Fregean paradigm (truth values and compositionality), and then tries to move smoothly towards other views which have been developed recently (even if the "meaning as proof" approach proposed by A. Ranta dates from the early 1990s, and the game semantics approach from the sixties!).

What changes the study of meaning is of course the evolution of logic. I think that it has not been well taken into account yet, and I hope it is an originality of this book to try to make this connection clear.

Let us then recall that in the history of logic, research on the conditions for a theoretical discourse to be coherent and rigorous, mainly initiated by G. Frege, was followed in the mid-twentieth century by a new approach, inspired by the computing revolution, focusing on the study of programs and their transformations. The view, according to which programs could be seen as the proofs of the statements expressing their correctness, made logical formalisms valuable objects of study for computer science. Among such logical frameworks, intuitionistic logic was considered an attractive system because of its constructiveness, which takes the form of a famous correspondence between programs and proofs, known as the Curry–Howard correspondence (Howard, 1980). After this discovery, the possibilities of extending such a correspondence were explored, and denotational semantics began to give more attention to the interpretation of proofs than to the interpretation of formulae.

Around the beginning of the 1990s, it became clear, mainly through the seminal work of A. Ranta based on Martin-Löf's type theory, that the semantics of proofs could provide the foundations for a view on natural language semantics going beyond the simple truth-conditional approach usually taken within the Montagovian tradition. Constructive type theory for instance turned out to be a suitable basis for an account of anaphora in so-called *donkey sentences*. Nevertheless, although the view of a sentence meaning as the set of its proofs allows for more fine-grained distinctions than a strict Fregean conception, it remains a *static* view. *Interaction*, which is the key point of *dialogue*, was still ignored. Pragmatics in type-theoretical frameworks remains concentrated on the agent's judgements, which are viewed as mainly *static*. The frameworks of linear logic and then ludics, developed by Jean-Yves Girard, have provided new ways of thinking about dynamic processes and interaction. In some respects, such an approach converges with previous game-theoretical views such as Hintikka's game-theoretical semantics and Lorenzen's dialogical logic, although it also differs from these in many respects.[1] Game-theoretical semantics (GTS)

[1]Particularly because the aims of these different frameworks are not identical: Lorenzen's dialogical logic was an attempt to show that intuitionistic logic was the most natural logic; Hintikka's GTS is also based on *a priori* rules but mainly aims at providing a basis

has helped us to study linguistic phenomena from a game viewpoint, thus leading to the emergence of a notion of *strategic meaning*, a notion which deserves much attention. Nevertheless, from our viewpoint, it lacks a *dynamical* dimension. As is claimed by Hintikka himself, game moves in GTS must never be seen as sequences of speech acts, but as mere games for "searching and finding". For instance, this theory reflects quite well the mental behaviour of a reader trying to find the reference of some proper name in a text. We could also say that GTS has mainly an *epistemic* perspective, and does not try to reflect the dynamics of the interpretation process. Compared to GTS, ludics is neither based on *a priori* rules, nor on some evaluation procedure which would lead to putting various weights on issues. It aims at showing that rules themselves are determined by more general considerations like *symmetry* and *orthogonality*, that is, *geometrical* considerations, so that it is possible to play with the rules of a game themselves, something important when confronted with dialogue situations.

The possibility of interpreting the same objects in two ways — as proofs, and as strategies in a game — is another important point which emerged with ludics, and allows us to combine the advantages of two approaches covered by this book: the "meaning as proof" theory and the game semantics theory.

Much of this book is based on work done within the context of the PRELUDE project, which ran from 2006 to 2009, with funding from the French National Agency for Research (ANR). I am indebted to many contributors to this programme, and especially to Myriam Quatrini (University of Marseilles), who is the co-author of Chapter 12, on ludics, and Christian Retoré (University of Bordeaux) who gave me many ideas and with whom I wrote many papers. And I thank Marie-Renée Fleury and Samuel Tronçon (University of Marseilles), without whom the PRELUDE project would not have been the success that it was. I am also indebted to Carl Vogel (Trinity College Dublin) for correcting many mistakes (though it is of course my own fault if there are still any mistakes remaining!). Special thanks are also addressed to Aarne Ranta who agreed to review part of Chapter 11 (but again, the mistakes which remain are my own fault). I address many thanks to all those who have made my life as a researcher so happy in this field, for now 25 years, particularly

for analyzing knowledge from this viewpoint; ludics is prior to any set of logical rules since it aims at providing new foundations for logic itself.

(in alphabetical order) Anne Abeillé, Michele Abrusci, Nicholas Asher, Karine Bashung, Gabriel G. Bès, Claire Beyssade, Philippe Blache, Claudia Casadio, Pierre Casteran, Francis Corblin, Marc Dymetman, Paul Egre, Christophe Fouqueré, Claire Gardent, Danielle Godard, Philippe de Groote, Jacques Jayez, Mark Johnson, Jean-Baptiste Joinet, Aravind Joshi, Ruth Kempson, François Lamarche, Pierre Livet, Jean-Marie Marandin, Michael Moortgat, Richard Moot, Glyn Morrill, Richard Oehrle, Guy Perrier, Pierre Pica, Carl Pollard, Shahid Rahman, Laurent Roussarie, Helge Rückert, Sylvain Salvati, Wilhelmijn Vermaat and to my former students Maxime Amblard, Houda Anoun and Sylvain Pogodalla. I address also my feelings of admiration to Jean-Yves Girard and my thanks to him for having allowed me to participate in the LIGC workshops that he initiated, where I met researchers from various fields (biology, computer science, physics...) including Giuseppe Longo, Jean-Louis Giavitto and Thierry Paul. I also address such feelings to Gérard Huet who brought a deep approach to the problem of natural language processing as an analyst of Sanskrit.

And I don't forget my colleagues at the University of Paris 8, who provided me with the opportunity to work, as a member of the UMR "Structures Formelles du Langage", in one of the most dynamic laboratories in this field in France. Among those not already mentioned, my particular thanks go to Joaquim Brandão de Carvalho, Brenda Laca, Léa Nash, Hans Obenauer, Sophie Wauquier and Anne Zribi-Hertz, with whom I have exchanged a lot during these last four years. The idea for this book came from a seminar given at the École Normale Supérieure of Paris within the framework of the master's programme "Linguistique Théorique et Descriptive" co-organized by the University of Paris 8 and by the Department of Cognitive Studies of ENS (DEC). I therefore thank the two co-directors of this master's programme, Léa Nash and Dominique Sportiche, who allowed me to conduct this seminar.

Contents

Part II. Logic

Part III. Proof Theory Applied to Linguistics

Part IV. Ludics

Chapter 1

Introduction

1.1 The Logical Space of Meaning

If the space of physics *is* mathematics, the space of meaning *is* logic. In the first case, mathematics produces objects and structures which are precisely those which make the external world understandable. It is only a small step to assume that this external reality *consists in* those mathematical objects. In support of this view, we must note that there is, for instance, no coherent formulation of quantum mechanics outside the equations of mathematics, so that this immaterial reality (the reality of mathematics) *exceeds* the reality of common sense. Mental entities are in many ways different from physical ones. It is not to say that mental entities cannot be thought of in physical terms in the last resort, but simply that so-called mental entities have distinct properties with regard to directly physical ones. Those properties *make* (in the strong sense) our mental life. According to Schrödinger (1958), the condition which gave birth to physics lies in that particular move of the mind extracting itself from reality, a process he named *elision*, and which preconditions the remote observation of physical objects, the price to pay being the rejection of subjectivity from science. In the case of mental objects, the subjective element cannot be constructively abstracted away because of *intrinsic reflexivity*: mental structures may always be applied to themselves. Nevertheless, a knowledge of them may come from a small gap introduced between the mental reality and itself, a gap which may be compared to *parallax*. Generally this reflection of thought on itself, where a discrepancy may occur, showing something of its structure, is provided by communication between people by means of language. This appears especially in dialogues, and it is no coincidence that the great works of our cultures are often based on them (if we think for instance of such different traditions as the Greek on one side, illustrated by Plato, and the Indian on the other side, as witnessed for instance by Nāgārjuna). Or as the famous cognitivist Steven Pinker (2007) said, "language emerges from interaction between human minds". Let us assume for instance a speaker A who claims *Every linguist knows an African language* to speaker B. B may respond, *I know* (acknowledgement), or *I'm not sure Paul, who is a linguist, knows any African language*, or she may answer, *Paul and Mark, who are linguists, don't know the same languages.* In the latter case, B shows that she had a slightly different comprehension from the two former cases. The feedback thus provided by B commits A to perceive that her sentence may have different meanings and realize that such a difference lies in the logical structure. Dialogue thus creates the opportunity for analyzing thought and meaning even if they cannot be remote from our mind. This aspect of meaning will be addressed mainly in the last part of this book, when more traditional ways of treating it will have been surveyed. In the small gap thus revealed between a thought and its reflection in another one, lies the possibility of logic, as a phenomenological approach to meaning.

On the other hand, as pointed out by F. Récanati (2008) among others, meanings may be assembled like syntactic constituents. We could make the assumption that human language is oriented towards dynamicity of meanings, which could not be connected without the help of syntactic structures. These syntactic structures form the core of syntax. It follows that it is not possible to study meaning without studying the syntax-semantics interface.

To study the space of meaning therefore implies that we have to address two domains simultaneously: logic and syntax. There is no guarantee that they can be harmoniously articulated. Like we shall see in the second part of this book, the study of logic is more and more oriented towards the discovery of deep symmetries, like those between the rules of linear logic in its sequent presentation, or between the two sides of a *sequent* $\Gamma \vdash \Delta$ either in classical logic or in linear logic. On the other hand, syntax seems to be a domain of massive anti-symmetry, at least if we follow the assumptions made by some of the most eminent linguists, like R. Kayne (1994) who stated the so-called *anti-symmetry* principle in generative syntax. There seems to be therefore a challenge in the idea of dealing simultaneously with both domains. That did not prevent many scholars from attempting to make a connection between them, above all since the seminal work by Jim Lambek (1958, 1961, 1988, 1993) who, in 1958, provided the foundations for the logical study of syntax in an authentic "syntactic" calculus. This work has been continued since the early 1980s, particularly by Glyn Morrill (1994) and Michael Moortgat (1988, 1997) and their followers. This led to a theory of language presented as *grammatical reasoning*. Such an enterprise is oriented to the search of *constants* which would play the same role with regard to language that constants, like *connectives* and *quantifiers*, play with regard to logic. On the path to this research, we meet recent evolutions of logic (Girard, 1987, 1995, 2001; Schroeder-Heister and Dosen, 1993) which show that by playing with the so-called structural rules in the sequent calculus (weakening, contraction and permutation), it is possible to enrich the spectrum of these connectives. It also appears that previously assumed primitive constants of logic (like the implicative connective) may actually be decomposed into more primitive symbols, thus allowing a more flexible management of formulae in the search for proofs.

Nevertheless, language is not logic. If we try to represent the more syntactic-semantic phenomena that we can in a logical system, probably we shall have to import into it devices which have little independent motivation within logic. This is the case for instance for *structural modalities*, which have been introduced since the 1980s, in the Lambek calculus in order to surpass its limitations (Kurtonina and Moortgat, 1997). Either we restrict the Lambek calculus **L** to a so-called "non-associative" Lambek calculus (**NL**), wishing to keep the notion of constituency which seems to be crucial in syntax, and so are obliged to introduce these devices to relax the constraints in some places, or we extend it, for instance, to accept non-peripheral extraction, but in this case we get an overgenerating system that we have to constrain if we don't wish to accept too many sentences. From our viewpoint, this reflects the conflict we mentioned between the fundamental symmetries of logic and the dissymmetries of syntax. In fact, strong arguments exist (see Girard, 2006) against non-associative logic. Non-associative logic would be contradictory with one of the main foundations of logic as they are provided by category theory.[1]

Our position in this book is not to hide the discrepancies between logic and syntax but to try to use the symmetries when possible, as much as possible. For instance, the use of symmetries in the computational applications

[1]Let us recall that a *category* is defined as a pair of sets, *objects* and *morphisms*, so that morphisms are supplied with an operation of composition which has a neutral element and is associative.

of classical logic will prove useful in representing elegantly some meaning phenomena, like scope ambiguities.

Another point which will be addressed in this book is a criticism of the widespread conception of meanings as sets of *truth conditions*, a view which dates from Frege and has been widely accepted in the studies of formal semantics by Montague (1974a,b) and Heim and Kratzer (1998), among others. The alternative view will be hinted at here, but developed more fully elsewhere.

1.2 The Aim of This Book

This book aims at presenting some of the main advances in natural language semantics and the interface between syntax and semantics in the last two decades, based on some of the most recent logical theories, mainly linear logic, and on precise analyses of computing due to advanced theories in theoretical computer sciences, like the extensions of the Curry–Howard isomorphism to classical logic which has led to several variants of calculi, like Parigot's $\lambda\mu$-calculus, Curien and Herbelin's $\lambda\mu\tilde{\mu}$-calculus, Wadler's dual calculus, the Lambek–Grishin calculus revisited by Moortgat and Bernardi and so on (Parigot, 1992; Curien and Herbelin, 2000; Herbelin, 2005; Wadler, 2003; Bernardi and Moortgat, 2010). The book surveys many solutions which have been proposed for the syntax-semantics interface, taking into account the specificities of the linguistic signs and fundamental mechanisms brought to light by the linguists and particularly the generativists. It ends with a presentation of ludics, a framework which allows us to characterize meaning as an invariant with regard to interaction between processes (Girard, 2001, 2003, 2006).

1.3 Starting from Traditional Formal Semantics

These advances described are not only technical: they are also philosophical. In this book, we start from a well-known stage of formal semantics, the one provided by Montague grammar, and we point out one of the problems it had to face: the question of scope construal. It has been well known for a long time that sentences with several quantifiers may be ambiguous. A question is whether such ambiguity is due to various potential syntactic analyses (as Montague and after him the most eminent linguists like

Chomsky seem to believe) or simply to various strategies for evaluating the logical meaning, as would seem to be the case if we took seriously the idea that this kind of ambiguity is properly semantic. Some linguists (like Culicover and Jackendoff (2005) when they called for a "simpler syntax") seem to call for a theory where the respective tasks of syntax and semantics would be quite separate and where, for instance, syntax would only deal with pure questions of assembling lexical signs, leaving to semantics the task of combining meanings. Of course, this trend still keeps close to the compositionality principle (Janssen, 1997), according to which the meaning of a sentence is a function of the meanings of its parts and the way they are combined, but there is perhaps a (later) stage in the interpretation (perhaps a "silent" one) where meanings are still combined without any syntactic counterpart. Some works based on continuation semantics (de Groote, 2001b) show that such a machinery is easily conceivable, by means of the particular reduction rules of the $\lambda\mu$-calculus. If it is so, we are confronted with an interesting question which regards the history of logic.

1.4 Semantics and the History of Logic (1): Intuitionism

1.4.1 *Curry–Howard*

One of the biggest discoveries in logic during the twentieth century was its *algorithmic content* (Howard, 1980). The Curry–Howard correspondence between formulae and types, and between proofs and programs revealed that proofs in intuitionistic logic may be coded by λ-terms, and that therefore if we represent, like in the Montagovian tradition, meanings by λ-terms, we have a way of computing meanings if we are able to display the analysis of a sentence as an intuitionistic proof. This had been exactly realized by Jim Lambek, several years before the "official" formulation of the Curry–Howard correspondence, in his seminal article (Lambek, 1958).

1.4.2 *Lambek and the substructural hypothesis*

The Lambek calculus is an intuitionistic framework since to recognize a sequence of words in that calculus precisely amounts to constructing a proof of the validity of a sequent with only one formula (or type) on its right-hand side (and, moreover, with a non-empty left-hand side, something which has actually nothing to do with logic, but with grammar, if that were not the case, we could insert "empty" categories with appropriate types anywhere, thus leading to strange sentences!). Of course, it is not a complete logic, since, for instance, it has no negation and no disjunction. Moreover it is a variant of so-called *substructural logics*, that is, logic where structural rules are missing. We recall here that a *structural rule* is a rule which allows the easy management of a sequence of formulae (or types). For instance the weakening rule is also known elsewhere as a monotonicity rule: it ensures that if something has been proven without a given formula A, it can be *a fortiori* proven if we add A to the set of premises. The contraction rule allows consideration of two occurrences of the same formula as the duplication of this formula and the permutation rule of course allows permutation of premises if needed in the course of a demonstration. Suppress one or more (eventually all) of these rules and you get what we call a *resource-sensitive logic*: exactly the kind of thing we seem to need in order to deal with "material" entities like *signs*. Of course, if you demonstrate that the sequence of words *Peter feeds the cat* is a correct sentence, by using the convenient types associated with the words, you would not be allowed to accept that the sequence *Peter Mary feeds the cat* is also a sentence. In the same way, the sentences *I call a cat* and *I call a spade a spade* are not similar in meaning, as would be the case if we had a contraction rule at our disposal!

Nevertheless, many linguistic phenomena are such that they need some violation of this strict *linearity* principle (we call *linearity* the property of a system based on a discipline of *consumption*: any object must be used once and only once), as is the case for anaphora (*he thinks that he is clever*) or parasitic gapping (*the book that I shelved without reading*).

1.5 Semantics and the History of Logic (2): Classicism

There have been a lot of works exploring the potential algorithmic content of classical logic (Girard, 1991; Vauzeilles, 1993; Danos *et al.*, 2003). In fact, it was shown, in particular by the categoricist Joyal (Lambek and Scott, 1986), that, strictly speaking, classical logic is not *constructive*! This is shown with category theory tools: if proofs are interpreted as morphisms in a category where formulae are objects, the constructive character of intuitionistic logic is revealed in its semantics in terms of a Cartesian closed category: each proof is an individual (modulo normalization), whereas in classical logic, the category obtained collapses into a simple Boolean algebra; that means that there is only one morphism corresponding to a provable formula, and this morphism only says that the formula is "true". Another aspect of the lack of constructivity of classical logic is provided by the fact that the normalization process in classical logic does not possess the Church–Rosser property: it is not confluent. It may happen that two different paths in the reduction lead to two different results. These negative facts did not prevent some researchers from dealing with classical logic, exploring the attempt to recover some determinism in the normalization process by *controlling* the strategy of reduction.

Computer scientists familiar with functional programming know that a symbolic form may be evaluated according to two different strategies: *call-by-value* and *call-by-name*. In call-by-value (CBV), each argument of a function must be evaluated (transformed into a *value*) before being passed to the function. In call-by-name (CBN), this is not required: arguments may be passed to the function before being evaluated. This gives an advantage when the same complex object is passed several times as an argument to a given function. We may summarize this opposition by saying that in CBV, *values* must be computed first, and that in CBN, *contexts* are computed first. In most programming systems, the two strategies lead to the same results (*confluence*), but it would be possible to extend those programming systems and then determinism would require following only one strategy. This is what is realized in the $\lambda\mu$-calculus. Nevertheless, for purposes other than strictly computational ones, we can imagine a calculus where two different strategies are equally usable, thus leading to different results. We claim that this is just what happens in language when we try to evaluate a sentence which contains several quantifiers.

If things are this way, it is reasonable to conclude that *semantic ambiguity* comes from the *classical* (and not *intuitionistic*) nature of language. However, there are still other tracks to follow. After all, these so-called *symmetric calculi* are fascinating but strange to manage: they are based on a permanent need for dissymmetrization. At each step, a formula must be distinguished (as an *active* one) and the story of activations and deactivations is stored in the stack of μ- (or $\tilde{\mu}$-)operators. Linear logic provides an entirely different way of making a symmetric calculus constructive.

1.6 Semantics and the History of Logic (3): Linear Logic

Removing the structural rules of weakening and contraction opens the field to another constructive logic: linear logic. In this case, the *functional* interpretation of logic is given up. Intuitionistic logic was based on a fundamental dissymmetry: several *inputs* for only one *output*. Now, in linear logic, we have a symmetry between inputs and outputs. Therefore the preferred interpretation is no longer in terms of functionality but in terms of interaction: if inputs and outputs are symmetric, it is possible to exchange them and to read a result either as a flow from inputs towards outputs or as a flow in the reverse direction. This suggests the interpretation of proofs as *processes*. Removing weakening and contraction moreover gives space for more connectives. The sequent calculus has two variants which coincide in the classical frame: the multiplicative and the additive ones. In general, we distinguish between the active formula (the one to which the rule is applied) and the inactive ones, which are its *contexts*. In an introduction rule for a given connective ($\wedge$ or $\vee$, on the left or on the right) we may formulate the rule either by mentioning disjoint contexts in the premises, or by mentioning identical ones: because of weakening and contraction, that does not matter. But in the absence of these structural rules, it matters, and the multiplicative and additive variants are no longer equivalent. Therefore we get *multiplicative* conjunction ($\otimes$) and disjunction ($\wp$), and *additive* conjunction ($\&$) and disjunction ($\oplus$). *Negation* is simply a duality operator: every formula may change its side in one sequent by being negated. *Linear implication* is a multiplicative connective. $A \multimap B$ is defined as $A^{\perp} \wp B$; it is this connective which expresses resource consumption: against A, you obtain B. As seen above, a natural interpretation of linear logic is in terms of processes. The linear implication gives an illustration of that: when you give A in order to obtain B, there are two entangled processes, one of giving A and one of receiving B in exchange. Therefore the multiplicative disjunction $\wp$ is naturally interpreted as a *parallelization* of processes, whereas $\otimes$ is interpreted as a sequentialization of processes ($A \otimes B$ is either A and then B or the other way round[2]). Game semantics interpretations have been provided very early on for linear logic, at least since the work of A. Blass (1992). In computational terms, a type could be regarded as a server from which a client can get, in one access, an element of that type; the client need not do anything more than show up. This kind of type is said to be simple because the client has nothing to do, but it is also possible to build *complex* types. In fact we can understand $A \& B$ as

[2] C. Retoré (1993) explored this in the context of non-commutative linear logic. For this purpose, he invented a connective $<$ such that $A < B$ is interpreted as A *before* B.

a *choice*: the resources you have can provide you with A as well as with B, therefore they provide you with a choice between two kinds of data. The complex type may be therefore interpreted as a *server* from which the client can get either A or B. Because $\oplus$ has a symmetric law with regard to &, it is natural to understand it as a dual choice: the client has nothing to do because the server makes the choice itself. Now, because of De Morgan's laws, *negation* may be interpreted as an exchange between the two roles, of server and of client. Blass still proposes a more elaborate interpretation by means of games. He says, "a play of $A \otimes B$ consists in interleaved runs of the two constituents A and B". Whenever it is the proponent's turn (= server's turn) to move, he or she must move in the same component in which the opponent (= client) last moved, while the opponent is free to switch components. Finally, $A \otimes B$ is a complex game with the impossibility for the proponent to make use of the information obtained from one component when playing in the other, while $A \wp B$ means such a game where he or she can.

This game interpretation is something that we get in the more recent *ludics* (Girard, 2001, 2003, 2006; Fleury and Quatrini, 2004; Curien, 2003; Lecomte and Quatrini, 2010). Ludics starts from the observation that proofs can always be represented as alternations of positive and negative steps (Andréoli, 1992), where a negative step is a reversible one (such that we may always make the reverse step because we never lose contextual information when making that step) and a positive one is non-reversible. Seeing attempts to find a proof as such successions of negative and positive steps allows us to see those attempts as games, where positive steps are moves by the proponent and negative ones are records of the expected answers by the opponent. Because the roles of proponent and opponent may be exchanged, we see that a similar attempt to build a proof exists from the opponent's viewpoint, but his or her attempt to build a proof has as its goal the negative sequent with regard to the sequent defended by the first player. Proofs are thus opposed to counter-proofs, in a space where both are coexisting (the space of *paraproofs*). An *orthogonality relation* may then be defined and in analogy to vector spaces, a null vector is introduced, which belongs to the proofs as well as to the counter-proofs: this paradoxical object is a *paralogism* that Girard calls the *daïmon*.

The "morality" of this story is that we are led to consider "true" proofs and "false" ones (that is, unsuccessful attempts) on the *same* level. What interests us above all is their interaction (called normalization) and not particularly their "truth" or their "falsity". In fact, this is exactly what is needed when we try to isolate various meanings that a sentence can have. A meaning is determined by the counter-meanings with which it interacts: there is no need to use truth values as primitive objects for representing that, as it suffices to have a formalization of the notion of interaction, something provided by ludics.

1.7 Presentation of the Book

This book may be read by beginners: it does not require much preliminary knowledge, except for some notions of predicate logic and λ-calculus that everyone can find in many textbooks (see for instance Gamut, 1991). It is particularly intended for students in computer science who wish to learn about applications of theoretical computer science to the understanding of natural language, for students in linguistics who wish to learn some logical and computer-scientist aspects of the study of language, and for students in philosophy who wish to have an overview of topics which are discussed in the area of language studies and which necessitate some technical background.

It is divided into 13 chapters (including this introduction and a conclusion). We start from a well-known state of the art: most formal semantics studies nowadays are based on two famous approaches, which share much in common, Montague's semantics and Heim and Kratzer's conception of semantics in generative grammar. That is not to claim our attachment to these frameworks, but to state things as they actually are. Nevertheless, this chapter states the connection between generative grammar and compositional semantics. We recall the main characteristics of the Montagovian approach to formal semantics, putting emphasis on the connection between syntactic and semantic rules and suggesting a move towards a more uniform treatment of compositional meaning. Heim and Kratzer's approach is also presented. It has the peculiarity with regard to Montague's of being directly based on a model-theoretic perspective. The question which arises from this chapter will be that of *uniformity*: how to deal in the most uniform way with all the problems which are posed from an analysis of scope construal?

Chapter 3 recalls Lambek's syntactic calculus, a calculus particularly suitable for getting a semantic translation from a syntactic derivation, seen as a proof in a fragment of an intuitionistic and resource-sensitive logic, but which is limited when considering for instance non-peripheral extractions or non-linear phenomena. Moreover, the pure Lambek calculus is not sensitive to the phrase structure of a sentence. In this chapter, a first comparison is attempted with *minimalist grammars*, which represent the current state of generative grammar (Stabler, 1997, 2003). Because a deeper study of such systems involves more knowledge in proof theory, Chapter 4 offers such a digression, presenting all the main tools that are nowadays used in this discipline, with some linguistic illustrations (like for *proof nets*).

The Curry–Howard correspondence is presented in Chapter 5.

Chapter 6 presents variants of the Lambek calculus, which are able to deal with the phrase structure (non-associative Lambek calculus). Various applications of this framework, as well as close variants of it are presented, like Hendricks' flexible types (Hendricks, 1993) and Jacobson's variable free semantics (Jacobson, 2000). Extensions are also shown, like the Lambek–Grishin calculus and Abrusci's non-commutative linear logic (Grishin, 1983; Abrusci, 1991).

Chapter 7 explores what Moortgat has called *grammatical reasoning*, that is, the view according to which complex linguistic analyses may be performed by logical means. These "logical" means include structural modalities (distinct from the usual modalities in modal logic, and therefore more or less tied to the linguistic purpose).

Chapter 8 goes back to the so-called Minimalist Program, proposing type-theoretical formulations for minimalist grammars, which would make the interface between syntax and semantics more tractable. In Chapter 9 other grammars are studied which propose similar but distinct views, like Pollard's convergent grammars (Pollard, 2007) and Anoun and Lecomte's linear grammars (Anoun and Lecomte, 2006).

Chapter 10 introduces the notion of *continuation*, a notion issued from computer science, which supports the management of *contexts* as well as of *terms* (or programs). It is shown that earlier solutions to some problems of the syntax-semantics interface, by Montague (like the systematic use of type raising), were already compatible with the idea of continuation: when an object of type e is seen as an object of lifted type $(e \rightarrow t) \rightarrow t$, it is transformed into a function which takes a continuation (something of type $e \rightarrow t$) as its argument. Barker (2002) and Shan (2004) were among the first to propose *continuized grammars* which precisely allow the kind of uniformity we are seeking. On the logical side, to systematically use continuations calls for a symmetric calculus where contexts and terms would be treated on a par. This chapter presents some variants of such a symmetric calculus, like the $\lambda\mu$- and $\lambda\mu\tilde{\mu}$-calculus. It shows how to directly apply these calculi to the framework of the Lambek grammars in order to overcome some of their limitations, in a direction which is currently explored by Bernardi and Moortgat (Moortgat, 2007; Bernardi and Moortgat, 2010). Chapter 11 is devoted to one of the most original contributions that proof theory can bring to semantics, that is, the constructive type theory approach, first advocated by Martin-Löf (1984) and then deepened by Ranta (1994). We try then to go further on to the dynamic aspect of language by taking (dialogical) *interaction* into account.

It is in Chapter 12 that we pose the main philosophical problems linked to the study of language: Is a non-denotational semantics possible? How to reconcile the *meaning as proof* paradigm (like it is formulated in the Brouwer–Heyting–Kolmogorov hypothesis, and used in natural language semantics by Ranta) with the game semantics paradigm developed by Hintikka and Sandu (1997) and Hintikka and Kulas (1983)? Ludics, we claim, provides an answer to these questions, and in this chapter, co-authored with Myriam Quatrini, the main concepts of ludics are presented, and linguistic applications are shown, concerning the notion of logical form, the semantics of questions and the study of discursive relations.

Finally, Chapter 13 tries to plan future works on the basis of the last part of this book, showing links with other perspectives like Kempson and Colleagues' dynamical syntax, Hamblin's perspective on fallacies, Groenendijk and Roelofsen's inquisitive semantics and works on semantics inspired by coherent spaces (Kempson *et al.*, 2003; Hamblin, 2004; Groenendijk and Roelofsen, 2010).

PART I

Truth-Conditional Meaning

Chapter 2

Compositional Approaches and Binding

2.1 Representing Logical Meaning: The Binding Issue

2.1.1 *The syntactic notion of binding*

One of the main problems to be elucidated by linguists when they study the structure of sentences is *binding*. How for instance does it come that a sentence like

Example 2.1. Who did he say Mary met?

Precisely means what it means, that is, something that we can schematize by

Example 2.2. For which x, he said that Mary met x

This kind of problem is known as the *binding problem*. We may introduce it by saying that, in some circumstances in sentences, the meaning (= the reference) of an item (a pronoun for instance) is not directly determined, like it is the case for ordinary, plain, lexical entries, or for phrases built from them, but is indirectly determined by the link it has at any distance with some particular other item. For instance, in Example 2.1, the meaning of *who* comes from the potential object x of the verb *met*. In standard formulations familiar to generativists, the syntactic representation of Example 2.1 is something like Example 2.3:

Example 2.3. $[_S$ who $[_S$ he said $[_{S'} t [_S$ Mary met $t']]]]$

Here *traces* are used for representing these links: t', as a trace, is bound by *who* (and by its trace in the lower embedded sentence). In principle, traces are co-indexed with their antecedent, that is, the elements which have moved to upper positions. In a generative analysis of Example 2.1, *who* moves from the embedded position presently marked by t', and successively occupies the position marked by t and (finally) the highest position (in a tree representation of the sentence). All these positions are therefore co-indexed. Because Move always displaces an element from the bottom to the top, it happens that not only t' and t are co-indexed with *who* but also that *who* *c-commands* its traces. Let us recall how the c-command relation is defined:

> In a tree, a node α c-commands a node β if and only if none of these two nodes dominates the other and the first branching node which immediately dominates α also dominates β.

There is therefore a possible "syntactic" definition of binding, the one which has been given by Chomsky, at least since his seminal book *Lectures on Government and Binding*. This theory may be summarized in a few notions. Chomsky (1981, p. 184) distinguishes between two notions of binding:

- operator binding (or $\overline{A}$-binding),
- antecedent binding (or A-binding).

This distinction refers to the opposition of A and $\overline{A}$ positions. The former are those where we can find any referential expression which may be an argument of a verb (that is, those positions where grammatical functions like subject of, object of, etc. may be assigned), while the latter are those which are not, that is, positions of adjunction, COMP and heads. For instance, in Example 2.1, *who* is in a COMP position, and therefore t is $\overline{A}$-bound, whereas in Example 2.4, the nominal *Peter* is in an A-position, and therefore t is A-bound.

Example 2.4. Peter seems [t to sleep]

Chomsky remarks that in Example 2.5, there are two traces, one introduced in the course of passivization and the other by the introduction of the operator *who*:

Example 2.5. Who [t was killed t']

Here, t is bound by *who* (and therefore $\overline{A}$-bound), and t' is bound by t (and therefore A-bound since t occupies a subject position). That amounts to saying that *who* is an *operator* which binds t and that t is the *antecedent* of t'. Elements bound by an antecedent are generally called *anaphors* (except for PRO, an empty element which intervenes in the case of control verbs like *to promise* or *to allow*), and Chomsky calls specifically *variables* the elements bound by an operator. The basic notions of the theory of binding are the following:

(1) (a) α is X-bound (that is, A or $\overline{A}$) by β if and only if α and β are co-indexed, β c-commands α, and β is in an X-position;
(b) α is X-free if and only if it is not X-bound;
(c) α is locally bound by β if and only if α is X-bound by β, and if γ Y-binds α then either γ Y-binds β or $\gamma = \beta$;
(d) α is locally X-bound by β if and only if α is locally bound and X-bound by β.

(2) α is a variable if and only if
(a) $\alpha = [_{NP} e]$ (empty NP),
(b) α is in an A-position,
(c) there is a β that locally $\overline{A}$-binds α.

By means of these definitions, the reader can easily see that in Example 2.6,

- only t is a variable, since it is in an A-position, and $\overline{A}$-bound by *who*,
- t' is A-bound and locally A-bound by t, and therefore $\overline{A}$-bound also by *who*,
- t'' is A-bound and locally A-bound by t', A-bound by t and $\overline{A}$-bound by *who*.

Example 2.6. Who $[_S t$ seemed $[_S t'$ to have been killed $t'']]$

Not only traces may be viewed as variables or anaphors; pronouns can also be viewed as such. For instance in Example 2.7, the pronoun *he* may be interpreted as co-referring to *John*. It plays the role of an anaphor, and if interpreted like this, it is A-bound by *John*. A correct representation would be Example 2.8 where co-indexing is made apparent.

Example 2.7. John thinks he's smart

Example 2.8. John_1 thinks $[\text{he}_1\text{'s smart}]$

By doing so, pronouns are treated exactly like variables. In other frameworks (see Montague below) they are dubbed *syntactic variables*. Because they must be indexed in order to be interpreted, we already understand why, in the Montagovian framework, they will be introduced into the analysis with an integer attached to them, so that we shall consider an infinity of such variables:

- $\text{he}_1, \text{he}_2, \ldots, \text{he}_n, \ldots, \text{she}_1, \text{she}_2, \ldots, \text{she}_n, \ldots$
- $\text{him}_1, \text{him}_2, \ldots, \text{him}_k, \ldots, \text{her}_1, \text{her}_2, \ldots, \text{her}_k, \ldots$

2.1.2 *The semantic notion of binding*

If there is a syntactic notion of binding, there is also a semantic one, which appears in the semantic translations of the sentences. Let us assume for instance that an operator is a kind of *quantifier* (like *every, each, a, some, most, all, etc.*). The notion of binding is then inherited from logic, where it is said in a formula like

$$\forall x\ boy(x) \Rightarrow (\exists y\ girl(y) \wedge kiss(x, y)) \tag{2.1}$$

that each occurrence of x is bound by $\forall$ and each occurrence of y by $\exists$. Our study of binding will therefore make explicit the similarity of the two notions of binding, one syntactic and the other semantic or logical. It is worth noticing that Chomsky qualifies only operator binding as "logical", the other being only "a syntactic notion relating to the syntax of LF" (LF = logical form). We may of course assume that antecedent binding is also logical in essence. It will be the case if we may characterize it by means of the logical formulae into which sentences are translated. Actually, we will assume that syntactic variables are translatable into logical ones. If τ is the translation function, that will amount to assuming

- $\tau(\text{he}_1) = x_1, \tau(\text{he}_2) = x_2, \ldots, \tau(\text{he}_n) = x_n, \ldots$
- $\tau(\text{him}_1) = y_1, \tau(\text{him}_2) = y_2, \ldots, \tau(\text{him}_k) = y_k, \ldots$

and then we may assume that α is bound if and only if $\tau(\alpha)$ is logically bound.

In any case, such an approach poses the question of translating sentences into the formulae of a rich, formal language, in a systematic and rigorous way. Moreover, there seems to exist no other way to achieve this goal than by using non-trivial compositionality: each word brings its own contribution to the entire meaning, which is itself derived step by step using the grammar rules to assemble meanings.

2.1.3 *The model-theoretic notion of binding*

Heim and Kratzer (1998) propose to directly interpret sentences as their truth values, via a procedure which directly assigns model-theoretic entities to syntactic structures, not passing through the formulae of a formal language. We will also study this conception. Let us here give simply a hint of their approach.

Suppose we have built a syntactic tree:

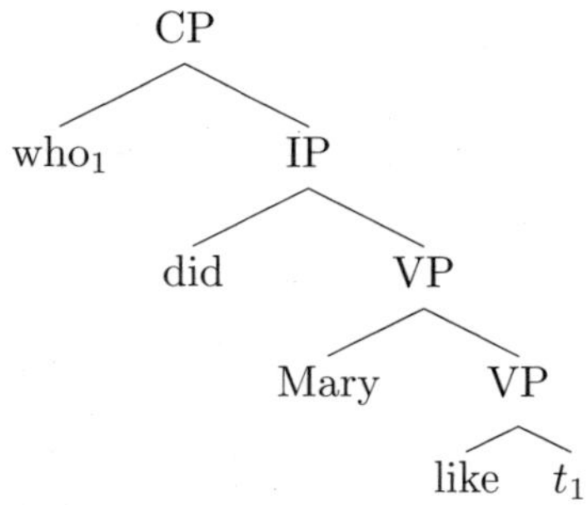

Intuitively speaking, this sentence has no *free* variable, but the subtree

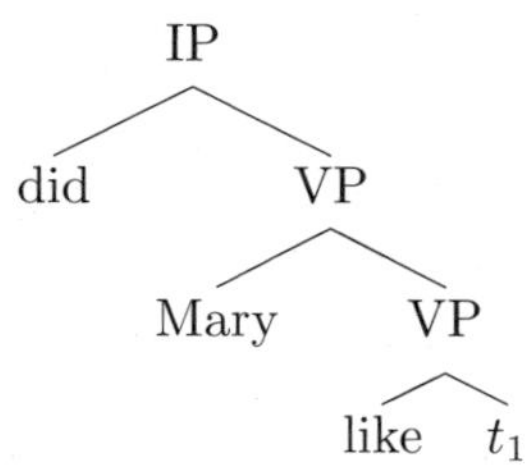

has a free variable, obtained by translating the trace t_1 into a logical variable x_1, so that the representation associated with this subtree is

$$liked(Mary, x_1) \tag{2.2}$$

a formula which may be interpreted with regard to some structure $M = \langle D, I \rangle$ only by means of an *assignment* a which gives x_1 a value. The CP is associated with a ground formula (that is, a formula, the truth value of which is independent of any assignment) and among the nodes it dominates, one is associated with a formula with a free variable and a second node is seen as a *binder* and the variable contained in the subtree is said to be bound by it.

In fact, the translation into formula (2.2) is unnecessary if we have a procedure that directly calculates truth values from trees, as we shall see in Section 2.5.1. Such a procedure involves assignments and it makes sense to say that the variable x_1 is free or not, inside a tree structure independent of any translation into a formula.

2.2 Syntactic Derivations and Semantic Composition

Semantic compositionality is based on the fact that sentences (and other phrases) may be decomposed into more elementary constituents. What brings to us such constituents if not syntax? We are therefore led to make use of syntactic structures. That presupposes we have established grammatical rules in order to generate those structures. The roots of this approach must be found in Montague (1974a,b), who showed in the 1970s how a grammar could be completed by a semantic component expressible in a formal language. Montague thought there was no difference between a human language and a formal one. However, we know that most human languages have never been consciously designed by humans,[1] in contrast with formal ones which are entirely designed by humans pursuing multiple goals like expressing mathematical theorems, formulating algorithms or representing knowledge in an abstract manner. Montague's viewpoint is therefore often criticized and it will not be defended here.

Nevertheless, the Montagovian viewpoint is a useful starting point for our purpose, since it proposes an operational and rigorous way of allowing the deduction of logico-semantic forms (or conceptual structures, if we prefer) from the analysis of sentences. Other viewpoints have followed Montague's, but always keeping this spirit of translating the natural language into a formal one.

2.3 Montague Grammar Revisited

2.3.1 *From rules to sequents*

It may help to situate oneself in the context of the years 1960–1970, when generative grammar was mainly viewed as a system of rewriting rules like

$$S \to NP\ \ VP; \quad VP \to V\ \ NP, etc.$$

Another way of expressing these rules consists in reading them from right to left, as

$$NP\ \ VP \vdash S$$

which amounts to seeing a grammar as a "bottom-up" process rather than a top-down one. An expression like $X\ Y \vdash Z$ is read as "from X and Y,

[1] Exception must be made for some artificial languages like Esperanto, Volapük or others, the "success" of which is not completely obvious.

we may deduce Z". Further on, we shall call such a kind of expression a *sequent.* It is interesting to note that if we have two sequents

$$NP\ \ VP \vdash S \quad \text{and} \quad V\ \ NP \vdash VP$$

we may deduce the sequent

$$NP\ V\ \ NP \vdash S$$

by using a very simple rule, that we shall call the "cut" rule:

$$\frac{\Gamma, X, \Delta \vdash Z \quad \Gamma' \vdash X}{\Gamma, \Gamma', \Delta \vdash Z}\,[cut] \tag{2.3}$$

where X and Z are syntactic categories and Γ, Γ' and Δ are sequences of categories. This rule may be viewed as a *transitivity axiom*: if a sequence of categories containing a category X may produce a category Z and if a second sequence of categories gives an X, then by putting the second sequence inside the first one in the place occupied by X, the new sequence so obtained will provide us with a Z. Otherwise, it is clear that we always have

$$X \vdash X \tag{2.4}$$

for every category X, something that we call an *identity axiom.*

In his grammar, of which we shall give here only a short overview, Montague was making use of a particular notion of syntactic category (that we shall deepen in Chapter 3, Sections 3.2 and 3.3), which uses the fact that, in principle, some words or expressions have a regular behaviour with regard to other words or expressions: they are "waiting for" some other words or expressions in order to become complete phrases. Thus we may consider, as a first approximation, that a determiner (DET) is an expression which must meet a common noun (CN) in order to give a nominal phrase. Thus, if CN is the category of common nouns and if T is the category of nominal phrases (or *terms* in Montague's terminology), we will be able to replace the symbol "DET" with the symbol "T/CN". The rule $T \to DET\ \ CN$, that we may rewrite in the form of a sequent $DET\ \ CN \vdash T$, therefore becomes the sequent

$$T/CN \quad CN \vdash T \tag{2.5}$$

We see then that if we have a very general rule, a kind of *rule scheme* or *meta-rule*, which applies to all X and all Y, we have the sequent

$$X/Y \quad Y \vdash X \tag{2.6}$$

We can dispense with the above particular rule: the "/" notation (that we name "slash") incorporates the particular syntactic rule into the category of determiners.

In the usual syntactic theory, the primitive categories are S, NP and CN; in Montague's theory, we have CN, T, t, VI, etc. The "/" operator will have later on a variant "\", the one denoting an expectation on the right, and the second on the left. We will have

$$Y\ Y\backslash X \vdash X \tag{2.7}$$

so that a French post-nominal adjective, like an adjective for nationality (for instance *américain*), will belong to the syntactic category $N\backslash N$ (cf. *écrivain Américain*, for *American writer*).

Given a category A, Montague denotes as P_A the set of expressions belonging to that category. The format of the syntactic rules is therefore the following:

> If $\alpha \in P_A$ and if $\beta \in P_B$, then (in some cases we have to enumerate) some function $F(\alpha, \beta)$ belongs to some set P_C.

Obviously, the case where $A = X/Y$ and $B = Y$ is a particular case of this general principle, where the function F amounts to concatenation and where $C = X$, but Montague wishes to deal with more subtle cases which do not always refer to mere concatenation. For instance, the negative morpheme in French wraps the verb (*regarde* gives *ne regarde pas*).

An example of a rule that he gives for English is the following[2] (which mixes phonological and syntactic considerations!):

> S_2: If $\alpha \in P_{T/CN}$ and if $\beta \in P_{CN}$, then $F_2(\alpha, \beta) \in P_T$, where $F_2(\alpha, \beta) = \alpha'\beta$ with $\alpha' = \alpha$ except if $\alpha = a$ and the first word of β begins with a vowel, in which case, $\alpha' = an$.

This rule allows us to store the expression *a man* as well as *an aristocrat* among the expressions of the T category.

[2]We follow more or less the numbering of rules which is given in Montague's (1974b) *The Proper Treatment of Quantification in Ordinary Language* (PTQ).

Another example of a rule is

> S_3: If $\alpha \in P_{CN}$ and if A $\in P_t$, then $F_{3,n}(\alpha, A) \in P_{CN}$, where $F_{3,n}(\alpha, A) = \alpha$ such that $A^\star$ where $A^\star$ is obtained from A by replacing every occurrence of $\mathbf{he}_n$ or $\mathbf{him}_n$ with resp. **he**, **she** or **it** or with **him**, **her** or **it** according to the gender of the first common noun in α (masculine, feminine or neutral).

Example: $\alpha =$ **woman**, $A = \mathbf{he}_1$ **walks**, $F_{3,1}(\alpha, A) =$ **woman such that she walks**. In this rule, the pronouns are dealt with as indices (Montague calls them *syntactic variables*), represented as integers, those indices being introduced along the discourse, according to their order of occurrence. This anticipates in some sense the future treatment of pronouns in discourse representation theory (as *reference markers*).

These rules are nevertheless for Montague mere scaffolding for introducing what is for him essential, that is, a rigorous and algorithmic way to build, step after step, compositionally, the "meaning" of a sentence. His method has two steps:

- First, to build formulae, well-formed expressions of a formal language (richer and richer, from ordinary first-order predicate logic to intentional logic).
- Then, to use a standard evaluation procedure on these formulae to deduce their truth value, with regard to a given model, assuming that the meaning of a sentence finally lies in its truth value.

To reach this goal, Montague crucially uses the λ-calculus.

Let us denote by $\tau(\alpha)$ the translation of the expression α, that is, its image by a function τ which, for every linguistic meaningful expression, associates a λ-term supposed to represent its meaning. We shall have, for instance,

$$\tau(\mathbf{unicorn}) = \lambda x.unicorn(x)$$
$$\tau(\mathbf{a}) = \lambda Q.\lambda P.\exists x(Q(x) \wedge P(x))$$

Comment: Rigorously, a function such as τ is defined on strings of characters that we shall denote, in conformity with the convention adopted by Montague, by expressions in bold.

τ will be extended further to the non-lexical expressions by means of a semantic counterpart associated with each rule, which indicates how to build the semantic representation of the resulting expression (the output of the rule) from the translations of its components (its inputs). Thus the rule S_2 will be completed in the following way:

> T_2: If α is of category T/CN and if β is of category CN, then $\tau(F_2(\alpha, \beta)) = \tau(\alpha)(\tau(\beta))$.

By using T_2 with $\alpha =$ **a** and $\beta =$ **unicorn**, we get

$$\tau(\textbf{a unicorn}) = \lambda P.\exists x(unicorn(x) \wedge P(x))$$

In what follows, we will mainly concentrate on simpler rules by trying to obtain from them the maximum of generalization. Let us therefore consider the following simplified rule:

> S: If $\alpha \in P_{X/Y}$ and if $\beta \in P_Y$, then $F(\alpha, \beta) \in P_X$, where $F(\alpha, \beta) = \alpha\beta$

which we saw has an easy translation in the form of the sequent X/Y $Y \vdash X$. We will adopt as its semantic counterpart

$$T : \text{If } \alpha \in P_{X/Y} \quad \text{and} \quad \text{if } \beta \in P_Y, \quad \text{then } \tau(\alpha\beta) = \tau(\alpha)(\tau(\beta)).$$

Symmetrically, by using "\",

$$T' : \text{If } \alpha \in P_Y \quad \text{and} \quad \text{if } \beta \in P_{Y\backslash X}, \quad \text{then } \tau(\alpha\beta) = \tau(\beta)(\tau(\alpha)).$$

Such rules may be simply given, in the sequent notation, by decorating the syntactic categories by means of the semantic translations of the expressions they belong to:

$$\alpha' : X/Y \;\; \beta' : Y \vdash \alpha'(\beta') : X$$

$$\alpha' : Y \;\; \beta' : Y\backslash X \vdash \beta'(\alpha') : X$$

where we simply write α' (resp. β') instead of $\tau(\alpha)$ (resp. $\tau(\beta)$).

2.3.2 *On relatives and quantification*

Among all the rules introduced by Montague, two of them will particularly attract our attention, because they pose non-obvious problems that we shall meet also in other formalisms. These rules are denoted as S_3, introduced in Section 2.3.1, which will be used to introduce expressions like "such that", and S_{14}, which is used for quantified nominal expressions.

S_3: If $\alpha \in P_{CN}$ and if $A \in P_t$, then $F_{3,n}(\alpha, A) \in P_{CN}$ where $F_{3,n}(\alpha, A)$ is defined by $F_{3,n}(\alpha, A) = \alpha$ such that $A^\star$ where $A^\star$ is obtained from A by replacing every occurrence of the pronoun $\mathbf{he}_n$ or $\mathbf{him}_n$ of type e with **he** or **she** or **it** or with **him**, **her** or **it** according to the gender of the first common noun in α (masculine, feminine or neutral), and if $\alpha' = \tau(\alpha)$ and $A' = \tau(A)$, then $\tau(F_{3,n}(\alpha, A)) = \lambda x_n.(\alpha'(x_n) \wedge A')$.

S_{14}: If $\alpha \in P_T$ and if $A \in P_t$, then $F_{14,n}(\alpha, A) \in P_t$ where

- if α is not a pronoun $\mathbf{he}_k$, $F_{14,n}(\alpha, A)$ is obtained from A by replacing the first occurrence of $\mathbf{he}_n$ or $\mathbf{him}_n$ with α and all the others respectively with **he**, **she** or **it** or with **him**, **her** or **it** according to the gender and case of the first *CN* or T in α,
- if α is the pronoun $\mathbf{he}_k$, then $F_{14,n}(\alpha, A)$ is obtained from A by replacing every occurrence of $\mathbf{he}_n$ or $\mathbf{him}_n$ respectively with $\mathbf{he}_k$ or $\mathbf{him}_k$,

and then $\tau(F_{14,n}(\alpha, A)) = \alpha'(\lambda x_n.A')$.

2.3.3 *Examples*

Let us study the case of some (pseudo)-sentences[3]:

Example 2.9. A woman such that she walks talks

Example 2.10. Peter seeks a woman

In the case of Example 2.9, the sentence $\mathbf{he}_1$ **walks** may be produced by means of the S_4 rule, which has not yet been mentioned. It consists in putting together an intransitive verb — or a *VP* — and its subject. This rule is simple. We must yet notice that, because from now on, nominal expressions have a higher-order type (see the result of an application of the S_2 rule), the semantic counterpart no longer amounts to applying

[3]We say "pseudo-sentences" in the sense that a sentence such as Example 2.9 may seem weird to a native speaker of English. In fact, Montague is not looking for a realistic grammar of English — or of any language — he simply tries to approximate by means of formal tools linguistic mechanisms such as relativization and quantification.

the semantics of VP to the meaning of T, but the other way round, the semantics of the subject T to that of VP. This makes us assume that all the nominal expressions (and not only the quantified ones like **a unicorn** or **every unicorn**) have such a higher-order type, that is, $(e \to t) \to t$. We have therefore the following representation for a proper name:

$$\tau(\textbf{Marie}) : \lambda X.X(marie)$$

As for a pronoun like $\textbf{he}_n$, it will be translated as

$$\tau(he_n) : \lambda X.X(x_n)$$

where we understand the role of the indices attached to pronouns: they allow a one-to-one correspondence to be maintained between the pronoun in a text and the enumerable set of variables we can use in semantics.

We may therefore use S_3 with $\alpha =$ **woman** and $A = \textbf{he}_1$ **walks**, given that their translations are respectively

$$\tau(\textbf{woman}) : \lambda u.woman(u)$$
$$\tau(\textbf{he}_1\textbf{walks}) : walk(x_1)$$

We obtain **woman such that she walks** of category CN, and of translation $\lambda x_1\ (woman(x_1) \wedge walk(x_1))$.

Then S_2 applies, which gives as a translation of **A woman such that she walks**

$$\lambda Q.\exists x\ woman(x)\ \wedge\ walk(x) \wedge Q(x)$$

By S_4, we get the final translation:

$$\exists x\ woman(x)\ \wedge\ walk(x)\ \wedge talk(x)$$

It is possible to represent this generation of a semantic form by the tree in Fig. 2.1, where we show the rule applications. As it may be seen, if S_2 and S_4 amount to standard applications, S_3 is not one. We could describe it semantically as a kind of coordination of two properties after the right-hand-side term (corresponding to **he walks**) has been abstracted over by using the variable associated with the pronoun (if not, that would not be a property but a complete sentence). It is this kind of irregularity that we shall try to avoid later on. In the case of Example 2.10, it is true that two generations are possible. The simplest uses the S_5 rule:

> If $\alpha \in P_{VT}$ (transitive verbs) and if $\beta \in P_T$ (terms), then $F_5(\alpha, \beta) \in P_{VI}$ (intransitive verbs), where $F_5(\alpha, \beta)$ is equal to $\alpha\beta$ if $\beta \neq \textbf{he}_n$ and $F_5(\alpha,\ \textbf{he}_n) = \alpha\ \textbf{him}_n$.

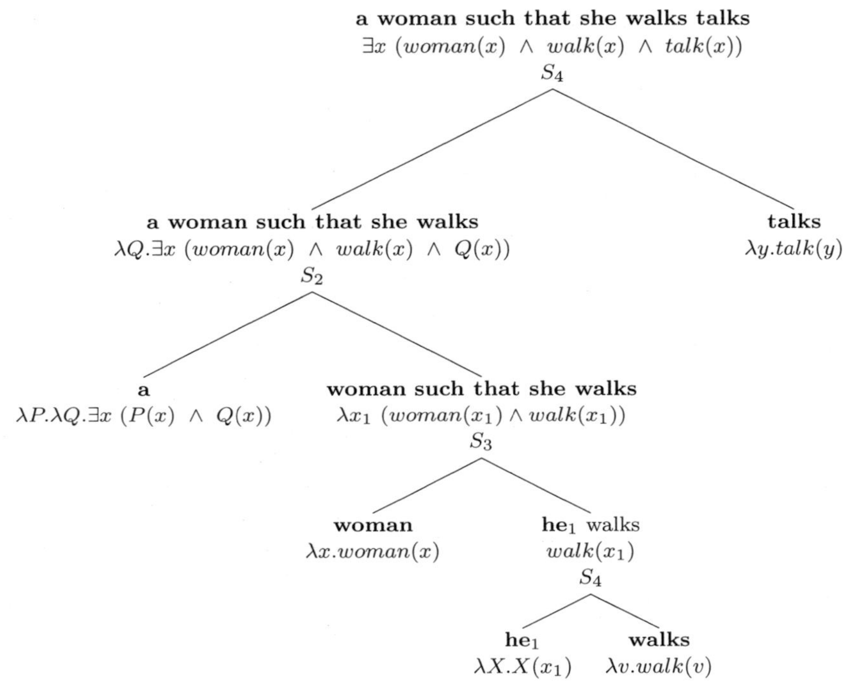

Fig. 2.1 **A woman such that she walks talks**

Since VT is of the form VI/T, this rule is a simple application rule and it results from it that, semantically, the translation of $F_5(\alpha, \beta)$ is $\alpha'(\beta')$. With that rule, we get the syntactic analysis in Fig. 2.2, where we let aside the semantic translation of **seek**. This analysis will give the *de dicto* analysis of the sentence. It would remain now to find the *de re* reading, for which the existential quantifier would be extracted from its embedded position as an object argument. Let us remark that this reading is the most familiar, the one which occurs for any ordinary transitive verb, if we accept for instance that the translation of *Peter eats an apple* is $\exists x\ (apple(x)\ \wedge\ eat(peter, x))$. At this stage, S_{14} must be used. This amounts to making the analysis and generation in Fig. 2.3. The λ-term at the root reduces according to

$$\lambda Q\ \exists x\ (woman(x)\ \wedge\ Q(x))(\lambda x_1 \lambda Y.Y(peter)(\text{seek}'(\lambda Z.Z(x_1)))) \longrightarrow$$

$$\exists x\ (woman(x)\ \wedge\ (\lambda x_1 \lambda Y.Y(peter)(\text{seek}'(\lambda Z.Z(x_1))))(x)) \longrightarrow$$

$$\exists x\ (woman(x)\ \wedge\ \lambda Y.Y(peter)(\text{seek}'(\lambda Z.Z(x))))$$

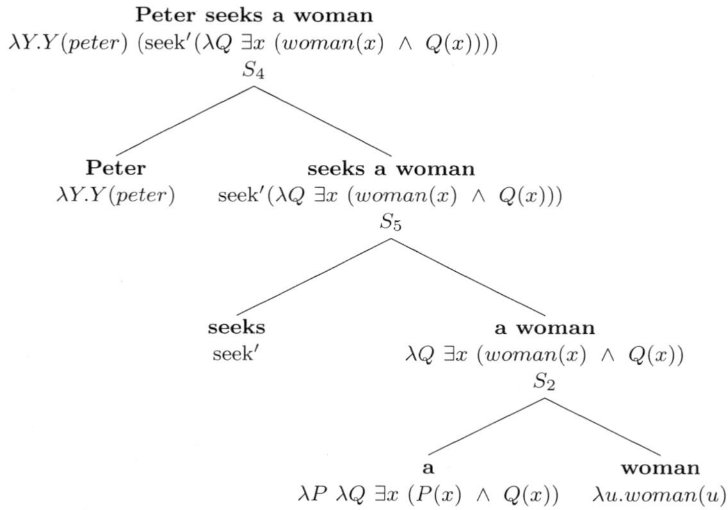

Fig. 2.2 **Peter seeks a woman**

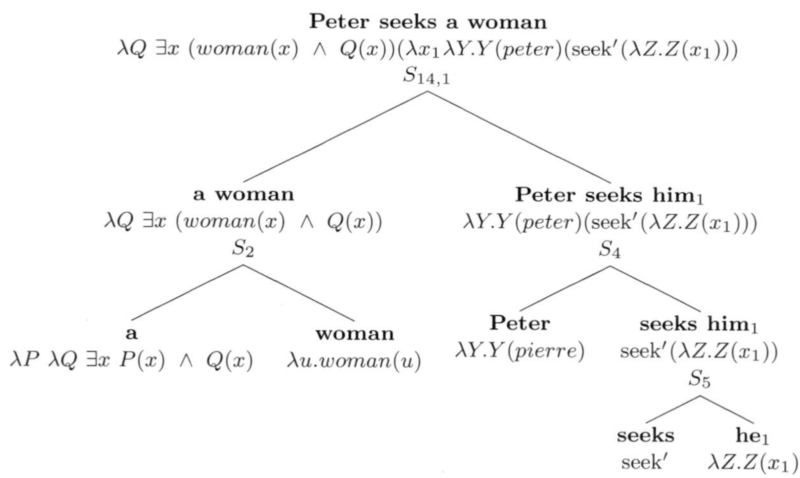

Fig. 2.3 **Peter seeks a woman** — *de re*

From these examples, we may draw the following conclusions:

- Different readings are obtained by means of various syntactic trees, thus making what are in principle "semantic" ambiguities (like scope ambiguities or *de re/de dicto* ambiguities) syntactic ones in reality, which does not seem to be satisfactory on the theoretical side. Let us notice that the type of syntactic analysis which uses the S_{14} rule strongly looks like the solution of *quantifier raising* in generative grammar, as it has been advocated by May and Fiengo. Thus, the "syntactic" approach to the semantic problem to scope ambiguities is not a drawback proper to Montague grammar, but that does not seem to be a reason to keep it!

- The rules of semantic construction contain steps which are "quasi-subliminal", consisting in abstracting just before applying. In the case of S_{14}, this abstraction uses the variable x_1 on the form $\lambda Y.Y$ $(peter)(\text{seek}'(\lambda Z.Z(x_1)))$, which corresponds to the sentence **Peter seeks him**, in the aim to transform it into a property.

2.3.4 *On binding*

Semantic representations *à la* Montague allow us to reformulate the question of binding seen in Section 2.1.2. Examples 2.9 and 2.10 involve binding in two different ways.

Example 2.9 is a relative and the pronoun **she** must be interpreted as co-referential with **a woman**, like the following representation shows it, with co-indexing:

$$[_S\ [_{DP}\ \text{a}\ [_N\ \text{woman}]_1\ \text{such that}\ [_S\ \text{she}_1\ \text{walks}]]\ \text{talks}]$$

A more "natural" sentence would be **A woman who walks talks**, the representation of which would be

$$[_S[_{DP}\ \text{a}\ [_N\ \text{woman}]_1\ \text{who}[_S t_1\ \text{walks}]]\ \text{talks}]$$

In this case, the relative pronoun **who** is a binder, like the expression **such that**. In Montague grammar, the fact of binding is rendered by the "subliminal" λ-abstraction which occurs during the application of the S_3 rule. It is by means of this abstraction that *woman* and — *walks* are applied to the very same individual.

Example 2.10 contains an (existentially) quantified expression (**a woman**) which serves as a binder. The use of S_{14} also contains an implicit λ-abstraction which allows the displaced constituent **a woman** to apply to a function of the individual variable x_1.

In each case, the scope of the binder is a formula with a free variable which is transformed into a function by means of λ-abstraction on this variable.

2.4 A Theory of Simple Types

Montague uses so-called *semantic types*. This means that every expression in Montague grammar is supposed to belong to (at least) one category (here, "type" means "category"). Let σ be the function which associates each expression with a set of types. It is assumed that expressions belonging to the same syntactic category also belong to the same set of semantic types, so that the σ function may be factorized through syntactic categories. Let *Typ* be the set of semantic types and *Cat* the set of syntactic categories, and μ the function which associates each expression with its syntactic category. Then there exists one and only one function τ from *Cat* to *Typ* such that $\sigma = \tau \circ \mu$.[4] These semantic types are built from a set of primitive types A according to the following rules:

$$\forall t \in A, \quad t \textit{ is a type } (t \in Typ) \tag{2.8}$$

$$\forall \alpha, \beta \in Typ, \quad (\alpha \rightarrow \beta) \in Typ \tag{2.9}$$

Originally, types were used in logic for eliminating paradoxes (like Russell's); the formulation we are using here is due to Church (from around the 1940s). For Church (1941), the set A consisted of only two primitive types: i (the type of individuals) and o (the type of propositions). For Montague and most applications in linguistic theory, the same types are denoted as **e** and **t**. As we remember, Russell's paradox is due to the fact that in Frege's framework, a predicate may apply to any object, even another predicate, thus giving a meaning to expressions like $\Phi(\Phi)$ where a predicate Φ is applied to itself. In order to avoid this pitfall, Russell suggested the use of types in order to make such an application impossible. The easiest way to do that is to associate each predicate with a type of the sort defined above and to stipulate that a predicate of type $\alpha \rightarrow \beta$ may only apply to an object of type α, thus giving a new object of type β. Curry and Feys (1958) took *functions* as primitive objects and Church used the λ-calculus to represent functions. Let us shortly recall that λ-terms are defined according to

$$M := x \mid (M\ M) \mid \lambda x.M \tag{2.10}$$

that is

- every variable is a λ-term,
- if M and N are λ-terms, then $(M\ N)$ is also one,

[4] $\circ$ denotes the functional composition.

- if M is a λ-term and x a variable, then $\lambda x.M$ is a λ-term,
- there is no way to build a λ-term other than the previous three clauses.

For instance $((M_1\ M_2)\ M_3)$ is a λ-term if M_1, M_2 and M_3 are. By assuming associativity on the left, this expression may be simply written as $M_1\ M_2\ M_3$.

β-conversion is defined as

$$(\lambda x.M\ N) \rightarrow M[x := N] \tag{2.11}$$

where the notation $M[x := N]$ means the replacement of x by N in M everywhere it occurs unbound.

η-conversion is the rule according to which

$$\lambda x.(f\ x) \rightarrow f \tag{2.12}$$

η-expansion is the reverse rule.

In the untyped λ-calculus, reduction may not terminate, like it is the case when trying to reduce

$$(\lambda x.(x\ x)\ \lambda x.(x\ x)) \tag{2.13}$$

λ-equality between terms is introduced via a set of specific axioms:

$$\lambda x.M = \lambda y.M[y/x] \quad y\ \textit{not free in}\ M$$

$$(\lambda x.M\ N) = M[x := N]$$

$$X = X \qquad \frac{X = Y}{Y = X} \qquad \frac{X = Y \quad Y = Z}{X = Z}$$

$$\frac{X = X'}{(Z\ X) = (Z\ X')} \quad \frac{X = X'}{\lambda x.X = \lambda x.X'}$$

Reduction always terminates in the *typed* λ-calculus, where the definition is enriched in the following way (it is supposed each type contains a set of variables):

- every variable of type α is a λ-term of type α,
- if M and N are λ-terms respectively of types $\alpha \rightarrow \beta$ and α, then $(M\ N)$ is a λ-term of type β,
- if M is a λ-term of type β and x a variable of type α, then $\lambda x.M$ is a λ-term of type $\alpha \rightarrow \beta$,
- there is no way to build a λ-term other than the previous three clauses.

In the typed λ-calculus, an equality is introduced for each type: "$=_\tau$" means the equality inside the type τ.

By means of these definitions, we can handle sequences of objects, like $a_1, a_2, \ldots, a_n$ which are interpreted as successive function applications:

$$((\ldots(a_1, a_2), \ldots, a_{n-1}), a_n)$$

These sequences may be reduced if and only if the sequences of their types may be reduced by means of the application rule

$$\frac{a : \alpha \to \beta \qquad b : \alpha}{(a\ b) : \beta}\ FA$$

More generally, we are interested in judgements of the form

$$x_1 : a_1, \ldots, x_n : a_n \vdash M : a \tag{2.14}$$

which mean the sequence of successive applications $(\ldots(x_1{:}a_1), \ldots, x_n{:}a_n)$ reduces to the object M of type a. But there is not only an *application* rule. For instance the following expression is a possible judgement:

$$x_1 : a_1 \vdash \lambda x.M : a \to b \tag{2.15}$$

which is obtained from $x_1 : a_1, x : a \vdash M : b$ by an application of the *abstraction* rule:

$$\text{if } A_1 : a_1, \ldots, A_n : a_n, x : a \vdash M : b$$

$$\text{then } A_1 : a_1, \ldots, A_n : a_n \vdash \lambda x.M : a \to b$$

This rule is precisely the one we use spontaneously when searching for the type of a function: if a function f applied to an object of type a gives an object of type b, then it is itself an object of type $a \to b$, which is denoted by $\lambda x^a.f$.

We may represent this abstraction rule also in the following way:

$$\frac{\begin{array}{c} A_1 : \alpha_1, \ \ldots, \quad A_1 : \alpha_n \quad [x : a] \\ \vdots \\ M : b \end{array}}{\lambda x.M : a \to b}$$

where the brackets denote a *hypothesis* that the rule discharges at the same time abstraction is performed on the resulting term M.

If after hypothesizing a value x of type a, the sequence of objects $A_1, \ldots, A_n$ of respective types $\alpha_1, \ldots, \alpha_n$ applied to this value gives a term M of type b, then, without the hypothesizing, the sequence $A_1, \ldots, A_n$ leads to the term $\lambda x.M$ of type $a \to b$.

In ordinary calculus, the same hypothesis may be used (that is, discharged) any number of times, even none. In our next applications, there will be a constraint according to which a hypothesis must be discharged once and only once. In that case, the type of calculus will be said to be *linear*, and we will employ the symbol $\multimap$ to denote the arrow.

All these points (rule systems and the linear calculus) will be studied in depth in the following sections of this book.

2.5 Heim and Kratzer's Theory

Montague grammar was conceived and built during the 1960s without much attention paid to works in the generative paradigm. Important syntactic notions like *projection*, *Move* and *control* were ignored by the logician and philosopher. Many similar phenomena (like extrapositions and questions, quantifiers in subject or object position) were treated in a non-uniform way. Heim and Kratzer, as generativists, have tried to overcome these difficulties. They have, moreover, tried to give semantic interpretations in terms of truth values independently of any intermediary formal language, aiming to "execute the Fregean program", as they declared in the second chapter of their book.

2.5.1 *Interpreting derivation trees*

As we know, a first-order language L is interpreted relative to some structure $M = \langle D, I \rangle$ where D is a non-empty set (the *domain* or *universe* of the interpretation) and I is an interpretation function defined at the beginning only on constants in L, and then extended to terms and formulae by means of a recursive procedure. In this frame, variables are interpreted by means of assignments, and the truth value of quantified formulae is obtained by making assignments vary. An assignment may be viewed either as a partial function defined in *Var* (the denumerable set of variables in L) with values in D or as a possibly infinite list of elements of $D(a_1, a_2, \ldots, a_n, \ldots)$, where for each i, a_i is the value assigned to the ith variable in L (with some a_i possibly replaced by $\perp$ for "undefined" when there is no value assigned to the ith variable).

Natural language is supposed to be interpreted similarly by Heim and Kratzer. What is required for that is simply a universe D and an interpretation function I which sends constant words to appropriate elements and sets definable from D and the set of truth values $\{0, 1\}$.

For instance, we shall assume

- for every proper noun c, $I(c)$ is an element of D,
- for every intransitive verb V, $I(V)$ is a function f_V from D to {0, 1},
- for every common noun N, $I(N)$ is a function f_N from D to {0, 1},
- for every transitive verb V, $I(V)$ is a function f_V from D to the set of functions from D to {0, 1}.

As an example, we may suppose D = {Ann, Mary, Paul, Ibrahim} and I defined as

I(**Ann**) = Ann
I(**Mary**) = Mary
I(**Paul**) = Paul
I(**Ibrahim**) = Ibrahim
I(**smokes**) = the function f: $D \longrightarrow \{0,1\}$ such that
  f(Ann) = 0, f(Mary) = 0, f(Paul) = 1, f(Ibrahim) = 1
I(**kisses**) = the function f: $D \longrightarrow \{0,1\}^D$ such that
  f(Ann) = the function f_{Ann}: $D \longrightarrow \{0,1\}$ such that
    f_{Ann}(Ann) = 0, f_{Ann}(Mary) = 0, f_{Ann}(Paul) = 1,
    f_{Ann}(Ibrahim) = 0
  f(Mary) = the function f_{Mary}: $D \longrightarrow \{0,1\}$ such that
    f_{Mary}(Ann) = 0, f_{Mary}(Mary) = 0, f_{Mary}(Paul) = 0,
    f_{Mary}(Ibrahim) = 1
  f(Paul) = the function f_{Paul}: $D \longrightarrow \{0,1\}$ such that
    f_{Paul}(Ann) = 1, f_{Paul}(Mary) = 0, f_{Paul}(Paul) = 0,
    f_{Paul}(Ibrahim) = 0
  f(Ibrahim) = the function $f_{Ibrahim}$: $D \longrightarrow \{0,1\}$ such that
    $f_{Ibrahim}$(Ann) = 0, $f_{Ibrahim}$(Mary) = 1, $f_{Ibrahim}$(Paul) = 0,
      $f_{Ibrahim}$(Ibrahim) = 0

Then, rules may be interpreted in order to define the denotation of a sentence in a small language defined on this lexicon. Let us assume the following grammar rules:

Rule 1: S → NP VP
Rule 2: NP → N
Rule 3: VP → VI
Rule 4: VP → VT NP
Rule 5: VI → **smokes**
Rule 6: VT → **kisses**
Rule 7: N → **Ann** | **Mary** | **Paul** | **Ibrahim**

Given a derivation tree τ in such a grammar, the yield of which is a sequence of words σ, we may interpret τ according to the following rules:

Rule 1: If τ has a root labelled by S, and two branches α, the root of which is labelled by NP and β, the root of which is labelled by VP, then $I(\tau) = I(\beta)(I(\alpha))$.

Rule 2: If τ has a root labelled by NP, and one branch α, the root of which is labelled by N, then $I(\tau) = I(\alpha)$.

Rule 3: If τ has a root labelled by VP, and one branch α, the root of which is labelled by VI, then $I(\tau) = I(\alpha)$.

Rule 4: If τ has a root labelled by VP, and two branches α, the root of which is labelled by VT and β, the root of which is labelled by NP, then $I(\tau) = I(\alpha)(I(\beta))$.

Rule 5: If τ has a root labelled by VI, and one branch α, simply labelled by **smokes**, then $I(\tau) = I(\textbf{smokes})$.

Rule 6: If τ has a root labelled by VT, and one branch α, simply labelled by **kisses**, then $I(\tau) = I(\textbf{kisses})$.

Rule 7: If τ has a root labelled by N, and one branch α, simply labelled by **Ann**, **Mary**, **Paul** or **Ibrahim**, then $I(\tau) = I(\textbf{Ann})$, $I(\textbf{Mary})$, $I(\textbf{Paul})$ or $I(\textbf{Ibrahim})$.

Let us consider the following sentence:

Example 2.11. Ann kisses Paul

Its derivation tree τ is

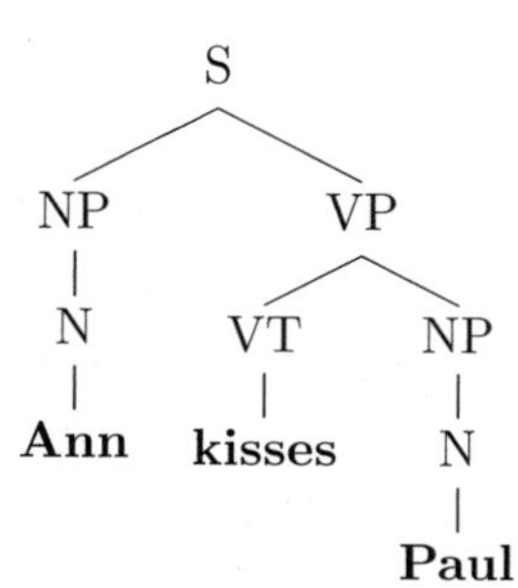

The interpretation of the tree is given by

$$I(\tau) = I([_{VP}\ [_{VT}\ \textbf{kisses}]\ [_{NP}\ [_{N}\ \textbf{Paul}]]])(I([_{NP}\ [_{N}\ \textbf{Ann}]]))$$

and we have

$$I([_{NP}\ [_{N}\ \textbf{Ann}]]) = I([_{N}\ \textbf{Ann}]) = I(\textbf{Ann}) = \text{Ann}$$

and

$$I([_{VP}\ [_{VT}\ \textbf{kisses}]\ [_{NP}\ [_{N}\ \textbf{Paul}]]]) = I([_{VT}\ \textbf{kisses}])(I([_{NP}\ [_{N}\ \textbf{Paul}]])) = I(\textbf{kisses})(I(\textbf{Paul}))$$

Applying I as defined on the constants of the language, we get

$$I(\textbf{kisses})(Paul) = f_{Paul}$$

and therefore $I(\tau) = [I(\textbf{kisses})(\text{Paul})](\text{Ann}) = f_{Paul}(Ann) = 1$.

Let us notice here that the interpretation of a transitive verb in this grammar is such that to each individual $x \in D$ is associated a function f_x from D to $\{0, 1\}$, such that for each individual $y \in D$, $f_x(y)$ is equal to 1 if and only if x is the object of the verb and y its subject and there is between x and y the relation denotated by the transitive verb.

Heim and Kratzer extend their notions and denote by [[.]] the function of interpretation that we can compute on each expression of the grammar (but we shall still use sometimes I to denote this function). An easier definition of the procedure amounts to defining it not on trees but on *nodes*. The interpretation of a node is then the previous interpretation of the (sub)tree of which it is the root. With this change of perspective, we can generally define the interpretation of any node, starting from terminal nodes and concluding that *all nodes in a phrase structure tree must be in the domain of the interpretation function* [[.]] (interpretability condition).

(1) Terminal nodes: If α is a terminal node, then α belongs to the domain of $[[\ .\ ]]$ if $[[\alpha]]$ is given by the lexicon.
(2) Non-branching nodes: If α is a non-branching node and β is its daughter, then α belongs to the domain of $[[\ .\ ]]$ if β belongs to it, and then $[[\alpha]] = [[\beta]]$.
(3) Branching nodes: If α is a branching node and β and γ are its daughters, then α belongs to the domain of $[[\ .\ ]]$ if β and γ belong to it and $[[\beta]]$ is a function defined on $[[\gamma]]$. In this case, $[[\alpha]] = [[\beta]]([[\gamma]])$.

In the previous example, for instance, let us take the node α labelled by the symbol VP: it is a branching node. β (labelled by VT) and γ (labelled by NP) are its daughters and it happens that $[[\beta]]$ is defined on $[[\gamma]]$. Therefore, $[[\alpha]]$ is defined and $[[\alpha]] = [[\gamma]]([[\beta]])$.

Heim and Kratzer also make use of the notion of type. Thus, a necessary condition for $[[\beta]]$ to be a function defined on $[[\gamma]]$ is that $[[\beta]]$ be of type $a \to b$ and $[[\gamma]]$ of type a, but cases may exist where this condition is not sufficient.

Let us call the *interpretability condition* the set of conditions expressed above. Of course the fact that $[[\beta]]$ (for instance in the case of a transitive verb) is a function defined on $[[\gamma]]$ may seem reminiscent of the θ-criterion:

> Every argument receives one and only one θ-role, and each θ-role is assigned to one and only one element.

We see nevertheless that the interpretability condition may sometimes be satisfied when the θ-criterion is not. For instance it is often admitted that a common noun assigns a θ-role, which is what happens in a sentence like *Paul is a student* where *Paul* receives the θ-role assigned by *student.* The sentence also satisfies the interpretability condition since a common noun is of type $e \to t$, and *Paul* is supposed to be of type e. But in the case of an NP like *the student*, the θ-role is not assigned whereas the condition is still satisfied: this time, it is the definite *the* which behaves as a function and "absorbs" the demand for an argument. In fact, *the* is of type $(e \to t) \to e$, so that *the student* is again of type e, just like any proper name (for the time being).[5]

[5]The function associated with the definite is called a *choice function* since it must *select* an individual in a set under some conditions.

2.5.2 *Predicate modification*

As the name indicates, modifiers modify items to which they apply. Nevertheless, application in this context must be *a priori* understood differently from the standard application of the semantics of an expression (say, a verbal phrase) to the semantics of another one (say, its subject). The modifier *in blue* may modify the predicate *man* in such a way that *man* becomes *man in blue*, but the two expressions have the same type (they are both of type $e \to t$). It seems to be regular to assign type $e \to t$ also to the modifier *in blue* since it can work as a predicate too. There are therefore cases where a node α is a branching node, but none of its daughters can apply to the other in the regular way (satisfying the interpretability condition). This requires a new semantic rule, that Heim and Kratzer named a *composition rule* (even if it does not correspond to what we usually name composition for two functions f and g). Such a rule takes two functions as inputs and returns a new function as output. For instance, the two inputs are

> f_{man} : the function $D \to \{0, 1\}$ such that for all $i \in D$, $f_{\text{man}}(i) = 1$ if and only if i is a man
> $f_{\text{in blue}}$: the function $D \to \{0, 1\}$ such that for all $i \in D$, $f_{\text{in blue}}(i) = 1$ if and only if i is in blue

and the output is

> $f_{\text{man in blue}}$: the function $D \to \{0, 1\}$ such that for all $i \in D$, $f_{\text{man in blue}}(i) = 1$ if and only if i is a man in blue

The easiest way to express this transformation rests on the use of λ-functions. Let us denote the functions f_{man} and $f_{\text{in blue}}$ by λ-expressions:

$$f_{\text{man}} : \lambda x \in D.x \text{ is a man}$$
$$f_{\text{in blue}} : \lambda x \in D.x \text{ is in blue}$$

The resulting function $f_{\text{man in blue}}$ is equal to $\lambda x \in D$. (x is a man) $\wedge$ (x is in blue), or also $\lambda x \in D$. (x is a man) $=$ (x is in blue) $=$ true.

Finally we get, as a new semantic rule,

Predicate modification: If α is a branching node and β and γ are its daughters, and $[[\beta]]$ and $[[\gamma]]$ are both of type $e \to t$, then $[[\alpha]] = \lambda x^e.([[\beta]](x) = [[\gamma(x)]] = 1)$.

2.5.3 *Variables and binding*

In the previous section, we had no need for variables since we only elaborated on phrase structure rules (or Merge operations). Things change as soon as we introduce displacements of constituents and therefore traces. A relative clause, for instance, **whom Mary met**, is obtained after a Move operation which displaces the object of **met** and leaves a trace behind. In a DP like **the man whom Mary met**, **man** and **whom Mary met** are two properties which are combined exactly like in predicate modification: **whom Mary met** is a modifier. Its denotation is the function

$$[[\textbf{whom Mary met}]] = \lambda x \in D.\textit{Mary met } x \tag{2.16}$$

Sometimes, following Heim and Kratzer, we shall denote this function by $\lambda x \in D.\ (\textit{Mary met } x = 1)$ but the reader knows that for a Boolean variable X, there is no distinction between the evaluations of X and of $X = 1$! ($X = 1$ is true if and only if X is true!)

It remains to know how the variable x is introduced. In fact the two questions, the syntactic one concerning Move and the semantic one concerning the introduction of a variable x, amount to the same mechanism.

Let us assume the following syntactic analysis of (2.16):

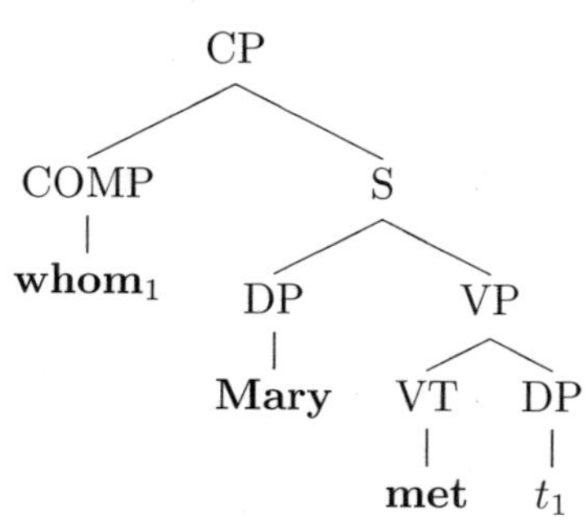

The point here is that we will translate the trace by a variable. It remains to interpret a node marked by a trace. Like the interpretation of formulae of first-order logic, this requires the use of *assignments*. From now on, trees (and nodes) will be interpreted not only relative to a structure $M = \langle D, I \rangle$, but also to an assignment $g : \mathbb{N} \rightarrow D$, and we will have, for any trace t_i (indexed by an integer $i \in \mathbb{N}$),

$$[[t_i]]^{M,g} = g(i)$$

It results that the interpretation relative to M, g of the tree representing **Mary met** t_1 (that we will denote as $\tau_{\text{Mary met } t_1}$) is calculated in the following way:

$I($ 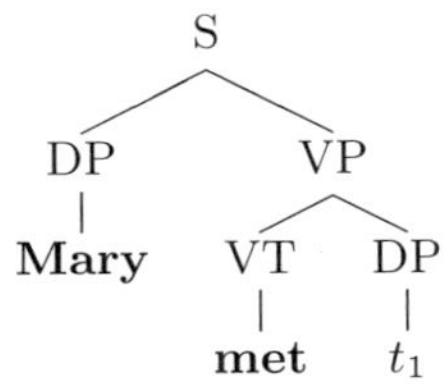 $)^{M,g} = I($ 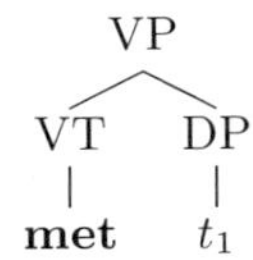 $)^{M,g}(I(\mathbf{Mary})^{M,g})$

with

$I($ 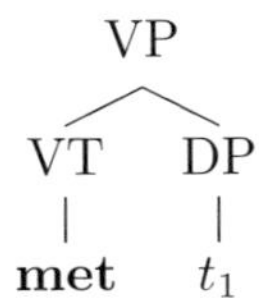 $)^{M,g} = I(\mathbf{met})^{M,g}(I(t_1)^{M,g})$

$$= [\lambda x.\lambda y.y\ met\ x](g(1)) = \lambda y.y\ met\ g(1)$$

so that, finally, $I(\tau_{\text{Mary met } t_1})^{M,g} = [\lambda y.y\ met\ g(1)](\text{Mary}) = \text{Mary}$ met $g(1)$, which clearly depends on the assignment g.

The next step consists in defining the role of **whom**. Heim and Kratzer introduced a new semantic rule: *predicate abstraction.*

Predicate abstraction: If α is a branching node whose daughters are a relative pronoun and $\beta(t_i)$, then $[[\alpha]]^M = \lambda x \in D.[[\beta(t_i)]]^{M,g[i:=x]}$.

In this rule, $\beta(t_i)$ designates a tree β containing a trace t_i and the notation $[[.]]^{M,g[i:=x]}$ means an interpretation relative to a structure M and some assignment g modified in i in order to assign a variable x to the integer i.

By using this rule we can compute the denotation of **whom Mary met**:

$$\begin{aligned} I(\tau_{\text{whom Mary met}})^{M,g} &= \lambda x \in D.I(\tau_{\text{Mary met } t_1})^{M,g[1:=x]} \\ &= \lambda x \in D.[Mary\ met\ g(1)]^{M,g[1:=x]} \\ &= \lambda x \in D.Mary\ met\ x \end{aligned}$$

In the last line, we get rid of g because the expression obtained no longer depends on it, and we get rid of M since all the constants which have to be interpreted with regard to I are already interpreted (**Mary** by the individual *Mary* and **met** by the function $\lambda x \in D.\lambda y \in D.y\ met\ x$). As we see it, the form which results is independent of any assignment, and so we can say it is closed. In the last step, we can interpret the expression **man whom Mary met** by predicate modification, thus obtaining

$$\lambda x \in D.(x\ is\ a\ man) \quad \text{and} \quad (Mary\ met\ x) \tag{2.17}$$

If moreover we interpret the definite **the** as a function from sets of individuals to individuals in the standard way, $I(\textbf{the})$ = *the function which associates to any subset E of D the unique element of E if E is a singleton, and is undefined for other subsets.*

Assuming that a function like (2.17) is the *characteristic function* of some set and that therefore we may also interpret **man whom Mary met** as a set (the subset of D whose elements are precisely the men whom Mary met and which is a mere singleton), by functional application, we obtain for the expression **the man whom Mary met** the denotation

$I(\textbf{the man whom Mary met}) =$

— the unique element m such that $m \in D$ and m is a man and Mary met m *if there is one and only one such individual,*

— and *undefined* if not.

Let us recall that, in a Montagovian setting, the translation of this nominal phrase is

$$\iota x.man(x) \wedge met(Mary, x) \tag{2.18}$$

using Russell's operator ι.

Let us summarize now Heim and Kratzer's rules for interpretation:

(1) Terminal nodes: If α is a terminal node, then α belongs to the domain of $[[\ .\ ]]$ if $[[\alpha]]$ is given by the lexicon.

(2) Non-branching nodes: If α is a non-branching node and β is its daughter, then α belongs to the domain of $[[\ .\ ]]$ if β belongs to it, and then, $[[\alpha]] = [[\beta]]$.

(3) Functional application: If α is a branching node and β and γ are its daughters, then α belongs to the domain of $[[\ .\ ]]$ if β and γ belong to it and $[[\beta]]$ is a function defined on $[[\gamma]]$. In this case, $[[\alpha]] = [[\beta]]([[\gamma]])$.

(4) Predicate modification: If α is a branching node and β and γ are its daughters, and $[[\beta]]$ and $[[\gamma]]$ are both of type $e \rightarrow t$, then $[[\alpha]] = \lambda x^e.([[\beta]](x) = [[\gamma(x)]] = 1)$.

(5) Predicate abstraction: If α is a branching node whose daughters are a relative pronoun and $\beta(t_i)$, then $[[\alpha]]^M = \lambda x \in D.[[\beta(t_i)]]^{M,g[i:=x]}$.

This allows some precise semantic definitions concerning *binding theory.*

Of course, that we consider traces as variables is coherent with the following sense we give to this notion of variable in linguistic theory:

Definition 2.1. (Variables and binders).

(1) A terminal symbol α is a variable if and only if there are at least two assignments a and a' such that $[[\alpha]]^a \neq [[\alpha]]^{a'}$.
(2) If α is a pronoun or a trace, a is an assignment defined on $i \in \mathbb{N}$, then $[[\alpha_i]]^a = a(i)$.
(3) A terminal node is a constant if and only if for any two assignments a and a', $[[\alpha]]^a = [[\alpha]]^{a'}$.
(4) An expression α is a variable binder in a language L if and only if there are trees β and assignments a such that
    (a) β is not in the domain of a,
    (b) there is a tree whose immediate constituents are α and β and this tree belongs to the domain of a.

For instance, if we go back to our previous example, it is obvious that there is an assignment f for which $[[\tau_{\text{Mary met } t_1}]]$ is not defined. It suffices to take the empty function $\emptyset$ as an assignment (or any partial function defined on $\mathbb{N}-\{1\}$) but since $[[\tau_{\text{whom Mary met } t_1}]]$ no longer depends on an assignment, it is defined even for the assignment $\emptyset$ (or any assignment not defined on 1), and therefore **whom** is a *variable binder*.

Now, to define the notions of *free* and *bound* variables poses the problem of cases where, in the same expression, some occurrences of a variable are bound and others free. We therefore define freeness and boundness for occurrences (Heim and Kratzer, 1998, p. 118):

Definition 2.2. (Freeness).

(1) Let α^n be an occurrence of a variable α in a tree β.
    (a) α^n is *free* in β if no subtree γ of β meets the following two conditions:
        (i) γ contains α^n.
        (ii) There is an assignment a such that α is not the domain of $[[.]]^a$, but γ is.
    (b) α^n is bound in β if and only if α^n is not free in β.

Of course, if there was some subtree γ of β (including β itself) containing α^n and such that γ would be defined for some assignment a which fails to give a value to α^n, that would amount to saying that γ contains a variable binder for α^n and α^n would be bound. The following theorem may be shown thanks to this definition.

Theorem 2.1. *α^n is bound in β if and only if β contains a variable binder which c-commands α^n.*

Proof (cf. Fig. 2.4): Suppose α^n is bound in β, then there is a subtree γ of β such that γ contains α^n and there is an assignment a such that α is not in the domain of $[[\ .\ ]]^a$, but γ is. Let in fact γ be the smallest subtree with these properties. γ has two daughters, δ, which contains α^n, and ϵ, which does not. Let a be an assignment such that γ but not α is in the domain of $[[\ .\ ]]^a$; because γ is the smallest subtree satisfying (i) and (ii), δ cannot satisfy them. Because (i) is true for δ, (ii) must therefore be false for it. Therefore δ is not in the domain of $[[\ .\ ]]^a$. This entails that ϵ is a variable binder, and it is precisely in a position to c-command α^n. Heim and Kratzer then defined the semantic notion of binding:

Definition 2.3. (Binding). Let β^n be a variable binder occurrence in a tree γ, and let α^m be a variable occurrence in the same tree γ, which is bound in γ. Then β^n *binds* α^m if and only if the sister node of β^n is the largest subtree of γ in which α^m is free.

We have now a good notion of a *binder*. But is that enough to deal with numerous questions like those Montague coped with, such as the interpretation of sentences with quantifiers? For a sentence like **Every boy met Mary**, the translation of which is $\forall x\ (boy(x) \Rightarrow met(x, Mary))$, a binding relation is assumed. Where does it come from? What is the binder?

As we have seen above, the translations of respectively *the boy whom Mary met* and *Mary met a boy* are rather similar: $\iota x.boy(x) \wedge met(Mary, x)$

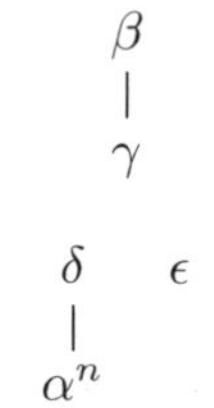

Fig. 2.4 ϵ binds α^n

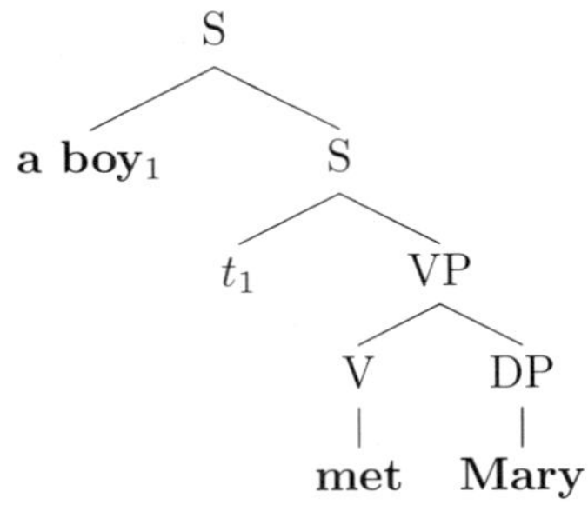

Fig. 2.5 A boy met Mary

and $\exists x.boy(x) \wedge met(Mary, x)$. For the nominal phrase, this is due to the use of the predicate modification rule, a use which is made possible by the fact that $boy(x)$ and $met(Mary, x)$ are two properties which are applied to the same individual x. In $met(Mary, x)$, the occurrence of the variable x comes from the trace in the object position of the verb **met**. From what trace comes the same variable x in the case of the sentence? Necessarily, it must come from a trace but we don't see what move could have left such a trace behind in this kind of sentence! From here comes the solution proposed very early on by May and Fiengo (May, 1977) (later on endorsed by Chomsky), according to which quantificational sentences have a logical form coming from the *quantifier-raising* transformation, which amounts to moving the quantificational phrase higher in the tree, so that it adjoins the S (or IP) node (see Fig. 2.5). Nevertheless, if we compare the tree in Fig. 2.5 with the one associated with the DP in Fig. 2.6, we may see that in the latter, **whom** is the binder, which is co-indexed with the trace, and not **boy**. It is the presence of **whom**, as a *variable binder*, which transforms $Mary\ met\ t_1$ into the closed term $\lambda x.met(Mary, x)$. If we wish to have a similar construct for the quantificational sentence, we must have a similar element in its tree, in such a way that we could have two λ-terms, $\lambda x.boy(x)$, and $\lambda x.met(Mary, x)$, and the quantifier **a** applying to both, thus giving $\exists x.boy(x) \wedge met(Mary, x)$. But what can play this role in this case?

Heim and Kratzer propose to insert a binder co-indexed with the trace for every move of a quantificational phrase, thus producing for instance the tree in Fig. 2.7. We must of course remember that we keep the same analysis of quantifiers as the one performed in Montague's frame, that is, a quantifier like **a** or **every** has the type $(e \to t) \to ((e \to t) \to t)$,

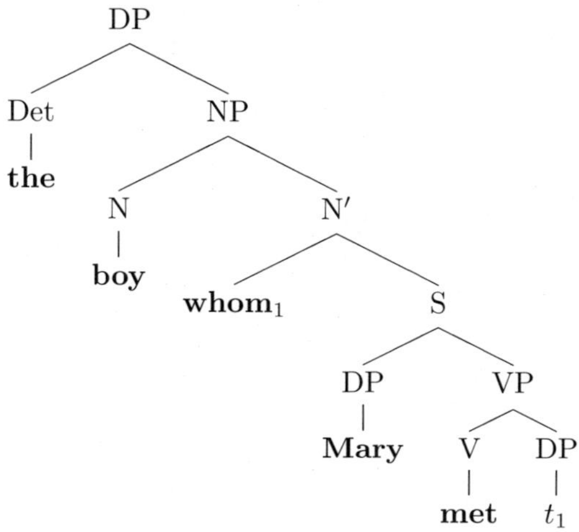

Fig. 2.6 The boy whom Mary met

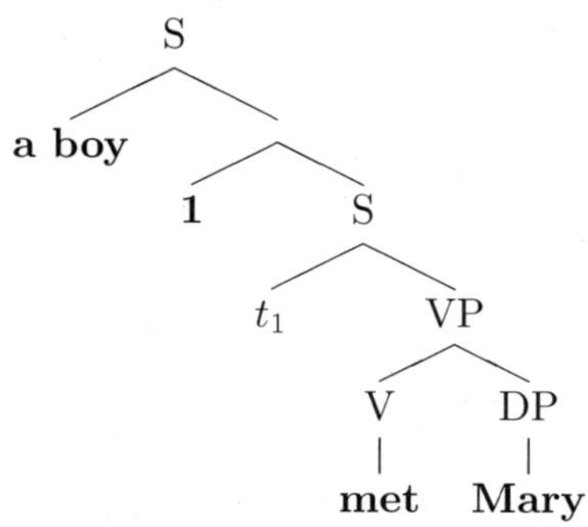

Fig. 2.7 The binder **1**

which compels us to have two $(e \rightarrow t)$ type predicates as arguments of the determiner. By taking this solution, we easily see how the meaning of such a quantificational phrase is obtained:

$I($ 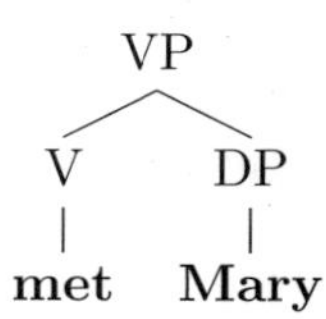 $) = \lambda y.((y\ met\ Mary) = 1)$

$I($
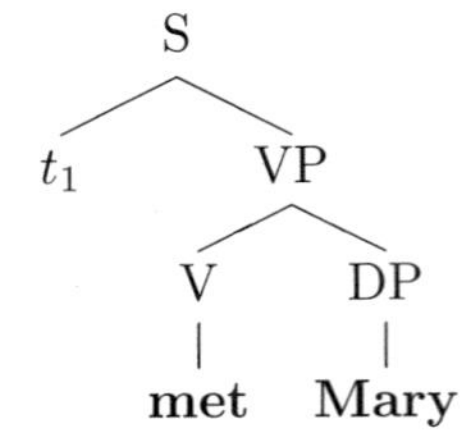

$) = (x_1\ met\ Mary) = 1$

By predicate abstraction:

$I($
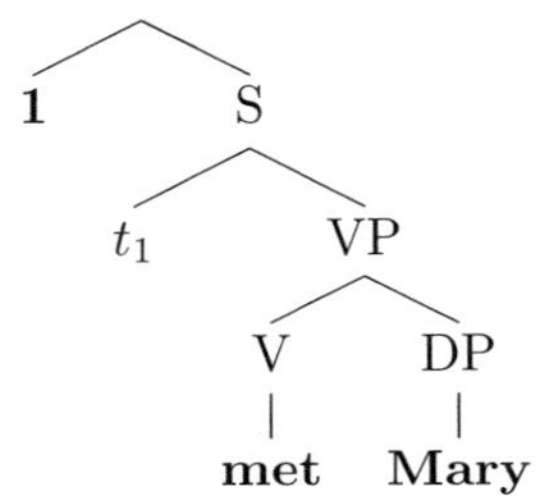

$) = \lambda x_1.((x_1\ met\ Mary) = 1)$

$I($
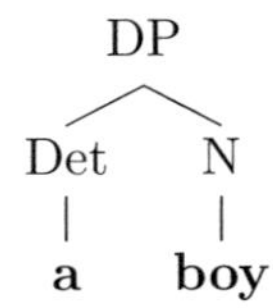

$) = \lambda g^{\langle e,t\rangle}.\exists x.(boy(x) = 1)\&(g(x) = 1)$

And finally, by functional application:

$$
\begin{aligned}
&I(\tau_{a\ boy\ met\ Mary})\\
&\quad = [\lambda g^{\langle e,t\rangle}.\exists x.(boy(x) = 1)\&(g(x) = 1)](\lambda x_1.(x_1\ met\ Mary = 1))\\
&\quad = \exists x.(boy(x) = 1)\&([\lambda x_1.(x_1\ met\ Mary) = 1)](x) = 1)\\
&\quad = \exists x.(boy(x) = 1)\&((x\ met\ Mary) = 1)
\end{aligned}
$$

Of course, predicate abstraction has been generalized:

Definition 2.4. (Predicate abstraction). If α is a branching node whose daughters are a binder which bears the index i and $\beta(t_i)$, then $[[\alpha]]^M = \lambda x \in D.[[\beta(t_i)]]^{M,g[i:=x]}$.

It is worth noticing here that such a solution has strong advantages. For instance, let us take a case where the quantificational phrase is in the object position. As we know, in this case, we have in principle a *type mismatch*:

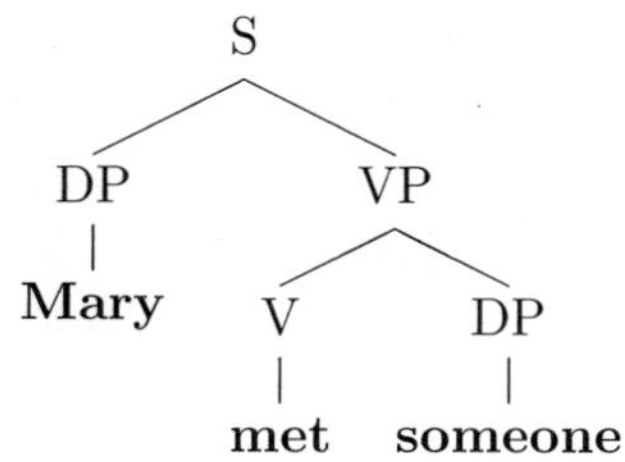

where **someone** is of type $(e \to t) \to t$ and **met** of type $e \to (e \to t)$, thus preventing any application of the functional application rule.[6] Quantifier raising mixed with this introduction of an "artificial" binder gives a correct solution, since we have now

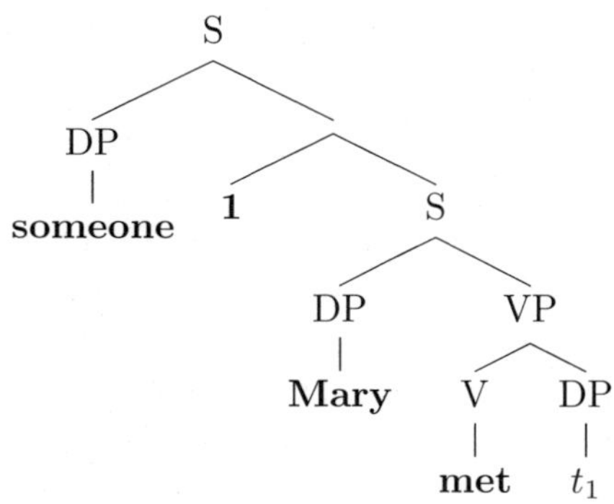

which yields the expected meaning (provided that, of course, we keep considering trace t as belonging to type e[7]).

There is one point which still deserves attention: when, exactly, may or must such moves, which have been apparent in the case of quantificational phrases, occur? Heim and Kratzer call Q-adjunction such a transformation and they say that the adjunctions may attach to nodes other than S (or IP), but also (why not?) to VPs, CPs and even DPs.

[6] Of course, functional composition could apply here, but this is another solution that we shall examine later on, in the categorial grammar framework.

[7] All the traces are not of this type; think of sentences where a constituent other than a DP is extracted, like a PP, or even a VP. But let us keep these issues aside for the time being.

It is worthwhile here to notice that Heim and Kratzer's viewpoint is based on a strict distinction of two "syntactic" levels:

- the level of surface structure (SS), which may be seen as the "syntactic" level, properly speaking,
- and the level of logical form (LF), which seems to be only legitimated by a transformation like QR, which makes a distortion with regard to the (phonetically) observed syntactic forms (since QR is only interpreted at LF and not at PF, the "phonetic form").

They show that binding may be defined at each level (syntactic binding for SS as opposed to semantic binding at LF) and they postulate, in their *binding principle*, that the two notions exactly correspond to each other.

> *Binding principle*: Let α and β be DPs, where β is not phonetically empty; then α binds β syntactically at SS if and only if α binds β semantically at LF.

Here, the notion of "semantic binding" is slightly different from the one we presented above (which only involves a variable binder and a trace or a pronoun). On the syntactic side, we know that co-indexing comes from the movement of a DP, which may be seen as previously indexed, and which leaves a trace with the same index behind it, at its original site, therefore a DP and a trace are co-indexed. That means that, syntactically, when looking at a quantificational sentence like **Mary met a man**, the DP **a man** and the trace it leaves are co-indexed: there is no variable binder at SS, and the representation is

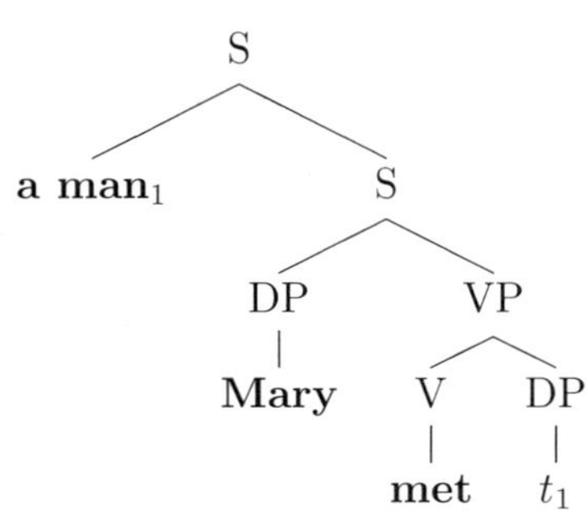

When passing to the logical form, we may of course keep this co-indexation, even if we add a supplementary node for the binder.

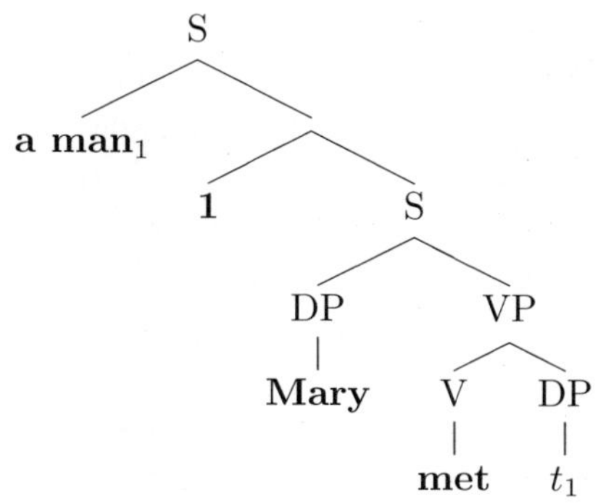

We may say in this configuration that the DP α (here **a man**) binds its trace t, even if it is, in fact, the binder **1** which does the job. This *abus de langage* is more clearly motivated if we now consider sentences with a reflexive, like **A man defended himself**, where the usual binding principles impose that **himself** be bound by **a man**. In this case, the logical form is

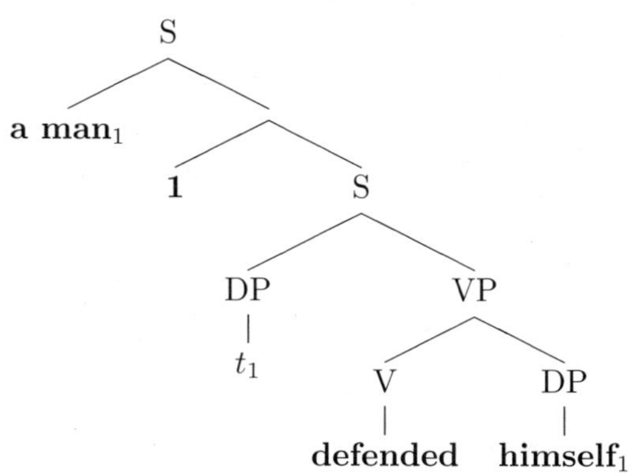

where it appears that **a man** and **himself** are bound by the same variable binder, which is the only way to give the expected reading. We can then propose a new definition of semantic binding between two DPs (instead of a DP and its trace like previously):

Definition 2.5. (Semantic binding). A DP α semantically binds a DP β if and only if β and the trace of α are semantically bound by the same variable binder.

This explains why in the principle above, we excluded traces for values of β. The need for QR would then be a consequence of this principle: the quantifier could bind a β only by putting the quantificational phrase α in a position where it c-commands β, therefore by moving it upward.

But what if we tend to dispense with such a distinction (like in the most modern formulations of Chomsky's Minimalist Program)? If the LF level is created only for the sake of some special transformations like QR, shouldn't we think it is *ad hoc*? Could it be possible to avoid the redundancy involved in *two* notions of binding, which cover the same cases in the end, and to keep only one universal notion, which could be based on the familiar use of binding in logic?

We shall tend in Section 3.4 towards a *monostratal* view, according to which a single level of syntactic analysis (and not three like in the former theory of deep structure (DS)/surface structure (SS) and logical form (LF)) is opposed to semantic evaluation.

Moreover, we arrive now at a stage where analogies with logical proofs may show advantages. We will therefore at first make apparent this analogy and provide more logical formulations of the theory, before giving solutions which help us to get rid of these distinct levels.

2.5.4 *Towards a proof-theoretic account of binding*

Suppose that, instead of inserting a new node **1** at the specifier node of S, we simply create a unary branch:

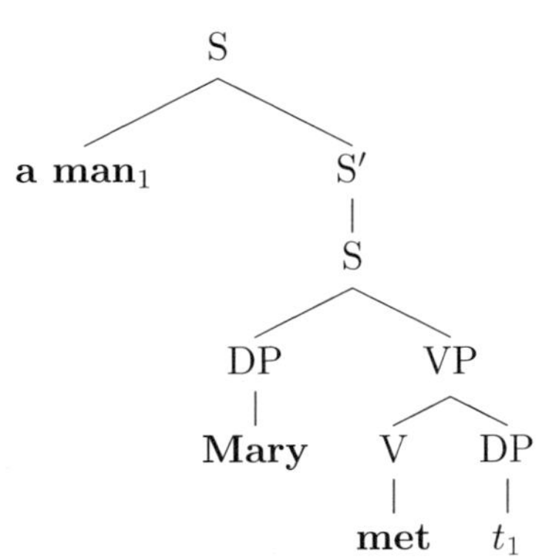

and suppose we rotate this tree by 180°, transforming it into a proof tree:

$$\dfrac{\dfrac{a\ man_1}{DP} \qquad \dfrac{\dfrac{\dfrac{Mary}{DP} \quad \dfrac{\dfrac{met}{V} \quad \dfrac{t_1}{DP}}{VP}}{S}}{S'}}{S}$$

Suppose then grammatical rules are expressed as deduction rules:

- Lexical ones:

$$\dfrac{Mary}{DP} \qquad \dfrac{met}{V} \qquad \dfrac{a\ man}{DP}$$

- Grammar rules:

$$\dfrac{V\ DP}{VP} \qquad \dfrac{DP\ VP}{S} \qquad \dfrac{DP\ S'}{S}$$

traces are seen as *hypotheses*, and there is a *discharge* rule which is semantically interpreted as the binding of the variable associated with some hypothesis (see Section 2.4):

$$\dfrac{\begin{array}{cc} A & [DP : x] \\ & \vdots \\ & B : \gamma \end{array}}{DP \multimap B : \lambda x.\gamma}$$

Then our inverted tree may be seen as a proof using these rules:

$$\dfrac{\dfrac{a\ man_1}{DP}\,Lex \qquad \dfrac{\dfrac{\dfrac{Mary}{DP}\,Lex \quad \dfrac{\dfrac{met}{V}\,Lex \quad \dfrac{t_1 : x}{DP}\,Hyp}{VP}\,R1}{S}\,R2}{S'}\,[Discharging]}{S}\,R3$$

where, in fact, $S' = DP \multimap S$. We use $\multimap$ in that system since, as stated in Section 2.4, exactly one DP hypothesis must be discharged. This opens the field to the logical approach to syntax and semantics, that we shall study in the following sections. One such approach is provided by categorial grammar, a formalism introduced very early on by Polish logicians after Husserl's philosophical work (Husserl, 1958) and Lesniewski's proposals (Lesniewski, 1929) with regard to the notion of semantic type.

By doing so, we obtain the binding principle for free since the surface structure and the logical form coincide by construction, and binding is simply associated with the use of a "discharge" rule, that is, what Moortgat calls a rule of *hypothetical reasoning.*

Modulo some generalization with regard to the strict consumption discipline, we could arrive at a proposition like

Definition 2.6. (Semantic binding (2)). A DP α semantically binds a DP β semantically represented by a variable x if and only if α applies to a λ-term $\lambda x.\Phi$, reflecting a proof, the last step of which is an introduction of the connective $\multimap$.

Chapter 3

Derivationalism

3.1 Introduction

Derivationalism is the linguistic perspective according to which syntactic units are obtained step by step simply using a rule system, with no step referring to some extra device, like the representation of the tree currently built. It is opposed to *representationalism*, a perspective which subordinates some steps to the checking of constraints on representations. It is sometimes said that, for instance, the passage, in Chomskyan theory, from the government and binding theory to the Minimalist Program is a shift from a representationalist to a derivationalist theory. In the minimalist theory, there are only two rules (called "generalized transformations" by Chomsky), and a sentence is generated by using them along a derivation which starts from an *enumeration* (of lexical items, represented by lists of features) and bifurcates at some point (called "spell out") towards a phonological form (PF) on one side and a logical form (LF) on the other side. Such a conception has its roots in former derivational theories which were invented long before. It is the reason why we begin this chapter by recalling elementary points on categorial grammars, which date from the 1930s. It is an opportunity to present the Lambek calculus in very simple terms (almost the same terms as those employed in Lambek's paper of 1958). We shall go back to this calculus after an excursion into logical systems in the next part of this book. The third section of this chapter addresses Stabler's formalization of the main minimalist ideas and thus shows how to give Chomskyan theory its full derivational meaning. A tentative attempt will be made further to bring the two frameworks (categorial grammar and minimalist grammar) closer.

This first approach of a deductive system (Lambek calculus) helps us to perceive the concrete meaning we can give to the view sketched at the end of the previous chapter in that it provides a way to answer the question: How to realize binding with no trace and no binder? If this is possible in the frame of a calculus like Lambek, how is it possible to export the solution to other systems, which are closer to the generative analyses most linguists are familiar with? Our first study of Stabler's minimalist grammars will show a first solution which is based on a particular procedure invented by Robin Cooper in the 1980s and named *Cooper storage.*

3.2 Categorial Grammars

Categorial grammar has been developed since the 1930s, starting from foundational works in logic and philosophy, which led to a method invented by the Polish logician Kasimir Ajdukiewicz (1935) for testing the syntactic well-formedness of a language expression. The method was called *syntactic connexity.* Bar-Hillel in the 1950s was the first to propose serious applications to the analysis of natural language, in a paper first published in 1953 in the volume *Language and Information* (Bar-Hillel, 1964). His main idea was to have "a simple quasi-arithmetic rule in order to calculate the well-formedness of any given linguistic sequence". For instance, the English sentence *Poor John sleeps* can be analyzed by assigning its words to the following categories:

poor	:	n/[n]
John	:	n
sleeps	:	s/(n)

In this notation, parentheses and brackets are used for directionality: a word of category A/[B] is expecting a word of category B on its right, whereas a word of category A/(B) is expecting a word of category B on its left. The reduction rules could be written as

$$A/[B]B \to A \tag{3.1}$$

$$BA/(B) \to A \tag{3.2}$$

In a more modern setting, two slashes are used, one which expects something on the right (/) and the other on the left (\). In this notation (due to Lambek), the resulting category (here A) is always *over* the argument category.[1]

The sentence *Poor John sleeps* gets thus what Bar-Hillel called an *index*, that is, the list of categories associated with its words:

$$\frac{n}{[n]} \; n \; \frac{s}{(n)} \tag{3.3}$$

A first reduction may then apply to the subsequence $n/[n]\ n$, giving the index's *first derivative*

$$n \; \frac{s}{(n)} \tag{3.4}$$

and finally a second reduction leads to the "second derivative"

$$s \tag{3.5}$$

thus ensuring that the sequence is well formed.

Let us call *functional application* rules the following two rules (in the Lambek notation):

FA forward:	$A/B\ B \to A$
FA backward:	$B\ B\backslash A \to A$

In the literature, such systems are named *AB-grammars* (for "Ajdukiewicz–Bar-Hillel"). Formally, an AB-grammar G is given by a vocabulary V and an assignment L of categories to the elements of V. G is said to be *rigid* if L is a function from V to the set of categories Cat. A sequence $\sigma \in V^*$[2] ($\sigma = \sigma_1, \sigma_2, \ldots, \sigma_n$ where $\forall i, 1 \leq i \leq n, \sigma_i \in V$) is said to be accepted by G if and only if the list of categories $L(\sigma) = L(\sigma_1), L(\sigma_2), \ldots, L(\sigma_n)$ reduces to a distinguished type s. If the grammar is not rigid (a case where L is a function from V to $\wp(Cat) - \{\emptyset\}$), σ is said to be accepted if and only if there is a list of categories belonging to $L(\sigma)$ which reduces to s. A language

[1]It must be noticed that other similar frameworks, like Steedman's, use a different convention; they still assume A/B expects B on the right, but the expectation on the left is expressed by A\B. In their case, the resulting category is always the first in a left-to-right reading of the sign, and the form of the slash simply indicates the orientation, something reminiscent of Bar-Hillel's notation, assuming that "/" is "/[]" and "\" is "/()".

[2]The "$\star$" here is the "Kleene star". Let us recall that, if E is a set provided with an associative law of composition "$\frown$", then $E^\star$ denotes the set of all the finite words on E, that is, all the sequences $a_1 \frown \ldots \frown a_n$ where $\forall i = 1 \ldots n, a_i \in E$ and where n may be 0, in which case the word is empty — an *empty word* — and is the neutral element of "$\frown$".

L is said to be accepted by G if and only if it is the set of all sequences in V^* accepted by G.

AB-grammars have been intensively studied and proved to be equivalent to context-free grammars, in the sense that for each language generated by a context-free grammar CG there exists an AB-grammar G which accepts it (with similar analysis trees), and conversely, for each language accepted by an AB-grammar G, there is a context-free grammar CG which generates it (with similar analysis trees, too).

Many considerations have led to a necessary enrichment of this "calculus". If we wish it may provide interesting (and not *ad hoc*) analyses of language expressions. For instance, there is the question of how to give an elegant description of phrases where some constituent has been displaced, like in Example 3.1.

Example 3.1. Mary read the book that she bought

We may assume here a quite intuitive assignment to the words of this sentence[3]:

read, bought	:	$(np\backslash s)/np$
Mary	:	np
she	:	$s/(np\backslash s)$
book	:	n
the	:	np/n
that	:	$(n\backslash n)/(s/np)$

Let us explain some of the assigned categories. *That* will give a noun modifier provided that it finds a sentence which lacks a nominal phrase on its right. The pronoun *she* receives a complex category (and not a simple one, like np) which takes into account the fact that it can occur only as a subject and not as an object. In this form, it says that *she* will give a sentence when provided with a sentence which lacks a subject.

If we display the list of categories which must reduce to the distinguished s, we have

$$np \quad (np\backslash s)/np \quad np/n \quad n \quad (n\backslash n)/(s/np) \quad s/(np\backslash s) \quad (np\backslash s)/np \tag{3.6}$$

[3]Using now the Lambek notation, where parentheses have their usual meaning, indicating in which order the arguments must be found.

and this sequence may not reduce since there is no np on the extreme right which would satisfy the verb expectation. The problem is solved *if other rules are added to the calculus*, like

Functional composition:

$$\begin{array}{ll} \textbf{FC forward}: & A/B\ B/C \rightarrow A/C \\ \textbf{FC backward}: & C\backslash B\ B\backslash A \rightarrow C\backslash A \end{array}$$

Actually, this leads to the history of the following reduction steps (with the output of the reduction below the horizontal line):

(1) FC forward:

$$np \quad (np\backslash s)/np \quad np/n \quad n \quad (n\backslash n)/(s/np) \quad \frac{s/(np\backslash s) \quad (np\backslash s)/np}{s/np}$$

(2) FA forward:

$$np \quad (np\backslash s)/np \quad np/n \quad n \quad \frac{(n\backslash n)/(s/np) \quad s/np}{n\backslash n}$$

(3) FA backward:

$$np \quad (np\backslash s)/np \quad np/n \quad \frac{n \quad n\backslash n}{n}$$

(4) FA forward:

$$np \quad (np\backslash s)/np \quad \frac{np/n \quad n}{np}$$

(5) FA forward:

$$np \quad \frac{(np\backslash s)/np \quad np}{np\backslash s}$$

(6) FA backward:

$$\frac{np \quad np\backslash s}{s}$$

What is surprising here is that functional composition has displaced a "hole" (the missing np) from the right periphery to the left until it may be absorbed by a category which precisely *licenses* this hole (here the complementizer *that*). Such missing arguments correspond to what are traces in the Chomskyan theory, but instead of having a moving constituent *np* leaving a trace behind it, it is the "hole", therefore the trace, which moves, and constituents remain in their own place.[4] For that reason, such a framework may be viewed as *displacement-free* (and with no empty categories like traces).

Moreover, this frame allows other rules, which will become valuable, like *type raising*. In a directional calculus, there are two such rules:

Type raising:

$$\begin{array}{ll} \textbf{TR forward}: & A \rightarrow B/(A\backslash B) \\ \textbf{TR backward}: & A \rightarrow (B/A)\backslash B \end{array}$$

They simply express the fact that any linguistic object of type A may be seen either as an object expecting on its right any object expecting on its left an object of type A, then getting the dominant type of the right object, or contrarily, as an object expecting on its left any object expecting on its right an object of type A, then getting the dominant type of the left object. This may be represented by Fig. 3.1. We may notice at this step that, as Lambek showed in his 1958 paper (see next section), those lifted types may be exemplified by means of the pronouns. Actually, our first intuition concerning the English pronouns *he* (or *she*, etc.) and *him* (or *her*, etc.) consists in providing them with an n type, like proper names (or *np* when differentiation is made between the two, something which is not made in the earlier stage of Lambek grammars), but then there would be neither any difference between

$$\frac{A \qquad A\backslash B}{B} \qquad\qquad \frac{B/(A\backslash B) \qquad A\backslash B}{B}$$

$$\frac{B/A \qquad A}{B} \qquad\qquad \frac{B/A \qquad (B/A)\backslash B}{B}$$

Fig. 3.1 Type lifting

[4]This analysis of dependencies between *fillers* and *gaps* across arbitrary intrasentential distances has been adopted in other frameworks, like HPSG (*head-driven phrase structure grammar*) (see Pollard and Sag, 1994).

- *He likes Mary*
- and **Mary likes he*

nor between

- *Mary likes him*
- and **Him likes Mary*

since all these expressions would reduce to s. What type then should we give to the pronouns *he* and *him*? Let us consider the sentence. *He sleeps.* The other assignment that would permit reduction of this sentence to s would consist in giving *he* the type $s/(n\backslash s)$, since in this case, we should have

$$\begin{array}{ll} he & sleeps \\ s/(n\backslash s) & n\backslash s \\ s & \end{array}$$

In a similar way, we may give *him* the type $(s/n)\backslash s$. We may notice that, in this case, we get the following reduction:

$$Mary\ likes\ him \tag{3.7}$$

$$n \quad ((n\backslash s)/n \quad (s/n)\backslash s) \tag{3.8}$$

$$(n \quad n\backslash(s/n)) \quad (s/n)\backslash s \qquad \text{(Assoc)}$$

$$s/n \quad (s/n)\backslash s \qquad \text{(FA backward)}$$

$$s \qquad \text{(FA backward)}$$

provided that we allow ourselves to use the rule according to which parentheses may be changed around the types in expressions like $n\backslash(s/n)$, that is, more generally speaking, the rule

$$(Assoc) \quad X\backslash(Y/Z) \leftrightarrow (X\backslash Y)/Z$$

But if now we want a derivation for

Example 3.2. He likes him

it seems we are stuck, unless we use the FC rules. In this case, we obtain the following reduction tree:

$$\dfrac{s/(n\backslash s) \qquad \dfrac{n\backslash(s/n) \quad (s/n)\backslash s}{n\backslash s}\ FC\ backward}{s}\ FA\ forward$$

Let us finally notice that, by TR, we have

$$n \rightarrow s/(n\backslash s) \qquad n \rightarrow (s/n)\backslash s$$

Therefore, any noun may be type-raised to the type of a pronoun, but

$$s/(n\backslash s) \not\rightarrow n \qquad (s/n)\backslash s \not\rightarrow n$$

That is, a pronoun cannot be type-lowered to the type of a noun, so that by assigning *he* (resp. *him*) the lifted type $s/(n\backslash s)$ (resp. $(s/n)\backslash s$) *in the lexicon*, we shall not get a lowering of this type to n, whereas by assigning type n to any noun, we shall be able to raise it in case of need.

But the time has come to bring some order into this set of rules.

3.3 The (Pure) Lambek Calculus

3.3.1 *The mathematics of sentence structure*

The 1958 paper by Jim Lambek ("The Mathematics of Sentence Structure") was the first attempt at exhibiting, under the notion of categorial grammar, a real logical system. The calculus proposed by Lambek is "logically complete" in two senses:

- It is possible to give it a semantics with regard to which it is complete (that is, the system permits the enumeration of all the "truths" according to these semantics).
- It provides us with a complete system of rules: to add other rules would make it degenerate.

Lambek's goal was really

> to obtain an effective rule (or algorithm) for distinguishing sentences from nonsentences, which works not only for the formal languages of interest to the mathematical logician, but also for natural languages such as English, or at least for fragments of such languages (Lambek, 1958).

Lambek rests on Bar-Hillel's idea, according to which grammatical categories may contain "/" as well as "\". Moreover, he tends to reject the notion of rigid assignment (only one type per word or expression), but the point is that the various types an expression may have must all be deduced in a logical way from some more elementary type.

3.3.2 *A categorical system*

Lambek shows that all the rule schemes we have seen above may be deduced as mere theorems inside a formal system comprising axioms and a few rules, where x, y, z denote syntactic types, and where we admit that if an expression A is of type x, an expression B is of type y, then the expression AB (obtained by simple concatenation) is of type xy, or $x \bullet y$ (which assumes that the calculus contains not only two operators, / and \, but also a third one, $\bullet$, that is a *non-commutative product* between types). Moreover, we write "$x \to y$"[5] to say that an expression of type x is also of type y.

(a)	$x \to x$		
(b)	$(x \bullet y) \bullet z \to x \bullet (y \bullet z)$	(b′)	$x \bullet (y \bullet z) \to (x \bullet y) \bullet z$
(c)	if $x \bullet y \to z$ then $x \to z/y$	(c′)	if $x \bullet y \to z$ then $y \to x\backslash z$
(d)	if $x \to z/y$ then $x \bullet y \to z$	(d′)	if $y \to x\backslash z$ then $x \bullet y \to z$
(e)	if $x \to y$ and $y \to z$ then $x \to z$		

The formulae (a), (b) and (b′) are obviously axioms: (a) expresses identity, (b) and (b′) express *associativity.* (c), (c′), (d), (d′) and (e) give rules: they show how new arrows may be added to the system. Rule (e) expresses the *transitivity* of the arrow. Readers aware of category theory (in the mathematical sense) recognize two axioms which are needed in order to get a category: identity and transitivity. In this case, formulae are interpreted as *objects* and arrows as *morphisms.* This kind of category is called a *deductive system.*

We shall go back to the Lambek calculus after we have studied deductive systems in general in more detail.

3.4 Minimalist Grammars

3.4.1 *Minimalist principles*

Minimalist grammars were proposed for the first time in 1996 by Stabler (1997). They are based on the main principles of Chomsky's Minimalist Pro-

[5]This arrow is the reduction arrow, something which will be later denoted by the turnstile "$\vdash$". It is not the arrow found in Section 2.4, which is better read as the implicative connective.

gram. These principles may be summarized according to Chomsky (1996) in the following way:

- A language L is taken to be a way of computing expressions that is a recursive definition of a set EXP.
- UG (universal grammar) postulates that
  - — the language faculty provides a set of features;
  - — the language faculty provides principles for assembling features into lexical items;
  - — the language faculty provides operations that apply successively to form syntactic objects of greater complexity ("call them CHL, the computational system for human language");
  - — the language faculty generates expressions $\langle PF, LF\rangle$ (phonological form, logical form) and thus correlates sound and meaning (which provide interface levels);
  - — a computation *converges at an interface level* if it is legible there, consisting solely of elements that provide instructions to the external systems; otherwise, it crashes at this interface. The computation converges if it converges at both interfaces.

Chomsky (2000) adds: "one might construe L literally as a step-by-step procedure for constructing EXPs, suggesting that this is how things really work" (that is, as a real property of the brain).

Stabler's minimalist grammars exactly try to follow these principles by introducing a new algebraic definition of grammars, mainly based on

- a general conception of lexical items as consisting of *list of features*;
- a very general notion of feature checking;
- two operations corresponding to what Chomsky called generalized transformations: *Merge* and *Move*.

All these notions may be defined only if we start from some elementary algebraic notions like *trees* and *precedence* and *domination* relations between nodes in a tree.

Let us then introduce some definitions.

We assume a finite tree given as a set of nodes endowed with an *immediate dominance* relation, $(N, \triangleleft)$. $x \triangleleft y$ means "x is a parent of y". The transitive closure $\triangleleft^*$ is the *ancestor* relation (assuming that every x is an ancestor of itself!). A second relation, $\prec$, is added to that structure, called *immediate linear precedence*, the transitive non-reflexive closure of which is $\prec^+$ (linear precedence) with the following properties:

- given two nodes x and y, either one is the ancestor of the other or one precedes the other;
- if a node x linearly precedes a node y and if it is an ancestor of a node w, then w linearly precedes all the nodes z dominated by y.

The concept of *projection* plays a central role in generative grammar: it has been introduced since the early days of X-bar theory. Let us recall that, according to this theory, each syntactic domain (nominal, verbal, prepositional or adjectival) has a similar architecture, based on several layers, distinguished by their number of bars (or primes) (X, X′, X″, ...). The lowest layer, X, is the *head*, while the highest one (generally supposed to be X″, or $\mathrm{X}^{\mathrm{max}}$) is called the *maximal projection* of X. Intermediary levels are intermediate projections. In the Minimalist Program, Chomsky sees Merge, the first generalized transformation, as an operation which combines two syntactic objects A and B, forming $\{A, B\}$, with one of the two items being the head. It is this item which gives its label to the final object: if $\{A, B\}$ has A (resp. B) for its head, it is an A (resp. B) too, or it belongs to the A-domain (resp. B-domain).

For this reason, Stabler introduces a third relation between nodes, $<$, such that $x < y$ means x *immediately projects over* y, the transitive but not reflexive closure $<^+$ being interpreted as x *projects over* y. This relation is introduced in such a way that when a node x is a parent of a node y, there exists a node u which immediately projects over every node z dominated by x. In other terms, when several nodes have the same parent, there is necessarily one of them which projects over the others.

These notions allow the definitions of

- the notion of head. x is a head of y if and only if
  - — either y is a leaf and $x = y$,
  - — or there exists a node z, the head of which is x which projects over all its sister nodes;
- the notion of maximal projection. The maximal projection of x is the smallest node for the relation $\triangleleft^*$ with head x;

- the notion of specifier. x is a specifier of the head y if and only if
  - — x is a maximal projection,
  - — x's parent has y for its head,
  - — $x \prec^* y$;
- the notion of complement. x is a complement of the head y if and only if
  - — x is a maximal projection,
  - — x's parent has y for its head,
  - — $y \prec^* x$.

The tree in Fig. 3.2 exemplifies these notions. In this tree, nodes are denoted by words on $\mathbb{N}$ according to an obvious convention: on the same level, labels are built by suffixing respectively $1, 2, \ldots$ to the word labelling the immediate parent. The root is labelled by ϵ (the empty word). 2 immediately projects over 1. 2.1 projects over 2.2 and all its daughters so that *house* at the node 2.1 is the head of this construction. 2.2.1 projects over 2.2.2 and its daughters: it is the head of the node 2.2 (that is, the prepositional phrase). The maximal projection of 2.1 is ϵ, and the maximal

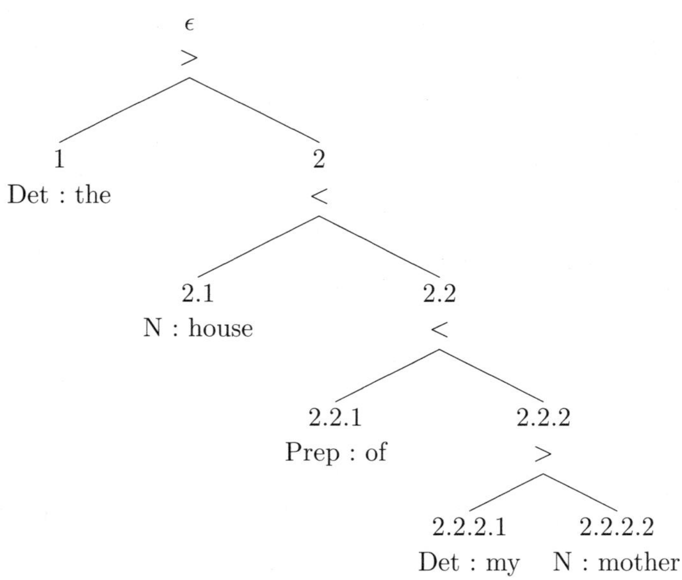

Fig. 3.2 An ordered tree with projection

projection of 2.2.1 is 2.2. 1 is the specifier of the head 2.1, and 2.2 is the complement of this same head, like 2.2.2 is the complement of the head 2.2.1.

3.4.2 *Features*

There are three kinds of features:

- phonetic features (which are pronounced according to the acoustic perceptive system), which constitute the set P;
- semantic features (which feed the conceptual system), which constitute the set I;
- formal features, which form the set C.

In the Minimalist Program, *formal features* are active in the sense that their role is to trigger specific operations.

C is partitioned into four subsets:

- basic features,
- selectors,
- licensors,
- licensees.

Basic features are basic syntactic categories: $\mathtt{v}, \mathtt{n}, \mathtt{d}, \mathtt{p}, \mathtt{c}, \mathtt{t}$, etc. *Selectors* express a demand for an expression of some definite syntactic category (that is, some basic feature), and they correspond to what are often called *c-selectors.* If α is a basic feature, then there is in the grammar a selector $= \alpha$. *Licensors* license a property to an expression which is in some specific relation (*specifier-head*). For instance, if an expression in a head position has the feature $+\mathtt{case}$, it will assign a case (nominative, accusative or oblique) to an expression which comes to occupy a specifier position with regard to it. In most instances, this assignment will trigger a Move operation. Moreover, the expression which is assigned the feature $+\mathtt{f}$ has the complementary feature $-\mathtt{f}$, which expresses the requirement for a value for the feature $\mathtt{f}$. For instance, in generative grammar, DP must receive a (structural) case. They are therefore endowed with the feature $-\mathtt{case}$.

As we may see, derivations in minimalist grammars are driven by a mechanism of resource consumption, exactly like in the case of categorial grammars. Each $= \mathtt{x}$ must match an $\mathtt{x}$ and each $+\mathtt{f}$ must match a $-\mathtt{f}$.

The necessity for matching a selector and a basic feature triggers Merge, whereas the necessity for matching a $+\texttt{f}$ and a $-\texttt{f}$ triggers Move.

Stabler (1997) distinguishes between weak and strong licensors, respectively denoted by $+\texttt{x}$ and $+\texttt{X}$. The former attract only the non-phonetic features and the later attract *all* the features of the subtree displaced during Move. This distinction allows us to differentiate between covert and overt movement. The variation weak/strong for a feature across various grammars may be seen as a *parameter*. Nevertheless, recent views by Chomsky reject this distinction because it is said to introduce a notion of *feature of a feature* which is not desirable at this stage.

We shall add selectors $= \texttt{C}$ and $\texttt{C} =$ for each basic category $\texttt{c}$ in order to implement *head movement* with right and left adjunction. When Merge will be performed between an expression t_1 with $= \texttt{C}$ (resp. $\texttt{C} =$) as its first feature and an expression t_2 with $\texttt{c}$, the phonetic feature of t_2 is concatenated with the phonetic feature of t_1 on the right (resp. on the left).

3.4.3 *Minimalist grammars*

A *minimalist grammar* is defined by a 4-tuple $\langle V, C, Lex, \Phi\rangle$ where

- V is a finite set of non-formal features,
- C is a finite set of formal features,
- Lex is a set of expressions built from V and C (the "lexicon"),
- Φ is a set of generative functions, which are partial functions from tuples of expressions to expressions.

Expressions are in fact, in that presentation, ordered, finite and binary trees with projection, with labels only on the leaves.

The language defined by such a grammar is the closure of Lex under the generative functions in Φ.

Labels on the leaves all respect the syntax given by the regular expression

$$selector^*(licensor)selector^*(base)licensee^*P^*I^* \qquad (3.9)$$

3.4.4 *Merge*

Let t be an expression of head $H(t)$. Let $t[\texttt{f}]$ be the expression obtained by prefixing the feature $\texttt{f}$ to the sequence of features of $H(t)$. Let $t(/\alpha/)$ be an expression of t such that the concatenation of its phonetic features is α, let $phon(t)$ be the phonetic feature of $H(t)$, let $+$ be the concatenation of phonetic features, then for all expressions t_1, t_2 and for all $\mathtt{c} \in Base$,

- $merge(t_1[=\mathtt{c}], t_2[\mathtt{c}]) = [^< t_1, t_2]$ if $t_1[=\mathtt{c}] \in Lex$
- $merge(t_1[=\mathtt{c}], t_2[\mathtt{c}]) = [^> t_2, t_1]$ if $t_1[=\mathtt{c}] \notin Lex$
- $merge(t_1[=\mathtt{C}], t_2[\mathtt{c}]) = [^< t_1(phon(t_1) + phon(t_2)), t_2(/\epsilon/)]$ if $t_1[=\mathtt{C}] \in Lex$
- $merge(t_1[\mathtt{C}=], t_2[\mathtt{c}]) = [^< t_1(phon(t_2) + phon(t_1)), t_2(/\epsilon/)]$ if $t_1[\mathtt{C}=] \in Lex$

The two first clauses of this definition say that when a lexical head selects something (t_2), this something becomes a *complement*: in the tree, it comes to the right of the head. But when t_2 is selected by a non-lexical unit, it becomes a *specifier*. Of course t_1 is still the head. The two last clauses rule define head movement: the first defines head movement with right adjunction, and the second with left adjunction.

3.4.5 *Move*

Let t^* be an expression which is a maximal projection of a head $H(t)$. Let $t\{t_1/t_2\}$ be the result of the substitution of t_2 to t_1 inside t^*. For all expressions $t_1[+\mathtt{f}]$ which contain one and only one maximal subtree $t_2[-\mathtt{f}]^*$, we define

- $move(t_1[+\mathtt{f}]) = [^> t_2^*, t_1\{t_2[-\mathtt{f}]^*/\lambda\}]$

where λ is the empty tree.

Using weak and strong licensors, we get also

- $move(t_1[+\mathtt{F}]) = [^> t_2^*, t_1\{t_2[-\mathtt{f}]^*/\lambda\}]$
- $move(t_1[+\mathtt{f}]) = [^> t_2(/\epsilon/)^*, t_1\{t_2[-\mathtt{f}]^*/t_2'\}]$ where t_2' is the t_2^* tree from which all the non-phonetic features have been deleted.

Example: Let *Lex* be the following list of expressions:

$=\mathtt{n}\ \mathtt{d}\ -\mathtt{case}\ -\mathtt{wh}$ */what/*	$\mathtt{n}$ */book/*	$\mathtt{d}\ -\mathtt{case}$ */you/*
$=\mathtt{c}\ =\mathtt{d}\ \mathtt{v}$ */think/*	$=\mathtt{t}\ \mathtt{c}$ */that/*	$\mathtt{d}\ -\mathtt{case}$ */Mary/*
$=\mathtt{d}\ +\mathtt{CASE}\ =\mathtt{d}\ \mathtt{v}$ */read/*	$=\mathtt{v}\ +\mathtt{CASE}\ \mathtt{t}$ */do/*	$=\mathtt{T}\ +\mathtt{WH}\ \mathtt{c}$
$=\mathtt{v}\ +\mathtt{CASE}\ \mathtt{t}$		

The possible steps of a derivation are

(1) Merge $=$`n d` $-$ `case` $-$ `wh` */what/* and `n` */book/*, resulting in

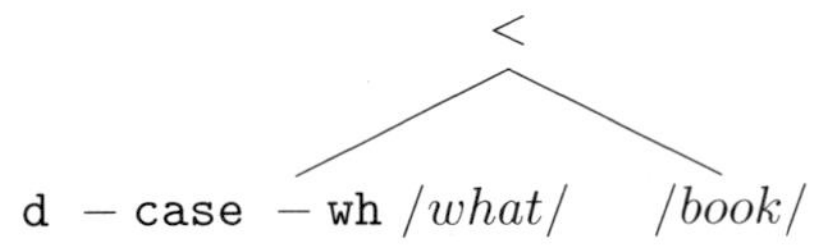

Comment: The item containing the selector $=$`n` belongs to the lexicon, therefore the second item is a complement and the first is the head.

(2) Merge $=$`d` $+$`CASE` $=$`d v` */read/*, resulting in

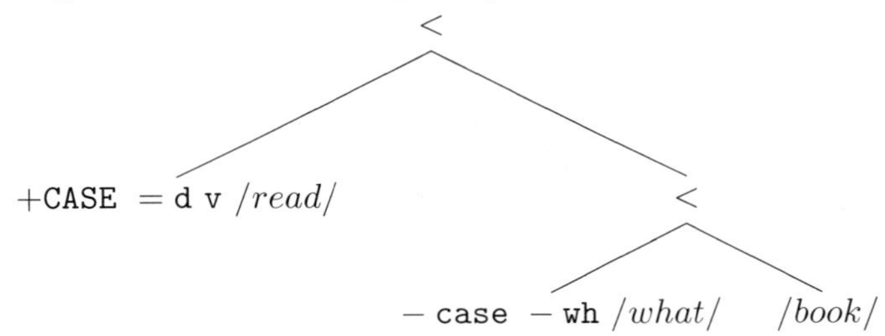

Comment: The item containing the selector $=$`d` is also lexical, therefore the expression coming from the previous step is also a complement, and the expression the selector belongs to is the new head.

(3) Move (since $+$`CASE` is the first feature of the head). The maximal subtree with a $-$`case` feature is the subtree of the head */what/*. Since `CASE` is strong, we get

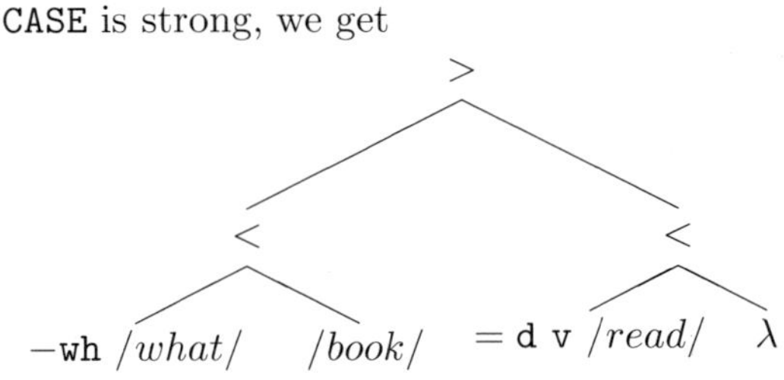

(4) The head of the previous tree has the feature $=$`d` as the first feature of its list, therefore we can merge it (as a non-lexical expression) with `d` $-$ `case` */Mary/*; the latter will occupy a *specifier*

position:

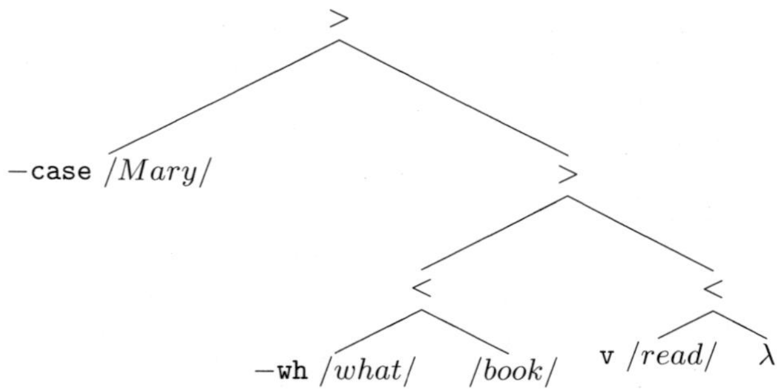

(5) Merge with = v + CASE t:

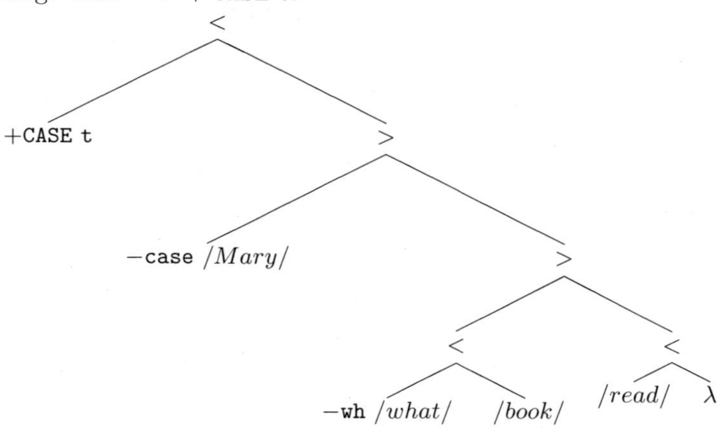

(6) Move *Mary*:

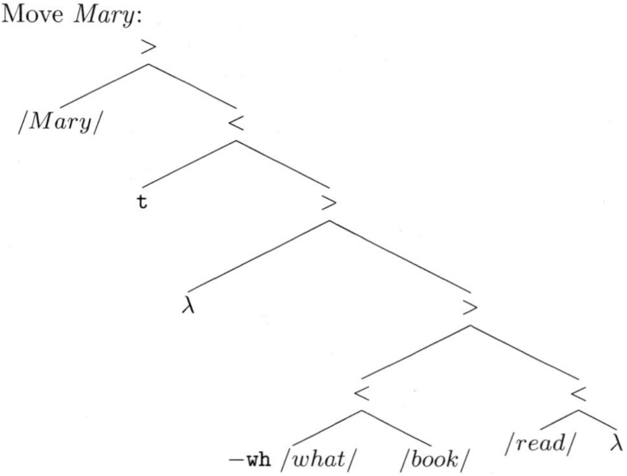

(7) Merge with = t c /*that*/:

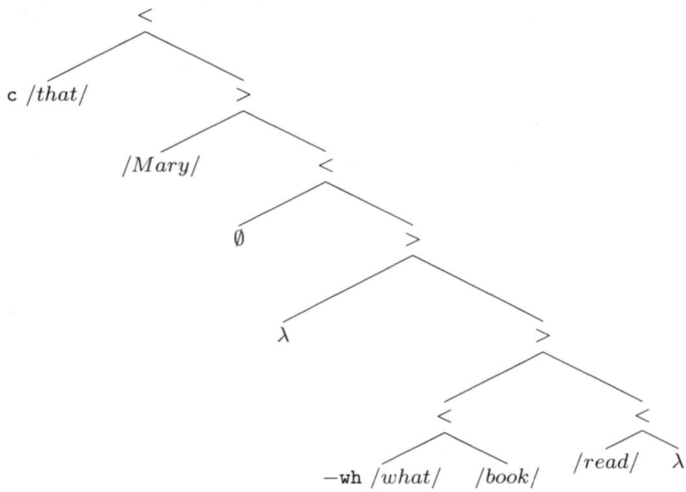

(8) Merge with = c = d v /*think*/:

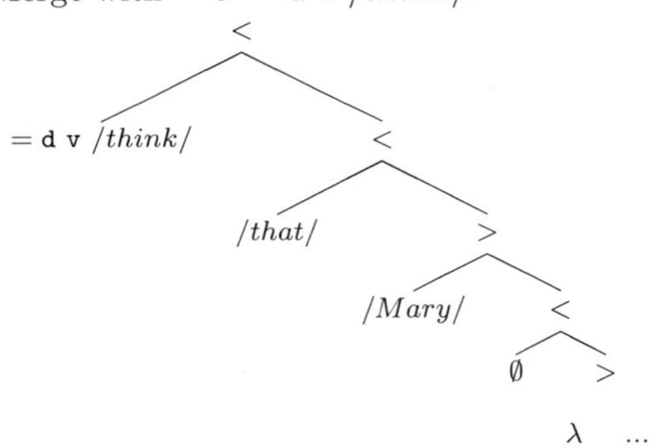

(9) Merge with d − case /*you*/:

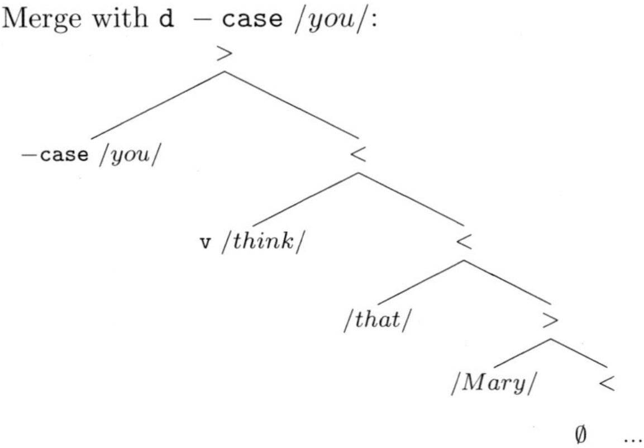

(10) (From now on, let us study only the upper part of the tree.) Merge with $=$ `v` $+$ `CASE` `t` */do/*:

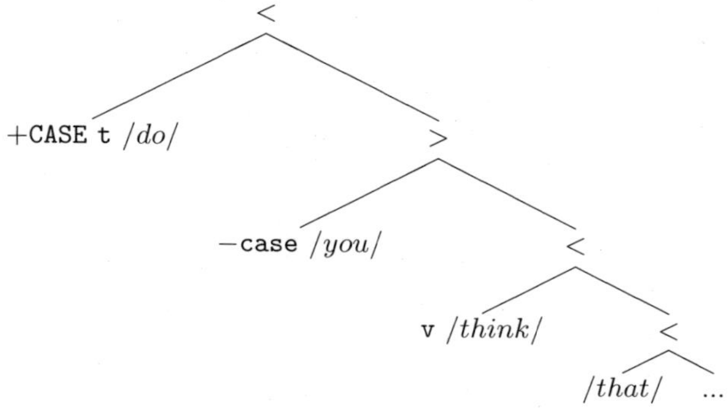

(11) Move *you*:

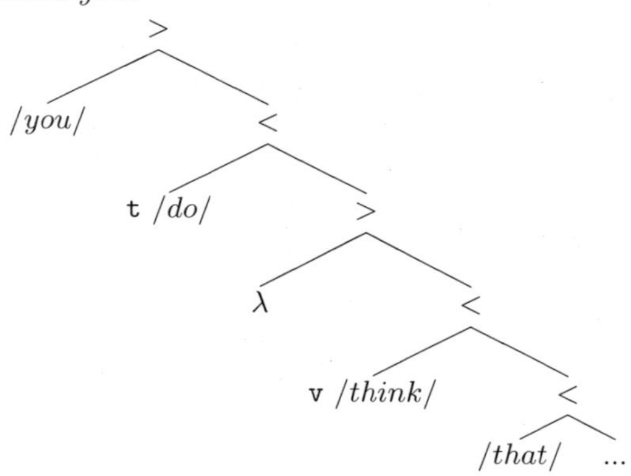

(12) Merge with $=$ `T` $+$ `WH` `c`, where $=$ `T` triggers head movement:

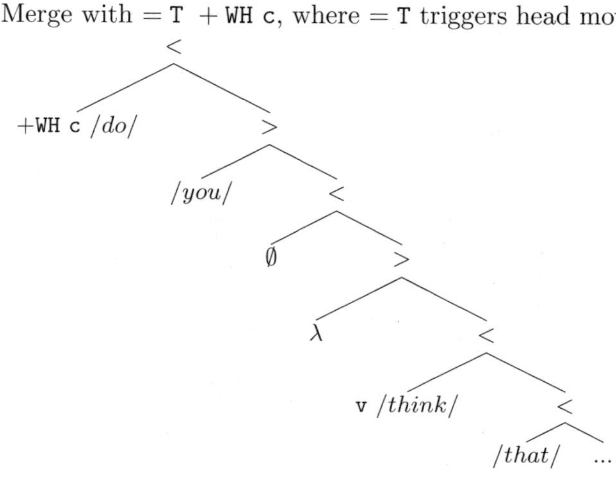

(13) Move *what book*:

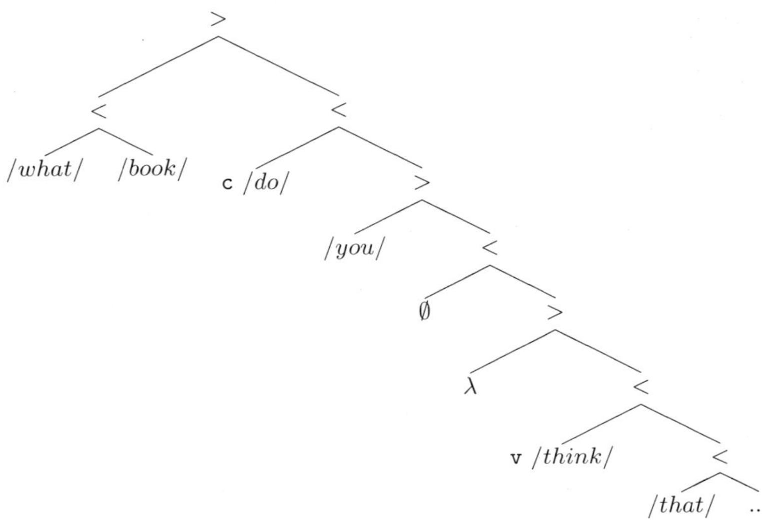

Another way to give the definitions of these partial functions amounts to presenting them as deductive rules which we will call *feature elimination rules*:

$$\frac{t_1[=\mathtt{c}] \qquad t_2[\mathtt{c}]}{[^< t_1, t_2]}\, merge\ [t_1 \in Lex] \qquad \frac{t_1[=\mathtt{c}] \qquad t_2[\mathtt{c}]}{[^> t_2, t_1]}\, merge\ [t_1 \notin Lex]$$

$$\frac{t_1[=\mathtt{C}] \qquad t_2[\mathtt{c}]}{[^< t_1(phon(t_1) + phon(t_2)), t_2(/\epsilon/)]}\, merge\ [t_1 \in Lex]$$

$$\frac{t_1[\mathtt{C}=] \qquad t_2[\mathtt{c}]}{[^< t_1(phon(t_2) + phon(t_1)), t_2(/\epsilon/)]}\, merge\ [t_1 \in Lex]$$

$$\frac{t_1[+\mathtt{F}](t_2[-\mathtt{f}]^*)}{[^> t_2^*, t_1\{t_2[-\mathtt{f}]^*/\lambda\}]}\, move \qquad \frac{t_1[+\mathtt{f}](t_2[-\mathtt{f}]^*)}{[^> t_2(/\epsilon/)^*, t_1\{t_2[-\mathtt{f}]^*/t_2'\}]}\, move$$

The derivation for *see a movie* is provided by

$$\cfrac{\cfrac{= \mathtt{d}\ + \mathtt{ACC}\ = \mathtt{d}\ \mathtt{v}\ /see/ \qquad \cfrac{= \mathtt{n}\ \mathtt{d}\ - \mathtt{case}\ /a/ \qquad \mathtt{n}\ /movie/}{[^< \mathtt{d}\ - \mathtt{case}\ /a/, /movie/]}\, merge\ [Lex]}{[^< + \mathtt{ACC}\ = \mathtt{d}\ \mathtt{v}\ /see/, [^< - \mathtt{case}\ /a/, /movie/]]}\, merge\ [\sim Lex]}{[^> [^< /a/, /movie/], [^< = \mathtt{d}\ \mathtt{v}\ /see/, \lambda]]}\, move\ (\mathtt{STRONG})$$

that we may contrast (w.r.t. weak and strong features) with

$$\dfrac{\dfrac{=\mathtt{d}\ +\mathtt{acc}\ =\mathtt{d}\ \mathtt{v}\ /see/ \qquad \dfrac{=\mathtt{n}\ \mathtt{d}\ -\mathtt{case}\ /a/ \qquad \mathtt{n}\ /movie/}{[^{<}\mathtt{d}\ -\mathtt{case}\ /a/, /movie/]}\ merge\ [Lex]}{[^{<}{+\mathtt{acc}}\ =\mathtt{d}\ \mathtt{v}\ /see/, [^{<}{-\mathtt{case}}\ /a/, /movie/]]}\ merge\ [\sim Lex]}{[^{>}\emptyset, [^{<}{=}\,\mathtt{d}\ \mathtt{v}\ /see/, [^{<}/a/, /movie/]]]}\ move\ (\mathtt{weak})$$

3.4.6 *Minimalist grammars and categorial grammars*

It seems obvious that there are many similarities between the two kinds of grammars. At first glance, for instance, a pair *selector, selectee* is exactly like a demand for some type with regard to this type. A difference may be that in categorial grammars, there is no autonomous feature $= \mathtt{x}$, which would be expressed as $/\mathtt{x}$ or $\mathtt{x}\backslash$: such features are always attached to a resulting type, like $\mathtt{y}/\mathtt{x}$ or $\mathtt{x}\backslash\mathtt{y}$. But in any case, the syntax of the MG expressions is such that there is also always a *basic* resulting type in those expressions. Another point of difference resides in that in MGs, selectors always select features which correspond, in CGs, to basic types (or 0-order types). Remember that in Lambek grammars, we have types corresponding to $A/(B\backslash C)$, where $B\backslash C$ is not a 0-order type, but a first-order one. In an MG, that would correspond to some expression like $= \{= \mathtt{z}\ \mathtt{y}\}\ \mathtt{x}$, which is not admitted. But this simply says that MGs take place in restricted CGs, for instance, Lambek grammars with at most types of order 1.

A more serious difference resides in the following remark by Lasnik and Uriagereka (2005, footnote 4, p. 55):

> [in CGs] adjuncts are typically analyzed as functions from X to X, in contrast with arguments:
>
> 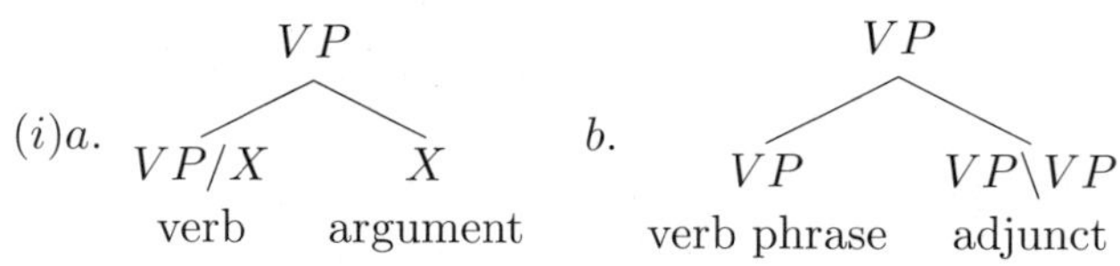
>
>
> If one were to say that VP/X is the head of VP in (i)a, one could not extend this claim to (i)b, where VP\VP is not the head of VP — instead, VP is.

The point is therefore that categorial grammars are confusing syntactic and semantic heads (we may assume that a semantic head is a function) and that, therefore, adjuncts are considered heads, which is at odds with most hypotheses in linguistic theory.

3.4.7 *Binding as "Cooper storage"*

Greg Kobele (2009) showed a way to conceive the syntax-semantics interface for minimalist grammars. It mainly relies on the theory of Cooper storage. What is *Cooper storage*? It was primarily developed (by Robin Cooper) in the context of Montague grammars for taking ambiguities into account. Let us assume a tree in the old style of generative grammars:

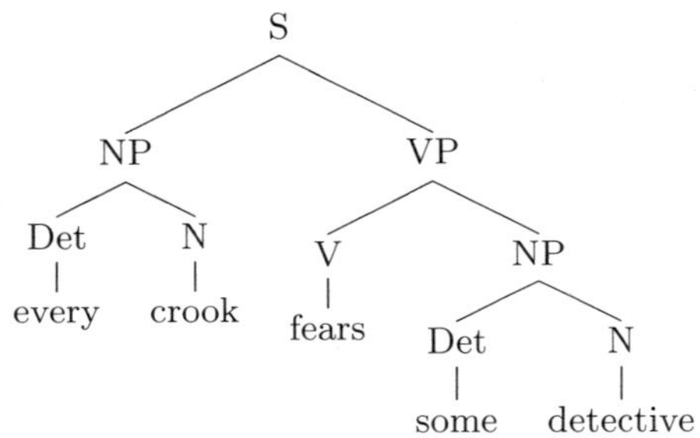

Each node in the tree is associated with two values:

(1) the semantic content,
(2) a quantifier store, which keeps still active quantifiers in memory.

A *quantifier store* is a pair $\langle i, M \rangle$ where i is an index and M a λ-term of type ((e, t), t). Let M_X and Δ_X be respectively the content and the quantifier store (QR) of the node X.

When semantically interpreting the tree, there is first a phase of *storing*, which consists, every time we merge two nodes B (a quantificational NP) and C to give a node A, in

- choosing a new index ι, associated with B;
- applying the content of the sister node C to the new variable x_ι, making the value of $(M_C(x_\iota))$ become A;
- making A's quantifier store become $\Delta_B \cup \Delta_C \cup \{\langle \iota, M_B \rangle\}$.

For *retrieving*, the procedure is the following:

- Let $\langle i, M' \rangle \in \Delta_A$;
- A may have the content $M'(\lambda x_i.M_A)$;
- In this case, its new QR will be $\Delta_A - \{\langle i, M' \rangle\}$.

Example:
Storing phase:

$$\begin{aligned}
&\textit{Every crook} &&:= \langle \lambda Q.\forall x.(crook(x) \Rightarrow Q(x)), \emptyset \rangle \\
&\textit{Some detective} &&:= \langle \lambda P.\exists y.(detective(y) \wedge P(y)), \emptyset \rangle \\
&VP\ \textit{node} &&:= fear(x_1), \{\langle 1, \lambda P.\exists y.(detective(y) \wedge P(y)) \rangle\} \\
&S\ \textit{node} &&:= fear(x_1)(x_2), \{\langle 1, \lambda P.\exists y.(detective(y) \wedge P(y)) \rangle, \\
& && \quad \langle 2, \lambda Q.\forall x.(crook(x) \Rightarrow Q(x)) \rangle\}
\end{aligned}$$

Retrieval phase:

$$\begin{aligned}
&fear(x_1)(x_2), \{\langle 1, \lambda P.\exists y.(detective(y) \wedge P(y)) \rangle, \langle 2, \lambda Q.\forall x.(crook(x) \Rightarrow Q(x)) \rangle\} \rightarrow \\
&(\lambda Q.\forall x.(crook(x) \Rightarrow Q(x)))(\lambda x_2.fear(x_1)(x_2)), \{\langle 1, \lambda P.\exists y.(detective(y) \wedge P(y)) \rangle\} \rightarrow \\
&(\forall x.(crook(x) \Rightarrow \lambda x_2.fear(x_1)(x_2)(x))), \{\langle 1, \lambda P.\exists y.(detective(y) \wedge P(y)) \rangle\} \rightarrow \\
&(\forall x.(crook(x) \Rightarrow fear(x_1)(x))), \{\langle 1, \lambda P.\exists y.(detective(y) \wedge P(y)) \rangle\} \rightarrow \\
&(\lambda P.\exists y.(detective(y) \wedge P(y)))(\lambda x_1.\forall x.(crook(x) \Rightarrow fear(x_1)(x))), \emptyset \rightarrow \\
&\exists y.(detective(y) \wedge (\lambda x_1.\forall x.(crook(x) \Rightarrow fear(x_1)(x)))(y)), \emptyset \rightarrow \\
&\exists y.(detective(y) \wedge (\forall x.(crook(x) \Rightarrow fear(y)(x)))), \emptyset
\end{aligned}$$

or the other way round:

$$\begin{aligned}
&fear(x_1)(x_2), \{\langle 1, \lambda P.\exists y.(detective(y) \wedge P(y)) \rangle, \langle 2, \lambda Q.\forall x.(crook(x) \Rightarrow Q(x)) \rangle\} \rightarrow \\
&(\lambda P.\exists y.(detective(y) \wedge P(y)))(\lambda x_1.fear(x_1)(x_2)), \{\langle 2, \lambda Q.\forall x.(crook(x) \Rightarrow Q(x)) \rangle\} \rightarrow \\
&(\exists y.(detective(y) \wedge (\lambda x_1.fear(x_1)(x_2))(y))), \{\langle 2, \lambda Q.\forall x.(crook(x) \Rightarrow Q(x)) \rangle\} \rightarrow \\
&(\exists y.(detective(y) \wedge (fear(y)(x_2)))), \{\langle 2, \lambda Q.\forall x.(crook(x) \Rightarrow Q(x)) \rangle\} \rightarrow \\
&(\lambda Q.\forall x.(crook(x) \Rightarrow Q(x)))(\lambda x_2.\exists y.(detective(y) \wedge (fear(y)(x_2)))), \emptyset \rightarrow \\
&\forall x.(crook(x) \Rightarrow (\lambda x_2.\exists y.(detective(y) \wedge (fear(y)(x_2))))(x)), \emptyset \rightarrow \\
&\forall x.(crook(x) \Rightarrow (\exists y.(detective(y) \wedge (fear(y)(x))))), \emptyset
\end{aligned}$$

3.4.8 *The interpretation of derivational trees*

Kobele proposes not to deal with the syntactic trees properly speaking, but with the derivation trees. There is a slight difference between the two objects. For us now, we have seen that the analogue of the traditional syntactic trees is actually *proof trees*, that is, the succession of the various steps in a proof. In a proof tree, nodes are syntactical objects resulting from the deduction rules or the operations (Merge and Move) that we have used. For instance, the tree in Fig. 8.2 is a proof tree in the natural deduction format.

A *derivation tree* is another kind of tree, the nodes of which are the *names of the rules* which have been applied, and the leaves of which are labelled by axioms (that is, in our view, lexical entries). Let us for instance look at the derivation tree for the (pseudo)-sentence *What book Mary does read?*

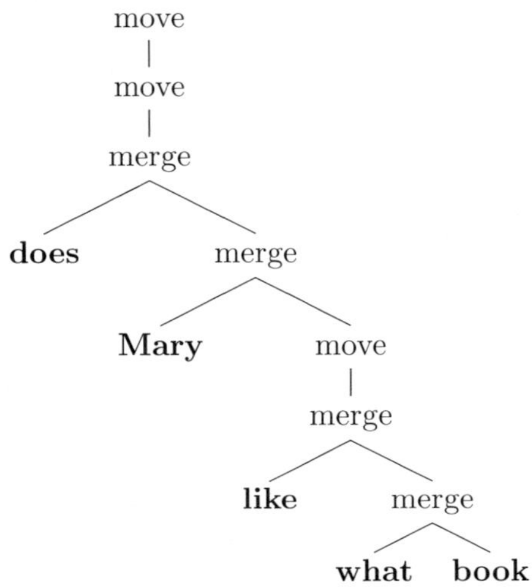

When we know the details of the lexical entries, and because of the *shortest move constraint* which prevents us from finding ambiguities in the determination of the application field of each operation, such a tree gives us one and only one syntactic tree, or proof tree.

Determining the meaning of a sentence may therefore also be done directly, starting from this tree. In this case, we simply have to define our semantic interpretation of each operation.

3.4.9 *The semantic interpretation of Merge*

Like Heim and Kratzer, Kobele proposes to directly map derivations to model-theoretic objects, that is, denotational entities and truth values, getting rid of a formulae level, but we shall allow ourselves to use formulae, as simpler devices.

The Merge principle is actually reminiscent of the *branching nodes* case:

> If α and β merge at some step of the derivation, then one of the two is of type $a \rightarrow b$ and the other of type b. Let us suppose α is of type $a \rightarrow b$; if β belongs to the domain of α, then $[[merge(\alpha, \beta)]] = [[\alpha]]([[\beta]])$ (the other way round if β is the function and α the object).

Let us look at for instance the following sentence:

Example 3.3. Every student sleeps

We assume here (provisionally) the following lexicon (where the third column gives the semantic type):

every	:=	**every**	$(e \to t) \to ((e \to t) \to (e \to t))$
student	:=	$\lambda x.$ **student**(x)	$e \to t$
sleeps	:=	$\lambda y.$ **sleep**(y)	$e \to t$

The story of the derivation is

(1) *merge*(every, student) $\to$ **every**$(\lambda x.$ **student**$(x))$
(2) *merge*(*merge*(every, student), sleeps) $\to$ **every**$(\lambda x.$ **student**$(x))(\lambda y.$ **sleep**$(y))$

And by assuming **every** $= \lambda P.\lambda Q.\forall z(P(z) \Rightarrow Q(z))$, we get by reduction

$$\forall z\ (\mathbf{student}(z) \Rightarrow \mathbf{sleep}(z))$$

But as is well known, if we transfer the quantificational DP to the object position, things turn wrong since we get a type mismatch. For instance, if we have

Example 3.4. John congratulated every student

where the semantic types are

John	:	e
congratulated	:	$e \to (e \to t)$

when merging the transitive verb with its object, we get a node α of type $e \to (e \to t)$ and another node β of type $(e \to t) \to t$ and none of them can apply to the other! Let us of course remark that we could in this case not *apply* one to the other, but *compose* the two, like it is feasible in categorial grammars, but that would entail changing the order of the variables in the formula expressing the transitive verb: $\lambda x.\lambda y.\texttt{congr}(y, x)$ should be changed into $\lambda x.\lambda y.\texttt{congr}(x, y)$. We already know that this problem has been the source of transformations and rules like quantifier raising and Montague's quantifying in.

3.4.10 *The semantic interpretation of Move*

How do we interpret Move? In fact some elements have already been brought to light at this stage. We mentioned earlier the similarity of Move with the logical procedure by which we make an assumption and then discharge it by an appropriate rule. When this rule is the introduction of *implication*, that results in a term we abstract over by means of the lambda abstractor: the term which corresponds to the proof just before discharging the assumption becomes a term of type $a \to b$ if it was of type b and if the variable we abstract over is of type a. In this case, the moved DP must necessarily be of a higher-order type if it must be applied to this term. It is as if, by syntactically moving, a DP was semantically type-raised, thus going, say, from type e to the higher-order type $(e \to t) \to t$. We may conclude that, semantically, moving a DP implies

(1) a previous Merge of a functional term ϕ with an object x of type e,
(2) a type raising of the displaced DP,
(3) and finally the application of this type-raised DP to the term $\phi(x)$ after an abstraction step.

The success of these operations of course depends on the fact that the object of type e is a *variable*, so that the object to which the type-raised DP applies can be a λ-abstraction $\lambda x.\phi(x)$.

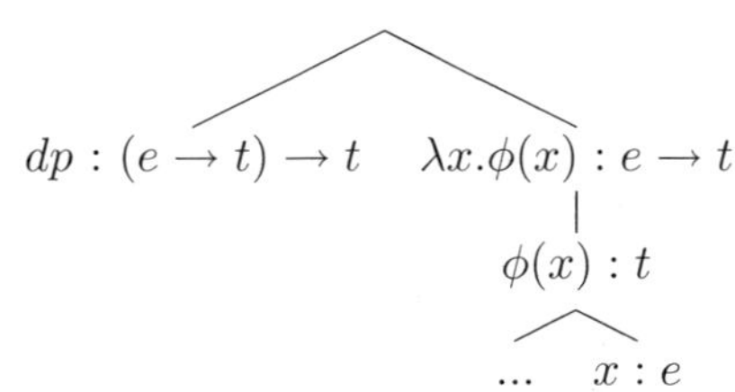

This analysis may be rendered by means of the Cooper-storage mechanism. We must in fact, as seen above, associate each node in the derivation not only with a semantic content, but also with a quantifier store. In the case of Merge, we have two modalities:

(1) Ordinary Merge:

syntactic operation	semantic content	Cooper store
$merge(\alpha, \beta)$	$[[\alpha]]([[\beta]])$ or $[[\beta]]([[\alpha]])$	$\text{store}(\alpha)\frown\text{store}(\beta)$

(2) Merge internal to Move:

syntactic operation	semantic content	Cooper store
$merge(\alpha, \beta)$	$[[\alpha]](x_i)$	$\text{store}(\alpha)\frown < i, [[\beta]] > \frown\text{store}(\beta)$

and for Move, we have

syntactic operation	semantic content	Cooper store
$move(\alpha)$	$[[\alpha]]$	$\text{store}(\alpha)$
$move(\alpha)$	$[[\alpha]](\lambda x_i.[[\gamma]])$	$\text{store}(\alpha) - \langle i, [[\gamma]]\rangle$

The first line concerns the case where the QR-store at node α does not contain any $\langle i, [[\gamma]]\rangle$ candidate, and (see below) the case where the constituent still has to move up (this can be known by still unchecked licensee features): in these cases, the meaning of Move is vacuous. The second line expresses the operation of retrieval (when the last licensee feature is checked). Let us take the following example:

Example 3.5. What book does Mary like?

Its derivation tree (with numbered nodes) is

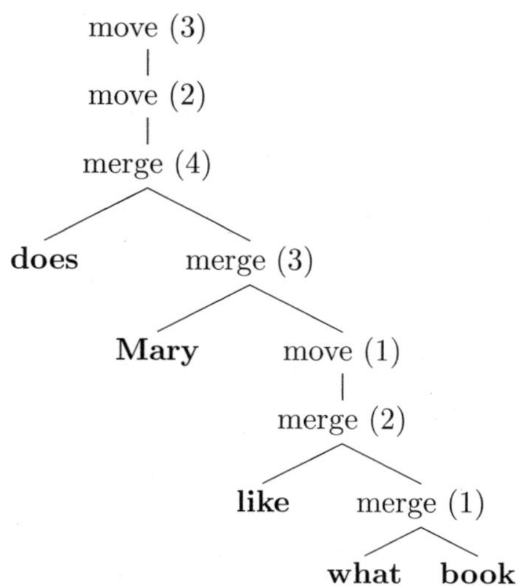

Let us look at the content of each node:

(1)

Operation	Content	Cooper-store
merge (1)	**what**(**book**)	$\emptyset$

(2)

Operation	Content	Cooper-store
merge (2)	$\mathbf{like}(x_1)$	$\{\langle 1, \mathbf{what(book)}\rangle\ \}$

(3)

Operation	Content	Cooper-store
move (1)	$\mathbf{like}(x_1)$	$\{\langle 1, \mathbf{what(book)}\rangle\ \}$

(4)

Operation	Content	Cooper-store
merge (3)	$(\mathbf{like}(x_1))(x_2)$	$\{\langle 1, \mathbf{what(book)}\rangle, \langle 2, \mathbf{Mary}\rangle\ \}$

(5)

Operation	Content	Cooper-store
merge (4)	$\mathbf{does}((\mathbf{like}(x_1))(x_2))$	$\{\langle 1, \mathbf{what(book)}\rangle, \langle 2, \mathbf{Mary}\rangle\ \}$

(6)

Operation	Content	Cooper-store
move (2)	$\mathbf{Mary}(\lambda x_2.\mathbf{does}((\mathbf{like}(x_1))(x_2)))$	$\{\langle 1, \mathbf{what(book)}\rangle\ \}$

(7)

Operation	Content	Cooper-store
move (3)	$\mathbf{what\ book}(\mathbf{Mary}(\lambda x_2.\mathbf{does}((\mathbf{like}(x_1))(x_2))))$	$\emptyset$

Comments:

- at step 2, Merge is performed with a variable since the constituent **what book** has still unchecked features $-$CASE and $-$WH; this entails the Cooper storage of an index associated with this variable and of the meaning of the constituent represented by this variable;
- at step 3, Move is vacuous since it is the step where $-$CASE is checked, but $-$WH is still unchecked;
- at step 4, Merge is still performed with a variable, but different from the previous one. Since it represents the moving constituent **Mary**, the semantic content of which is of a high order (like in Montague semantics: $\lambda Q.Q(\mathbf{Mary})$), the store is enriched. It now contains two "waiting" meanings to apply, but we cannot confound them since they are associated with different still unchecked features ($-$WH for 1 and $-$CASE for 2);
- at step 6, **Mary** may be applied: it is the stage where its last unchecked feature is checked;
- at step 7, **what book** may be applied since it is the step where its $-$WH feature is checked.

3.5 Concluding Remarks

This chapter presented two derivational systems into which a derivational account of grammar may be taken. The Lambek calculus, that we shall study further, has the advantage of providing a direct construction of a logical form by means of what is known as the Curry–Howard correspondence, but it suffers from several limitations with regard to the syntactic forms that may be derived (for instance, extraction may be accounted for but only from peripheral positions). Minimalist grammars are supposed to provide more flexibility (they are proven to be *mildly context-sensitive*[6] (Michaelis, 2001), that is, they have the same generative capacity as Joshi's tree-adjoining grammars) but they lack an easy and uniform way to extract a semantic representation.

[6]Let us recall here that languages (in the mathematical sense of sets of strings on an alphabet A) are hierarchized according to a classification originally made by Chomsky and Schützenberger in the 1950s. This hierarchy corresponds to different degrees of complexity concerning the problem of recognizing a string as belonging to a given language. Finite languages aside, "linear" languages — nothing to do with *linearity* in the logical sense — are the most easily recognizable — by finite states automata. They are strictly included in context-free languages, which constitute a class itself strictly included in the class of context-sensitive languages. There are also languages which are not context-sensitive but are still semi-decidable; they are called Turing languages. Undecidable sets are out of this classification. The question arose whether natural languages coincided with one of these classes. It appeared, after many works, among them Joshi's, that natural languages exceeded the generative power of context-free grammars but that they are strictly included in the class of context-sensitive languages. It is the reason why Joshi defined an intermediary class called *mildly context-sensitive languages*.

PART II
Logic

Chapter 4

Deductive Systems

4.1 Fitch's Natural Deduction System

Deductive systems are sets of rules allowing proofs. One of the systems which corresponds best to our usual understanding of proofs in mathematics is the one attributed to Frederic Fitch, who himself improved a system originally due to the Polish logician Stanislaw Jaśkowski. This *style* of proofs consists in displaying formulae which are progressively deduced in a row, with premises at the top and the conclusions at the bottom (on the schema in Fig. 4.1, there are i premises and one conclusion). The main deduction is the space on the immediate right of the first vertical bar.

It may happen that we are led to make assumptions during a proof. In this case, the formulae which are deduced from these assumptions are of course valid only under them. When making an assumption (or *hypothesis*), the rule is therefore to inscribe it in the current deduction with a slight shift on the right and then to open a new space for a deduction (strictly embedded into the previous one). This new deduction is called a *sub-deduction*. It starts at the most recent hypothesis which has been made. Of course we can make new hypotheses in the space of any sub-deduction, thus leading to a hierarchy of sub-deductions. For instance, the schema in Fig. 4.2 has two premises: A and B, then a hypothesis C opens a sub-deduction, then in that sub-deduction, a new hypothesis D is made. A deduction is therefore a sequence (vertically displayed):

$$(\phi_1, \ldots, \phi_{n1}, \delta_1, \phi_{n1+1}, \ldots, \phi_{n2}, \delta_2, \phi_{n2+1}, \ldots, \phi_{nk}, \delta_k, \phi_{nk+1}, \ldots, \phi_{n_{k+1}})$$

where ϕ_i are formulae and δ_j deductions. Of course k may be 0, and any subsequence of subformulae $\phi_{np+1}, \ldots, \phi_{nq}$ may be empty. Every ϕ_i is a premise, an axiom or some formula coming from the application of

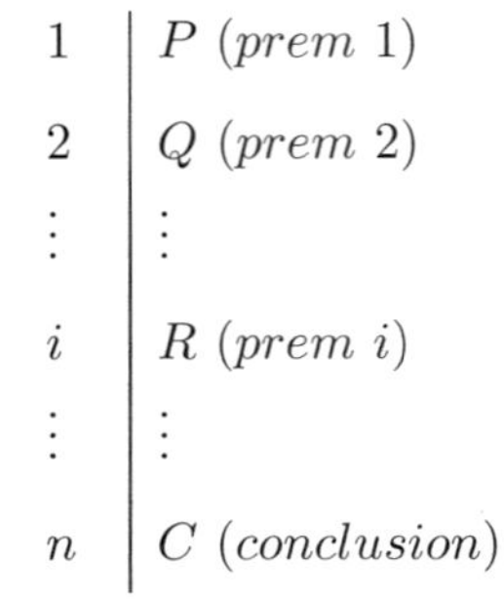

Fig. 4.1 Form of a proof in natural deduction

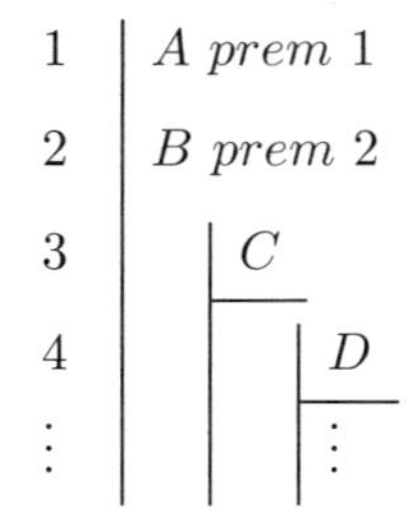

Fig. 4.2 Deduction with hypotheses

deduction rules to previous formulae and sub-deductions in the sequence. The first formula of each δ_j is a formula called a *hypothesis*. A formula is said to belong to a deduction δ if and only if it is one of the ϕ_i. Each δ_j is said to be an immediate sub-deduction in δ.

Given two deductions δ and γ, we write $\gamma \prec \delta$ to say that γ is an immediate sub-deduction of δ, that is, a deduction starting from a hypothesis made just after a formula belonging to δ. Let $\prec^*$ be the transitive closure of $\prec$; more generally, γ is said to be a sub-deduction of δ if and only if $\gamma \prec^* \delta$.

It is always possible to repeat a formula (a previous conclusion or a premise or a hypothesis) provided that this repetition is made in the appropriate space, that is, in the same row including the formula which is repeated or in a sub-deduction which has been opened after the inscription of this formula. For that, the *repeat* rule is used (see Fig. 4.3).

⋮	⋮	
m	A	
⋮	⋮	
n	A	R, m

⋮	⋮	
1	A	
⋮	⋮	
i	│ B	
⋮	│ ⋮	
m	│ A	R, 1
⋮	│ ⋮	

Fig. 4.3 The *repeat* rule

There are two kinds of logical rules properly speaking, and they are associated with each connective (or *particle*). Using a premise, an axiom or an already proven formula is equated with *eliminating* its main connective, while using the structure of a formula we aim at proving is equated with *introducing* its main connective. A natural deduction system therefore contains

- elimination rules,
- introduction rules,

with one pair for each connective.

4.1.1 *Conjunction*

Let us begin with conjunction. The elimination and introduction rules are respectively given in Figs. 4.4 and 4.5. The elimination rule has two variants depending on the "side" of the conjunction on which we project.

⋮	⋮	
m	$A \wedge B$	
⋮	⋮	
n	A	$\wedge$E, m

⋮	⋮	
m	$A \wedge B$	
⋮	⋮	
n	B	$\wedge$E, m

Fig. 4.4 Conjunction elimination

$$
\begin{array}{l|ll}
\vdots & \vdots & \\
m & A & \\
\vdots & \vdots & \\
n & B & \\
\vdots & \vdots & \\
p & A \wedge B & \wedge\mathrm{I},\ m,\ n
\end{array}
$$

Fig. 4.5 Conjunction introduction

$$
\begin{array}{l|ll}
\vdots & \vdots & \\
m & A \Rightarrow B & \\
\vdots & \vdots & \\
n & A & \\
\vdots & \vdots & \\
p & B & \Rightarrow\mathrm{E},\ m,\ n
\end{array}
$$

Fig. 4.6 Elimination de l'implication

4.1.2 *Implication*

Let us immediately look at the case of implication, our most useful connective for the applications we have in mind.

The elimination rule (Fig. 4.6) is actually the most well-known rule in logic, that is, *modus ponens*, which says that from $A \Rightarrow B$ and A, we may always deduce B. It is also called the *detachment* rule.

The introduction rule (Fig. 4.7) also reproduces a proof process we are accustomed to: proving $A \Rightarrow B$ exactly amounts to *making the assumption* of A and then deducing B in the sub-deduction opened by the hypothesis A. This provides our first case of so-called *hypothetical reasoning*.

Disjunction will provide us with another case of hypothetical reasoning.

$$\begin{array}{l|lll}
1 & & \multicolumn{1}{|l}{A} & \\ \cline{3-3}
\vdots & & \multicolumn{1}{|l}{\vdots} & \\
n & & \multicolumn{1}{|l}{B} & \\
n+1 & A \Rightarrow B & & \Rightarrow\text{I, } 1\text{–}n
\end{array}$$

Fig. 4.7 Implication introduction

4.1.3 *Disjunction*

Introducing disjunction is something symmetrical with regard to eliminating conjunction. We have

$$\begin{array}{l|ll}
\vdots & \vdots & \\
m & A & \\
\vdots & \vdots & \\
n & A \vee B & \vee\text{I, } m
\end{array}
\qquad\qquad
\begin{array}{l|ll}
\vdots & \vdots & \\
m & B & \\
\vdots & \vdots & \\
n & A \vee B & \vee\text{I, } m
\end{array}$$

Elimination is more particular. It rests on the idea that $A \vee B$ allows the inference of C only if A as well as B allows us to infer it. Therefore in order to infer C, I first make hypothesis A, check that C may be deduced from it, and then I do the same thing with B. Hence the following rule:

$$\begin{array}{l|lll}
\vdots & \vdots & & \\
i & A \vee B & & \\
i+1 & & \multicolumn{1}{|l}{A} & \\ \cline{3-3}
\vdots & & \multicolumn{1}{|l}{\vdots} & \\
j & & \multicolumn{1}{|l}{C} & \\
j+1 & & \multicolumn{1}{|l}{B} & \\ \cline{3-3}
\vdots & & \multicolumn{1}{|l}{\vdots} & \\
k & & \multicolumn{1}{|l}{C} & \\
l & C & & \vee\text{E, } i,\ i+1\text{–}j,\ j+1\text{–}k
\end{array}$$

4.1.4 *Negation*

Rules for negation may be derived if we introduce a special constant $\perp$ interpreted as *absurdity* (or always false), and if we assume that something is not the case if and only if it would imply such an absurdity. This leads us to assert the following equivalence:

$$\neg A \equiv (A \Rightarrow \perp)$$

$\perp$ has an elimination rule but no introduction rule (it would consist in introducing an absurdity into a demonstration):

$$\begin{array}{l|l}
\vdots & \vdots \\
i & \perp \\
i+1 & Q
\end{array}$$

This rule is precisely the one that medieval logicians called *ex falso sequitur quodlibet*: from something false, or contradictory, any formula may be derived. Negation has then the two following rules:

$$\begin{array}{l|l l}
\vdots & \vdots & \\
m & A & \\
\vdots & \vdots & \\
n & \neg A & \\
n+1 & \perp & \neg\text{E}, m, n
\end{array}
\qquad
\begin{array}{l|l l}
1 & \begin{array}{|l} A \\ \hline \vdots \\ \perp \end{array} & \\
\vdots & & \\
m & & \\
m+1 & \neg A & \neg\text{I}, 1\text{–}m
\end{array}$$

As we see, $\perp$ may only occur in a deduction via the elimination rule of $\neg$, but applying this rule is precisely the case when we have a contradiction $(A \wedge \neg A)$, hence $\perp$ has really the sense of a contradiction.

Is that system sufficient for giving all the proofs of classical logic? The answer is no. In particular, it lacks the law of *excluded middle* and the law of *double negation.* This is not astonishing. Proving the excluded middle $(A \vee \neg A)$ would be possible only by using the rule $\vee\, I$, but this rule assumes that we know either A or $\neg A$ when this rule is essentially used when we don't know which one is the case (e.g. *either the universe is flat or it is not*). It is otherwise possible to show that as soon as we introduce the double negation law into the system, the excluded middle may be derived, where this law is expressed by

$$\begin{array}{l|ll} \vdots & \vdots & \\ n & \neg\neg A & \\ n+1 & A & \neg\neg\text{E}, n \end{array}$$

If we don't add explicitly this rule, we get a subsystem of classical logic: *intuitionistic logic*, a denomination justified by the fact that intuitionist mathematicians (led by Brouwer) rejected the law of excluded middle from their deductions.

We shall come back to natural deduction in intuitionistic logic in the next section in order to show that proofs in this logic have the nice property of being displayable as trees, something appreciated by linguists, who emphasize the algebraic structure of trees.

4.2 Natural Deduction in Intuitionistic Logic

4.2.1 *Tree format*

Another way of formulating the previous rules is the following, where premises are at the top, and for each rule, the conclusion is at the bottom of every scheme. These schemes are elementary trees, and when combined they will give trees of any depth.

Introduction rules:

$$\frac{A \quad B}{A \wedge B}\,[\wedge\ I]$$

$$\frac{A}{A \vee B}\,[\vee_1\ I] \qquad \frac{B}{A \vee B}\,[\vee_2\ I]$$

$$\frac{\begin{array}{c}[A]\\ \vdots \\ B\end{array}}{A \Rightarrow B}\,[\Rightarrow\ I]$$

$$\frac{\begin{array}{cc}[A] & [A]\\ \vdots & \vdots \\ B & \neg B\end{array}}{\neg A}\,[\neg\ I]$$

Elimination rules:

$$\frac{A \wedge B}{A}\,[\wedge_1\ E] \qquad \frac{A \wedge B}{B}\,[\wedge_2\ E]$$

$$\frac{A \vee B \quad \begin{matrix}[A] \\ \vdots \\ C\end{matrix} \quad \begin{matrix}[B] \\ \vdots \\ C\end{matrix}}{C}\,[\vee\ E]$$

$$\frac{A \Rightarrow B \quad A}{B}\,[\Rightarrow\ E]$$

$$\frac{\neg A \quad A}{B}\,[\neg\ E]$$

4.2.2 *Normalization*

A proof in natural deduction may contain useless detours. Let us assume for instance that the rules [⇒ I] and [⇒ E] are successively applied, with the same formula A introduced as a hypothesis and then discharged at the first step, and then "eliminated" at the second one, that would result in a useless detour. It could be suppressed in an obvious way, like it is shown in Fig. 4.8. Technically, such a simplification is called a normalization step. If we want to find a "canonical" proof for a theorem, that is, the simplest one (using no "detours"), then we must transform our original proof by applying a sequence of such steps. That amounts to *normalizing* the proof and brings a new dynamic perspective into the world of proofs. Moreover,

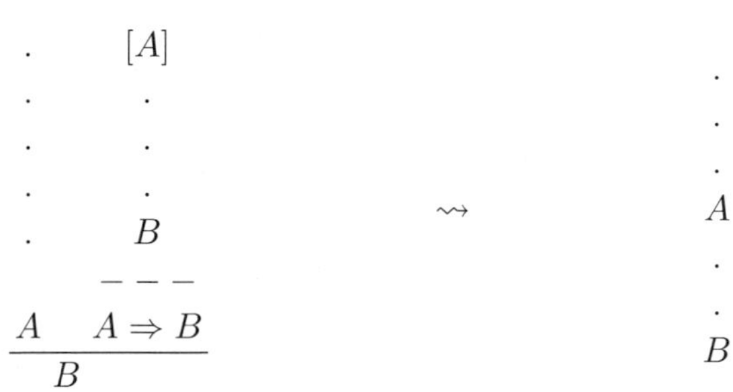

Fig. 4.8 A normalization step

like we shall see soon (Chapter 5), such a normalization process mirrors the *reduction* process at work in the λ-calculus.[1]

4.2.3 *Sequent format*

Another presentation of natural deduction is due to Gentzen. It rests on the notion of *sequent*. A natural deduction sequent (*ND- sequent* for short) is a deduction relation between a set of formulae (on the left-hand side) and only one formula (on the right-hand side).

Axiom:

$$\Gamma \cup \{A\} \vdash A$$

Weakening:

$$\frac{\Gamma \vdash A}{\Gamma \cup \Delta \vdash A}$$

Introduction rules:

$$\frac{\Gamma \vdash A \quad \Gamma \vdash B}{\Gamma \vdash A \wedge B}\,[\wedge\ I]$$

$$\frac{\Gamma \vdash A}{\Gamma \vdash A \vee B}\,[\vee_1\ I] \qquad \frac{\Gamma \vdash B}{\Gamma \vdash A \vee B}\,[\vee_2\ I]$$

$$\frac{\Gamma \cup \{A\} \vdash B}{\Gamma \vdash A \Rightarrow B}\,[\Rightarrow\ I]$$

$$\frac{\Gamma, A \vdash B \quad \Gamma, A \vdash \neg B}{\Gamma \vdash \neg A}\,[\neg\ I]$$

Elimination rules:

$$\frac{\Gamma \vdash A \wedge B}{\Gamma \vdash A}\,[\wedge_1\ E] \qquad \frac{\Gamma \vdash A \wedge B}{\Gamma \vdash B}\,[\wedge_2\ E]$$

$$\frac{\Gamma \vdash A \vee B \quad \Gamma \cup \{A\} \vdash C \quad \Gamma \cup \{B\} \vdash C}{\Gamma \vdash C}\,[\vee\ E]$$

[1] For a proof of the normalization theorem, we send the reader to the excellent *Proofs and Types* (Girard *et al.*, 1989, Chapter 4).

$$\frac{\Gamma \vdash A \Rightarrow B \quad \Gamma \vdash A}{\Gamma \vdash B} [\Rightarrow E]$$

$$\frac{\Gamma \vdash A \quad \Gamma \vdash \neg A}{\Gamma \vdash B} [\neg E]$$

4.3 Intuitionistic Sequent Calculus

Let us study now a slightly different system, which contains another kind of rule: the *sequent calculus* properly speaking (to differentiate from what we have called ND-sequent calculus). The interest of this new formulation is that it provides a *decision procedure* for intuitionistic logic (something that we of course have for classical logic, by means of truth tables, but truth tables don't work for intuitionistic logic since we know that all the classical tautologies are not theorems of this logic, for instance $A \vee \neg A$). The sequent calculus will contain in fact only *introduction rules*, but they differ in the side of the sequent the connective is introduced on. Reading a deduction in the sequent calculus from the bottom to the top (that is, starting from the formula to prove) will therefore show only eliminations of connectives (except when applying the cut rule or an axiom, but in the latter case, the proof stops and in the former one, as we shall see, instances of the cut rule may be eliminated). The rules (where Γ is a finite set of formulae) are the following:

4.3.1 *Structural rule*

$$\frac{\Gamma \vdash A}{\Gamma \cup \Delta \vdash A} [Af]$$

This rule is said to be structural because it doesn't concern any logic connective and only plays on the context of proof of A. It is needed in order to indicate that each time we add new hypotheses to the left-hand side, the conclusion is preserved. It may also be seen as a *monotonicity* rule saying that the set of conclusions drawn from a set of assumptions never decreases when the size of the set of hypotheses grows. We call it the *weakening* rule.

4.3.2 *Identity rules*

Axiom:

$$\Gamma, A \vdash A$$

Cut:

$$\frac{\Gamma \vdash A \quad \Gamma, A \vdash C}{\Gamma \vdash C}\,[cut]$$

4.3.3 *Logical rules*

Negation:

$$\frac{\Gamma \vdash A}{\Gamma, \neg A \vdash \bot}\,[\neg L] \qquad \frac{\Gamma, A \vdash \bot}{\Gamma \vdash \neg A}\,[\neg R]$$

Conjunction:

$$\frac{\Gamma, A \vdash C}{\Gamma, A \wedge B \vdash C}\,[\wedge_1 L] \qquad \frac{\Gamma, B \vdash C}{\Gamma, A \wedge B \vdash C}\,[\wedge_2 L]$$

$$\frac{\Gamma \vdash A \quad \Gamma \vdash B}{\Gamma \vdash A \wedge B}\,[\wedge R]$$

Disjunction:

$$\frac{\Gamma, A \vdash C \quad \Gamma, B \vdash \bot}{\Gamma, A \vee B \vdash C}\,[\vee_1 L]$$

$$\frac{\Gamma, A \vdash \bot \quad \Gamma, B \vdash C}{\Gamma, A \vee B \vdash C}\,[\vee_2 L]$$

We may remark that, given the rules for negation, in the case of $[\vee_1\ L]$ for instance, $\Gamma', B \vdash \bot$ is equivalent to $\Gamma' \vdash \neg B$. Thus, this rule simply formulates the scheme according to which if we have $A \vee B$, then, if the context provides us with $\neg B$, it will be necessary, to prove C, that C be deducible from A, and similarly for $[\vee_2\ L]$ where B will have, in the given context, to produce C if this context leads to $\neg A$.

$$\frac{\Gamma \vdash A}{\Gamma \vdash A \vee B}\,[\vee_1 R] \qquad \frac{\Gamma \vdash B}{\Gamma \vdash A \vee B}\,[\vee_2 R]$$

Implication:

$$\frac{\Gamma \vdash A \quad \Gamma, B \vdash C}{\Gamma, A \Rightarrow B \vdash C}[\Rightarrow L] \qquad \frac{\Gamma, A \vdash B}{\Gamma \vdash A \Rightarrow B}[\Rightarrow R]$$

4.3.4 *An example of a proof in intuitionistic logic*

The following deduction represents a proof of the sequent $A \Rightarrow (A \Rightarrow B) \vdash A \Rightarrow B$ in the sequent calculus: which shows an alternation of the left rule and of the right rule.

$$\dfrac{\dfrac{A \Rightarrow A \qquad \dfrac{\dfrac{}{A \Rightarrow A}[Ax] \quad \dfrac{}{A, B \Rightarrow B}[Ax]}{A, A \Rightarrow B \vdash B}[\Rightarrow L]}{A, A \Rightarrow (A \Rightarrow B) \vdash B}[\Rightarrow L]}{A \Rightarrow (A \Rightarrow B) \vdash A \Rightarrow B}[\Rightarrow R]$$

The particularity of this calculus is its *additive* use of contexts (the sets of formulae in uppercase Greek letters), which is in contrast with their *multiplicative* use that we will see very soon: the very same context Γ occurs on top (on both sides, left and right) and at the bottom of the $[\Rightarrow L]$ rule, for instance. Moreover, when applying this rule, this context remains unchanged. We could say the same thing for any rule. Concerning this example of a proof, this results in formula A, which gives an A in the left premise of the lower $[\Rightarrow L]$ application, being still there to give a new A in the left premise of the upper one. In this deduction, therefore, a formula, like A, may be used twice (or more) during the deduction. When we present this proof in a tree format, we get

$$\dfrac{\dfrac{\dfrac{A \Rightarrow (A \Rightarrow B) \quad [A]^1}{A \Rightarrow B}[\Rightarrow E] \qquad [A]^1}{B}[\Rightarrow E]}{A \Rightarrow B}[\Rightarrow I](1)$$

The re-use of the same formula A (as emphasized above) is shown by the fact that, here, the same hypothesis A is made at the first and at the second step, but both occurrences will be discharged together at the last step. This puts the accent on something important in this style of proofs. Each hypothesis made corresponds to a *package* of occurrences, of which we may discharge one item, several or none at the same time. When none are discharged, that can be the case of an empty package. For instance, the following deduction is correct:

$$\frac{A}{B \Rightarrow A}[\Rightarrow\ I]$$

No hypothesis has been made (therefore we have an empty package) but this does not prevent us from using the $[\Rightarrow\ I]$ rule. We simply obtain an equivalent of the tautology $A \Rightarrow\ (B \Rightarrow\ A)$, which was taken as an axiom by Frege. In the natural deduction sequent format, we would have

$$\cfrac{\cfrac{\cfrac{A \vdash A \quad \vdash A \Rightarrow (A \Rightarrow B)}{A \vdash A \Rightarrow B} \qquad A \vdash A}{A \vdash B}}{\vdash A \Rightarrow B}$$

Here the set-theoretical nature of contexts plays an important role. We find $A \vdash B$ by applying the *modus ponens* rule to the two sequents $A \vdash A \Rightarrow B$ and $A \vdash A$. A is thus the left context which is common to both. When *modus ponens* is applied, these two contexts, instead of being juxtaposed, "unify", that is, their union is made and redundant items are suppressed. Like it is normal in the set union, there is still one unique A in the resulting sequent.

4.3.5 *The cut rule*

Among the rules given above, one deserves particular attention: the *cut rule*. In some sense it has a sort of "negative" face with regard to the identity axiom since it also links two instances of the same formula but in such a way that this formula disappears when passing from the top of the rule to its bottom. The cut rule is in fact a mere generalization of a transitivity axiom: if by means of Γ, you can prove A, and if by means of A, you can prove B, then you can directly prove B from Γ. The cut rule is therefore a device we can use to assemble proofs, forming a bigger one. It is also a very useful device because it allows us to re-use results already proven, exactly like we do in the case of *lemmas*, that we may re-use at wish in the demonstration of various theorems.

Of course, when we use lemmas for demonstrating something, we know that we could as well demonstrate that thing directly, but this would amount to still re-proving the lemma inside the main proof. Nevertheless, we use lemmas because we are guaranteed that it does not alter the correctness of the demonstration. In logic, this important property must be proven: it is the result known as Gentzen's *Hauptsatz*, or *cut-elimination theorem*.

4.3.6 *Lists and sets*

Let us note that the previous system would have been slightly modified if we had used *lists* instead of *sets* in its formulation. In computer science, the use of the list structure is widespread. A list is a finite succession of items (possibly empty) where some may be duplicated; what counts is an occurrence (or a token) of each item. For instance $[A, B, A, A, C]$ is a list, and it can be reduced neither to $[A, B, A, C]$ (by a kind of contraction, the pair A, A being reduced to a single A) nor to $[A, B, C]$ (by suppressing all the reduplications of the same symbol), and it is not equal to any of its permutations (except of course the identity), for instance $[A, B, A, A, C] \neq [B, A, A, A, C]$. The structure of a list is more material (or "locative")[2] than the structure of a set. In the set structure the precise location of an item is immaterial, which is not the case for the list structure where the location counts. Of course, we may simulate sets by means of lists, but that implies novel "rules": one for allowing contraction (several occurrences of the same item are supposed to be reducible to only one), another for allowing permutation. It is only by means of these rules that we may approximate sets. For instance, by means of permutation, $[A, B, A, A, C]$ would be equal to $[B, A, A, A, C]$; by contraction, it would be equal to $[B, A, C]$, and therefore there would be no difference between giving the list or the set $\{A, B, A, A, C\}$, since this set is in fact the set $\{B, A, C\}$. Hence the new *structural rules*, as given below.

4.3.7 *Structural rules*

(Where Γ is a list and the comma between Γ and Δ represents their juxtaposition.)

[2]If we adopt the terminology introduced by Girard, a set is a *spiritualist* entity, whereas a list is a *locative* one. This opposition refers to the difference of levels of these entities. Locative entities are identified by their addresses, whereas spiritualist ones may be defined modulo isomorphisms.

Permutation:

$$\frac{\Gamma, A, B, \Delta \vdash C}{\Gamma, B, A, \Delta \vdash C}\,[P\ L]$$

Contraction:

$$\frac{\Gamma, A, A \vdash C}{\Gamma, A \vdash C}\,[C\ L]$$

The weakening rule should be formulated as

$$\frac{\Gamma, \Delta \vdash C}{\Gamma, A, \Delta \vdash C}\,[W\ L]$$

but because of permutation, we can content ourselves with

$$\frac{\Gamma \vdash C}{\Gamma, A \vdash C}\,[W\ L]$$

4.4 Classical Sequent Calculus

It remains to see how to complete the previous system in order to obtain the whole classical logic. An idea which comes to mind is to make the calculus symmetrical by considering that on the right-hand side as well as on the left-hand side of any sequent, we have a *list* of formulae. This requires us to extend the structural rules to the right-hand side.

4.4.1 *Structural rules*

Permutation:

$$\frac{\Gamma, A, B, \Delta \vdash \Theta}{\Gamma, B, A, \Delta \vdash \Theta}\,[P\ L] \qquad \frac{\Gamma \vdash \Delta, A, B, \Theta}{\Gamma \vdash \Delta, B, A, \Theta}\,[P\ R]$$

Weakening:

$$\frac{\Gamma \vdash \Theta}{\Gamma, A \vdash \Theta}\,[W\ L] \qquad \frac{\Gamma \vdash \Theta}{\Gamma \vdash A, \Theta}\,[W\ R]$$

Contraction:

$$\frac{\Gamma, A, A \vdash \Theta}{\Gamma, A \vdash \Theta}\,[C\ L] \qquad \frac{\Gamma \vdash A, A, \Theta}{\Gamma \vdash A, \Theta}\,[C\ R]$$

4.4.2 *Identity rules*

Axiom:

$$A \vdash A$$

Cut:

$$\frac{\Gamma \vdash A, \Delta \quad \Gamma', A \vdash \Delta'}{\Gamma, \Gamma' \vdash \Delta, \Delta'}\,[cut]$$

4.4.3 *Logical rules*

Negation:

$$\frac{\Gamma \vdash A, \Delta}{\Gamma, \neg A \vdash \Delta}\,[\neg L] \qquad \frac{\Gamma, A \vdash \Delta}{\Gamma \vdash \neg A, \Delta}\,[\neg R]$$

Conjunction:

$$\frac{\Gamma, A \vdash \Delta}{\Gamma, A \wedge B \vdash \Delta}\,[\wedge_1 L] \qquad \frac{\Gamma, B \vdash \Delta}{\Gamma, A \wedge B \vdash \Delta}\,[\wedge_2 L]$$

Variant:

$$\frac{\Gamma, A, B \vdash \Delta}{\Gamma, A \wedge B \vdash \Delta}\,[\wedge L]$$

Multiplicative version:

$$\frac{\Gamma \vdash A, \Delta \quad \Gamma' \vdash B, \Delta'}{\Gamma, \Gamma' \vdash A \wedge B, \Delta, \Delta'}\,[\wedge R]$$

Additive version:

$$\frac{\Gamma \vdash A, \Delta \quad \Gamma \vdash B, \Delta}{\Gamma \vdash A \wedge B, \Delta}\,[\wedge R]$$

Disjunction:

Multiplicative version:

$$\frac{\Gamma, A \vdash \Delta \quad \Gamma', B \vdash \Delta'}{\Gamma, \Gamma', A \vee B \vdash \Delta, \Delta'}\,[\vee L]$$

Additive version:

$$\frac{\Gamma, A \vdash \Delta \quad \Gamma, B \vdash \Delta}{\Gamma, A \vee B \vdash \Delta}[\vee L]$$

$$\frac{\Gamma \vdash A, \Delta}{\Gamma \vdash A \vee B, \Delta}[\vee_1 R] \qquad \frac{\Gamma \vdash B, \Delta}{\Gamma \vdash A \vee B, \Delta}[\vee_2 R]$$

Variant:

$$\frac{\Gamma \vdash A, B, \Delta}{\Gamma \vdash A \vee B, \Delta}[\vee R]$$

Implication:

Multiplicative version:

$$\frac{\Gamma \vdash A, \Delta \quad \Gamma', B \vdash \Delta'}{\Gamma, \Gamma', A \Rightarrow B \vdash \Delta, \Delta'}[\Rightarrow L] \qquad \frac{\Gamma, A \vdash B, \Delta}{\Gamma \vdash A \Rightarrow B, \Delta}[\Rightarrow R]$$

Additive version:

$$\frac{\Gamma \vdash A, \Delta \quad \Gamma, B \vdash \Delta}{\Gamma, A \Rightarrow B \vdash \Delta}[\Rightarrow L]$$

When looking at structural rules, it immediately appears that the comma (“,”) cannot have the same interpretation on the right-hand side as on the left. $[W\ L]$ obviously means that if Γ allows us to deduce Θ, then *a fortiori*, Γ *and* A allow us to deduce the same conclusion Θ. $[W\ R]$ cannot mean that from $\Gamma \vdash \Theta$ we may deduce that Γ gives Θ *and* A! If it were the case, anything could be deduced. Nevertheless, from $\Gamma \vdash \Theta$ we may deduce that Γ gives Θ *or* A, which is actually a way to weaken a result. Hence the fact that the comma on the right must be interpreted as an “or”. The $[\wedge\ L]$ and $[\vee\ R]$ rules confirm this interpretation.

Negation rules are interpreted easily as side changes for the formula which is negated. Let us note that we get easily the excluded middle principle by means of them (thus obtaining a classical calculus).

$$\dfrac{\dfrac{\dfrac{}{A \vdash A}[Ax]}{\vdash A, \neg A}[\neg\ R]}{\vdash A \vee \neg A}[\vee\ R]$$

The additive and multiplicative versions are equivalent, like we show easily in the case of conjunction.

First, let us prove that the additive version of conjunction implies its multiplicative version. By means of as many applications of the weakening rule as necessary, we prove

$$\frac{\Gamma \vdash A, \Delta}{\Gamma, \Gamma' \vdash A, \Delta, \Delta'}$$

and

$$\frac{\Gamma' \vdash B, \Delta'}{\Gamma, \Gamma' \vdash B, \Delta, \Delta'}$$

We use then the additive rule to prove

$$\frac{\Gamma, \Gamma' \vdash A, \Delta, \Delta' \quad \Gamma, \Gamma' \vdash B, \Delta, \Delta'}{\Gamma, \Gamma' \vdash A \wedge B, \Delta, \Delta'}$$

Second, let us prove that the multiplicative version implies the additive. We have

$$\frac{\Gamma \vdash A, \Delta \quad \Gamma \vdash B, \Delta}{\Gamma, \Gamma \vdash A \wedge B, \Delta, \Delta}$$

by means of the multiplicative rule. Then, we use permutation and contraction as many times as necessary to obtain

$$\frac{\Gamma, \Gamma \vdash A \wedge B, \Delta, \Delta}{\Gamma \vdash A \wedge B, \Delta}$$

4.5 Some Properties of the Sequent Calculus

4.5.1 *Subformula property*

A peculiarity of the sequent calculus, which makes it so helpful, is that for each rule, except for the cut rule, the formulae in the premises which are not in the contexts (which always remain passive) are *subformulae* of the new formula we get in the conclusion.[3] We can conclude that a *cut-free* proof of a formula uses only its subformulae. When applied to only the

[3]If we extend the calculus to predicates, the formula $A[t/x]$, which results from the substitution of t in A where x occurs freely, is seen as a subformula of a quantified formula $\forall x A$ or $\exists x A$.

propositional part of the calculus, this provides a *decision procedure*, and thus gives intuitionistic logic such a procedure.[4] We shall use this result intensively in the research of proofs (that is, syntactic derivations) in the Lambek calculus.

4.5.2 *Cut-elimination*

If the subformula property proves useful for getting a decision procedure, we know it only applies to cut-free proofs. So what do we do in the case of a proof with cuts? The answer is easy: cuts can always be eliminated. This is a result which corresponds, in the sequent calculus, to the normalization theorem in natural deduction. As we mentioned above, it is known as Gentzen's *Hauptsatz*, and can be formulated in the following way:

Theorem 4.1. *The cut rule is redundant in the sequent calculus.*

For a demonstration of this theorem, we refer the reader to Girard *et al.* (1989). It is important to notice that the proof of this theorem is *constructive*, that is, it provides at the same time an algorithm for eliminating the cuts in a demonstration.

4.6 Linear Logic

In the previous section, we stated the equivalence of additive and multiplicative versions of the classical sequent calculus. It rests on the use of structural rules of weakening and contraction. Suppose we suppress those rules, then this equivalence is lost, which means that the additive version and the multiplicative one define different connectives. This leads to linear logic, invented by Jean-Yves Girard in the mid-1980s. The additive rules define the additive connectives & and $\oplus$, while the multiplicative rules define the multiplicative connectives $\wp$ and $\otimes$.

4.6.1 *Identity rules*

Axiom:

$$A \vdash A$$

[4]Of course this does not work with the predicate calculus because of an infinity of subformulae.

Cut:

$$\frac{\Gamma \vdash A, \Delta \quad \Gamma', A \vdash \Delta'}{\Gamma, \Gamma' \vdash \Delta, \Delta'} [cut]$$

4.6.2 *Logical rules*

Negation:

$$\frac{\Gamma \vdash A, \Delta}{\Gamma, A^{\perp} \vdash \Delta} [^{\perp}L] \qquad \frac{\Gamma, A \vdash \Delta}{\Gamma \vdash A^{\perp}, \Delta} [^{\perp}R]$$

Tensor (multiplicative conjunction):

$$\frac{\Gamma, A, B \vdash \Delta}{\Gamma, A \otimes B \vdash \Delta} [\otimes L] \qquad \frac{\Gamma \vdash A, \Delta \quad \Gamma' \vdash B, \Delta'}{\Gamma, \Gamma' \vdash A \otimes B, \Delta, \Delta'} [\otimes R]$$

"with" (additive conjunction):

$$\frac{\Gamma \vdash A, \Delta \quad \Gamma \vdash B, \Delta}{\Gamma \vdash A \& B, \Delta} [\& R]$$

$$\frac{\Gamma, A \vdash \Delta}{\Gamma, A \& B \vdash \Delta} [\& L]_r \qquad \frac{\Gamma, B \vdash \Delta}{\Gamma, A \& B \vdash \Delta} [\& L]_r$$

Disjunction:

"par" (multiplicative disjunction):

$$\frac{\Gamma, A \vdash \Delta \quad \Gamma', B \vdash \Delta'}{\Gamma, \Gamma', A \wp B \vdash \Delta, \Delta'} [\wp L] \qquad \frac{\Gamma \vdash \Delta, A, B, \Delta'}{\Gamma \vdash \Delta, A \wp B, \Delta'} [\wp R]$$

"plus" (additive disjunction):

$$\frac{\Gamma, A \vdash \Delta \quad \Gamma, B \vdash \Delta}{\Gamma, A \oplus B \vdash \Delta} [\oplus L]$$

$$\frac{\Gamma \vdash A, \Delta}{\Gamma \vdash A \oplus B, \Delta} [\oplus_1 R] \qquad \frac{\Gamma \vdash B, \Delta}{\Gamma \vdash A \oplus B, \Delta} [\oplus_2 R]$$

Linear implication:

$$\frac{\Gamma \vdash A, \Delta \quad \Gamma', B \vdash \Delta'}{\Gamma, \Gamma', A \multimap B \vdash \Delta, \Delta'} [\multimap L] \qquad \frac{\Gamma, A \vdash B, \Delta}{\Gamma \vdash A \multimap B, \Delta} [\multimap R]$$

It may be showed that, in analogy with classical logic, we have

$$A \multimap B \equiv A^{\perp} \wp B$$

Moreover, we may prove several equivalences (we write $A \equiv B$ for the conjunction of $A \vdash B$ and $B \vdash A$):

De Morgan's rules:

- $(A \otimes B)^{\perp} \equiv A^{\perp} \wp B^{\perp}$
- $(A \wp B)^{\perp} \equiv A^{\perp} \otimes B^{\perp}$
- $(A \& B)^{\perp} \equiv A^{\perp} \oplus B^{\perp}$
- $(A \oplus B)^{\perp} \equiv A^{\perp} \& B^{\perp}$

Distributivity:

- $A \otimes (B \oplus C) \equiv (A \otimes B) \oplus (A \otimes C)$
- $A \wp (B \& C) \equiv (A \wp B) \& (A \wp C)$

It is worth noticing that these two distributivity laws dictated the choice by Girard of these particular symbols. In each family (respectively $\{\otimes, \oplus\}$ and $\{\&, \wp\}$), a symbol intervening in a distributivity law is a 180° rotation of the other.

4.6.3 *Exponentials*

Linear logic is not only a resource-sensitive logic, it also allows encoding of classical and intuitionistic logics, provided that we get the permission to reintroduce locally the power of structural rules, which have been abandoned at the global level. This is done by introducing new unary connectives, that Girard named *exponentials* because of their behaviour which is analogous to the mathematical function of the same name, which translates an additive structure into a multiplicative one.

The exponential "!" is such that the resource affected by it may be used freely as many times as we want (including none). The exponential "?" is its dual; in other terms, it is defined by

$$\begin{aligned} (?\,\mathrm{A})^{\perp} &= \,!(\mathrm{A}^{\perp}) \\ (!\,\mathrm{A})^{\perp} &= \,?(\mathrm{A}^{\perp}) \end{aligned}$$

We have

Introduction of "!":

$$\frac{\Gamma, A \vdash \Delta}{\Gamma, !A \vdash \Delta}\,[!\ L] \qquad \frac{!\Gamma \vdash A, ?\Delta}{!\Gamma \vdash !A, ?\Delta}\,[!\ R]$$

Structural rules:

$$\frac{\Gamma \vdash \Delta}{\Gamma, !A \vdash \Delta}[!\ W] \qquad \frac{\Gamma, !A, !A \vdash \Delta}{\Gamma, !A \vdash \Delta}[!\ C]$$

Introduction of "?":

$$\frac{\Gamma \vdash A, \Delta}{\Gamma \vdash ?A, \Delta}[?\ R] \qquad \frac{!\Gamma, A \vdash ?\Delta}{!\Gamma, ?A \vdash ?\Delta}[?\ L]$$

Structural rules:

$$\frac{\Gamma \vdash \Delta}{\Gamma \vdash ?A, \Delta}[?\ W] \qquad \frac{\Gamma \vdash ?A, ?A, \Delta}{\Gamma \vdash ?A, \Delta}[?\ C]$$

The encoding of intuitionistic logic rests on the following equivalence, which makes a bridge between our usual implication and the linear one.

$$A \Rightarrow B \equiv !A \multimap B$$

The translation of intuitionistic logic into linear logic is then recursively defined:

$$\begin{array}{lcll} (A)^o & = & A & \text{if } A \text{ is an atom} \\ (A \Rightarrow B)^o & = & !A^o \multimap B^o & \\ (A \wedge B)^o & = & A^o \& B^o & \\ (A \vee B)^o & = & !(A^o) \oplus !(B^o) & \\ (\neg A)^o & = & !A^o \multimap \bot & \end{array}$$

(see below the use of $\bot$).

Other equivalences may be shown:

- $!(A\&B) \equiv !A \otimes !B$
- $?(A \oplus B) \equiv ?A \wp ?B$

and other deduction relations:

- $!A \vdash 1 \& A \& (!A \otimes !A)$
- $!(A\&B) \equiv !A \otimes !B$

The relation $!(A\&B) \equiv !A \otimes !B$ gives its name to the exponential "!" since it transforms an additive structure into a multiplicative one.

4.6.4 *Constants*

We may add to the classical calculus two constants $\top$ (for "true") and $\bot$ (for "false"), the introduction rules of which are

Left introduction of $\top$:

$$\frac{\Gamma \vdash \Delta}{\Gamma, \top \vdash \Delta}$$

Right introduction of $\bot$:

$$\frac{\Gamma \vdash \Delta}{\Gamma \vdash \bot, \Delta}$$

Axioms:

$$\Gamma, \bot \vdash \Delta \qquad \vdash \top$$

Comments: The two rules are due to the fact that $\top$ and $\bot$ are respectively the neutral elements of $\vee$ and $\wedge$. The first axiom interprets the famous *ex falso sequitur quodlibet*. It is a kind of left introduction rule of $\bot$ but without premises. The second axiom interprets the classical principle that "true" is always true, and it is a kind of right introduction of $\top$.

When transposed to linear logic, these constants have also their variants. Let us keep the notation $\top$ for &, and let us introduce **1** for $\otimes$, but let us give the notation $\bot$ for the neutral element of $\wp$, and the notation **0** for the additive disjunction $\oplus$. The left introduction rule of $\bot$ becomes the following multiplicative rule:

$$\frac{\Gamma \vdash \Delta}{\Gamma, 1 \vdash \Delta}$$

while the axiom of truth becomes

$$\vdash 1$$

Otherwise, the right introduction rule of $\bot$ remains unchanged:

$$\frac{\Gamma \vdash \Delta}{\Gamma \vdash \bot, \Delta}$$

By duality with the axiom for **1**, we get

$$\bot \vdash$$

Let us admit then that **0** is absorbing for the tensor:

Axioms:

$$\Gamma, 0 \vdash \Delta$$

and by duality:

$$\Gamma \vdash \top, \Delta$$

We may then show easily

- $A \otimes \mathbf{0} \equiv \mathbf{0}$
- $A \wp \top \equiv \top$
- $A \multimap \top \equiv \top$
- $\mathbf{0} \multimap A \equiv \top$
- $A \otimes \mathbf{1} \equiv A$
- $A \wp \bot \equiv A$
- $\mathbf{1} \multimap A \equiv A$
- $A \multimap \bot \equiv A^{\bot}$
- $!\top \equiv \mathbf{1}$

NB: The last of these equivalences reinforces the analogy between the unary connective "!" and the exponential function ($\exp(0) = 1$).

4.6.5 *The one-sided calculus*

It is now obvious that, by means of negation (which allows transfer of formulae from the left to the right), the linear calculus may be expressed only by right rules. Hence, the following one-sided calculus:

Axiom:

$$\vdash A, A^{\bot}$$

Cut:

$$\frac{\vdash A, \Gamma \quad \vdash A^{\bot}\Gamma'}{\vdash \Gamma, \Gamma'}\,[cut]$$

Multiplicative part:

$$\frac{\vdash A, \Gamma \quad \vdash B, \Gamma'}{\vdash A \otimes B, \Gamma, \Gamma'}\,[\otimes R] \qquad \frac{\vdash \Gamma, A, B, \Gamma'}{\vdash \Gamma, A \wp B, \Gamma'}\,[\wp R]$$

Additive part:

$$\frac{\vdash A, \Gamma \quad \vdash B, \Gamma}{\vdash A\&B, \Gamma}[\&R]$$

$$\frac{\vdash A, \Gamma}{\vdash A \oplus B, \Gamma}[\oplus_1 R] \qquad \frac{\vdash B, \Gamma}{\vdash A \oplus B, \Gamma}[\oplus_2 R]$$

Exponentials:

$$\frac{\vdash A, ?\Gamma}{\vdash !A, ?\Gamma}[!\ R] \qquad \frac{\vdash A, \Gamma}{\vdash ?A, \Gamma}[?\ R]$$

Structural rules:

$$\frac{\vdash \Gamma}{\vdash ?A, \Gamma}[?\ W] \qquad \frac{\vdash ?A, ?A, \Gamma}{\vdash ?A, \Gamma}[?\ C]$$

NB: Negation, in such a system, is meta-linguistic, that is, it is recursively defined by stating

$$\begin{array}{lcl}
(\mathrm{A}^\perp)^\perp & = & \mathrm{A} \\
(A \otimes B)^\perp & = & \mathrm{A}^\perp \wp \mathrm{B}^\perp \\
(A \wp B)^\perp & = & \mathrm{A}^\perp \otimes \mathrm{B}^\perp \\
(A\&B)^\perp & = & \mathrm{A}^\perp \oplus \mathrm{B}^\perp \\
(A \oplus B)^\perp & = & \mathrm{A}^\perp \& \mathrm{B}^\perp \\
(!A)^\perp & = & ?(\mathrm{A}^\perp) \\
(?A)^\perp & = & !(\mathrm{A}^\perp)
\end{array}$$

It is of course always possible to add one-sided rules for constants, knowing that $\mathbf{1}^\perp = \perp$, $\perp^\perp = \mathbf{1}$, $\top^\perp = \mathbf{0}$ and $\mathbf{0}^\perp = \top$:

$$\frac{}{\vdash 1}[1] \qquad \frac{\vdash \Delta}{\vdash \perp, \Delta}[\perp] \qquad \frac{}{\vdash \top, \Delta}[\top]$$

There is no rule for **0** in the one-sided calculus since it is already expressed via its dual $\top$.

4.6.6 *Intuitive interpretation*

The *multiplicative* management of contexts corresponds to a cumulative process: if two disjoint contexts (Γ, Δ) (Γ on the left, Δ on the right) and

(Γ', Δ') may give separately A and B respectively, then, by putting them together, we get at the same time A *and* B.

In contrast, the *additive* management corresponds to superposition: the same context is used for producing (in the case of &) several different formulae. If we look at the $[\& R]$ rule, we can see that if the context (Γ, Δ) may be used as well for producing A as for producing B, then it may be used to produce $A\&B$. The best interpretation we can have for & is that of a *choice*. Let us imagine a vending machine serving drinks which, in exchange for 50 c, gives at our choice a coffee or a tea. Then we have

$$\frac{50\ c \vdash COFFEE \quad 50\ c \vdash TEA}{50\ c \vdash COFFEE\ \&\ TEA}$$

In other words, for 50 c, you have the choice between a coffee and a tea.

The $\oplus$ case is dual. Here, let us imagine the machine gives tea or coffee only, depending on what it possesses at the time. If it gives coffee, it also gives coffee $\oplus$ tea; if it gives tea, it also gives the same combination of coffee $\oplus$ tea. In other terms, a constant situation (coffee $\oplus$ tea) is realized when it provides any of the two hot drinks, but the point is that this time, that is its own choice and not ours! If we recall here the De Morgan laws

$$(A\&B)^{\perp} \equiv A^{\perp} \oplus B^{\perp}$$
$$(A \oplus B)^{\perp} \equiv A^{\perp}\&B^{\perp}$$

we see that negation makes us pass from an *active* viewpoint to a *passive* one, and *vice versa*. If we admit formulae have from now on a *polarity* (positive or negative), and if we interpret a negative formula as the process of giving something and a positive one as the process of accepting it, the above equivalences can be interpreted in the following way:

- to give an active choice between A and B amounts to accepting either to give A or to give B, according to the choice of our partner,
- to give a passive choice between A and B amounts to accepting to give A or B according to our own choice.

The polarized pair of *acceptance* and *gift* may be of course replaced by another, like the pair of *demand* and *offer*. What is on the right side is interpreted as a demand and what is provided on the left side is interpreted as an offer which must correspond to the demand. In the coffee & tea (or coffee $\oplus$ tea) example, the user asks for coffee or tea and she offers a corresponding amount of money. By duality, the vending machine asks for an amount of money and in exchange offers a hot drink.

With regard to the linear implication, let us observe that old classical theorems are no longer theorems in linear logic. Let us look at for instance the formula $A \Rightarrow (B \Rightarrow A)$ again, which in linear logic would be written $A \multimap (B \multimap A)$; an attempt to prove it in LL leads to

$$\dfrac{\dfrac{A, B \vdash A}{A \vdash B \multimap A}}{\vdash A \multimap (B \multimap A)}$$

but, because of the lack of weakening, $A, B \vdash A$ is not provable. We may notice here in passing that the absence of such a theorem goes in the direction desired by authors like Belnap and Anderson (1975) who have criticized this kind of "law", arguing of the "non-relevance" of B in the classical or the intuitionistic proof: the hypothesis B may be introduced freely but nothing in the proof comes from it, as we may observe when looking at the deduction in the Fitch presentation.

$$\begin{array}{l|lll}
1 & \quad\vert\; A & & \\
2 & \quad\vert\quad\vert\; B & & \\
3 & \quad\vert\quad\vert\; A & & \text{R, 1} \\
4 & \quad\vert\; B \Rightarrow A & & \\
5 & A \Rightarrow (B \Rightarrow A) & &
\end{array}$$

If it is not possible to obtain this result in LL, it is because a formula introduced as a hypothesis must be, in this system, used exactly once by a logical rule. In the same way, $(A \multimap (A \multimap B)) \multimap (A \multimap B)$ is also not provable: in the classical sequent calculus, its proof needs the contraction rule.

$$\dfrac{\dfrac{\dfrac{\dfrac{A \vdash A \quad \dfrac{A \vdash A \quad B \vdash B}{A \Rightarrow B, A \vdash B}}{(A \Rightarrow (A \Rightarrow B)), A, A \vdash B}}{(A \Rightarrow (A \Rightarrow B)), A \vdash B}}{(A \Rightarrow (A \Rightarrow B)) \vdash (A \Rightarrow B)}}{\vdash (A \Rightarrow (A \Rightarrow B)) \Rightarrow (A \Rightarrow B)}$$

Without the contraction rule, this theorem would therefore be not provable. In other words, there must be in the context as many occurrences of formulae as needed to make the proof, but there must not be more than necessary! In the end, the occurrences of formulae (or *tokens*) behave exactly like *resources* and a proof is like a process which consumes all the resources which are provided to it and only them. This explains why there are so many "economic" examples in the literature.

For instance, if with 50 c, I may get a coffee or a tea, then I shall have the formulae

$$50\ c \multimap COFFEE \qquad 50\ c \multimap TEA$$

but obviously not

$$50\ c \multimap COFFEE \otimes TEA$$

since the 50 c consumed for the coffee is no longer available for the tea, and reciprocally.

This situation has also been compared with the one which prevails in quantum mechanics, a well-known principle of which is that it is not possible to have a precise measure at the same time of the location and the moment of a particle. We have therefore

$$measure \multimap POS \qquad measure \multimap MOM$$

but not

$$measure \multimap POS \otimes MOM$$

The precise measure of one of the two quantities destroys the possibility to make the other.

If we consider now exponentials, it looks as if

- "!A" meant: "A may be used (or consumed or provided) any number of times (including 0)",
- "?A" meant: "there is a *potentiality* of any number of A (the ability to demand or to ask for it, for instance)".

A simple potentiality does not cost anything, which explains why it may be given for free, and thus any number of times. Asking for $?A$ is therefore asking for something which has no cost, like access to the location where a resource was stored (precisely the resource A). In computer terms, and after an interpretation by Lafont, $!A$ may be viewed as the *storage* of the resource A, with the possibility to use this resource as long as we desire, while $?A^{\perp}$ means the possibility to activate this potentiality, by really using it, or by *erasing* it or by *duplicating* it (it is called *costorage*). When a client S_1 asks a server S_2 to give her an indefinite amount of the good A, from S_2's viewpoint, it is as if it offered to S_1 the right to solicit it an indefinite number of times to provide this good. So, by duality (negation), the request for $!A$ becomes the acceptance of $?A^{\perp}$.

The restaurant example (also due to Lafont) is also well known:

Hors d'oeuvre
Green salad
or
Terrine
***
Main dish
T-bone steak
French fries at wish
***
Dessert
According to what is furnished:
Banana
or
Pineapple
***
PRICE
15 euros

It gives rise to the formula

$$15\ E \multimap (GS \& T) \otimes (STEAK \otimes !\ FRENCH\ FRIES) \otimes (BAN \oplus PIN)$$

where the exponential means "at your wish". The formula means that against 15 euros, you receive, in sequential order, a green salad or a terrine at your choice, a T-bone steak with as many French fries as you want (theoretically infinitely many!), and at the choice of the owner, either a banana or a pineapple.

It may be noticed that this gives the viewpoint of the consumer. The one of the restaurant's owner is expressed by

$$15\ E \otimes (GS^{\perp} \oplus T^{\perp}) \wp (STEAK^{\perp} \wp ?(FRENCH\ FRIES^{\perp})) \wp (BAN^{\perp} \& PIN^{\perp})$$

that is, the restaurant's owner receives 15 euros, and she prepares herself to provide in parallel a green salad or a terrine according to the choice of the consumer, a steak, infinitely many French fries (depending on the demand from the consumer) and according to her own decision a banana or a pineapple.

The "economic" interpretation makes us always assume two sides of goods, as expected: one side corresponding to *demand* and the other to *offer*. In a transaction (like in the case of the restaurant's owner), the two roles are exchanged. When the merchant asks for an amount of money, the consumer asks for the goods, and reciprocally when the first one offers goods, the second offers money. With regard to constants, that amounts to saying that if $\mathbf{1}$ is the neutral element of $\otimes$ and $\bot$ the neutral element of $\wp$, they correspond to each other; if one asks for $\mathbf{1}$, she may give in exchange only $\bot$, and reciprocally. We may therefore interpret the pair ($\mathbf{1}$, $\bot$) as an empty transaction, as if somebody gave something perfectly free for nothing. With this interpretation, we respect the laws, for instance,

$$\frac{\Gamma \vdash \Delta}{\Gamma, 1 \vdash \Delta}$$

since a null value which is provided for consumption does not change anything in a transaction process. For the dual law concerning $\wp$, it is the same: adding something with no value to an amount of goods does not change this amount from an economic viewpoint. $\bot$ costs nothing!

The interpretation of multiplicatives is therefore easy; it is not exactly the same with regard to the additives. What do $\mathbf{0}$ and $\top$ mean from this transactional viewpoint? $\mathbf{0}$ is the neutral element of $\oplus$ but also an absorbing element for $\otimes$, while $\top$ is neutral for & and absorbing for $\wp$. We suggest $\mathbf{0}$ could be interpreted as something *never available*, an *absence* of goods. Of course if you have a passive choice between A and such an absence, you know that you will get at the end A. Nevertheless, the relation $A \otimes \mathbf{0} = \mathbf{0}$ seems strange. It is as if the cumulation of a good A and an absence of goods collapsed into this absence. You will never get at the same time A and an absence of goods, even if you may get A alone! And of course the sequent $\Gamma,\ \mathbf{0} \vdash \Delta$ means that if you can use these lacking goods — which is of course impossible — then you will be able to do anything: exactly the economic counterpart of *ex falso sequitur quodlibet*.

There is of course the dual problem with $\top$. What is $\top$? It is worth noting that we have $\Gamma \vdash \top, \Delta$ for any Γ and Δ, that is, we can get $\top$

against *any* amount of money! Thus, $\top$ is what satisfies *any* offer. That can be interpreted by *to get it, give me what, you want* including nothing and what is impossible to give (something reminiscent of Lacan's aphorism: "to love is to give what we have not"!). The fact that $!\top = \mathbf{1}$ may then be explained: *you may have as much of this good that you want... it is always in fact of null value.* And if you have the active choice $A\&\top$, from a "rational" viewpoint, you will choose A since for the other goods, you may obtain it for free with any amount of money!

4.7 Back to the Lambek Calculus

4.7.1 *The Lambek calculus as non-commutative linear logic*

Like Girard remarks in his *Le Point Aveugle* (2006, Vol. I, Section 11.1.3), the Lambek calculus, of which we gave a short introduction in Section 3.3, is not a complete logic in the strong sense, since as connectives, it uses only the two arrows ($/$ and $\backslash$) and, in the version with product, the non-commutative tensor ($\bullet$). To have the product without the disjunction, the implications without the possibility of expressing their equivalences with disjunctive formulae (which would involve negation) appear to be at odds. Transforming the Lambek calculus into an authentic logic would entail the introduction of all these ingredients, starting from linear logic. In fact, at first sight, the passage from linear logic to the Lambek calculus rests on the suppression of the last structural rule: permutation. If we wish to keep the idea that $A \multimap B \equiv A^{\perp} \wp B$, it is of course necessary to add $\wp$, which, exactly like $\otimes$, must be non-commutative.

In this case, the direction of the implication (from left to right or from right to left) will only result from the non-commutativity of $\wp$. We shall have

$$A\backslash B \equiv A^{\perp} \wp\, B \qquad B/A \equiv B \wp A^{\perp}$$

but if we need a disjunction, symmetrical with regard to the product, we must necessarily stay inside a classical calculus, that is, a calculus with formulae on the right and on the left of the sequents. This way, we obtain the following rules for $\wp$ (on the left) and for $\bullet$ (on the right):

$$\frac{\Gamma, A \vdash \Delta \quad B, \Gamma' \vdash \Delta'}{\Gamma, A\wp B, \Gamma' \vdash \Delta, \Delta'}[\wp\, G] \qquad \frac{\Gamma \vdash \Delta, A \quad \Gamma' \vdash B, \Delta'}{\Gamma, \Gamma' \vdash \Delta, A \bullet B, \Delta'}[\bullet\, D]$$

We can actually retrieve the usual rules of the Lambek product by taking Δ, δ' as empty (notice that the left context of the left formula in the product concatenates on the left with the similar context of the right formula and that, symmetrically, the right context of A, which is on the left of A, stays on the left in the concluding sequent, while the right context of B, which is on the right of B, stays on the right).

$/$ and $\backslash$ may then be defined as mentioned above, as illustrated by the two following deductions:

$$\frac{np, np^{\perp} \vdash \quad s \vdash s}{np, np^{\perp} \wp\, s \vdash s} \qquad \frac{s \vdash s \quad np^{\perp}, np \vdash}{s \wp\, np^{\perp}, np \vdash s}$$

which translate into

$$\frac{np \vdash np \quad s \vdash s}{np, np\backslash s \vdash s} \qquad \frac{np \vdash np \quad s \vdash s}{s/np, np \vdash s}$$

Let us notice on the other hand that in the Lambek calculus, $(C/B)/A \equiv C/(A \bullet B)$. Let us for instance imagine some verb which would take two ordered complements (like in *Mary calls her son Kim*) and that would have the type $(s/np_1)/np_2$, which would mean that it waits on its right at first np_2, and then np_1, in other words the product $np_2 \bullet np_1$. In disjunction terms (and using associativity) that type would give as well

$$s \wp (np_1^{\perp} \wp\, np_2^{\perp}) \quad as \quad s \wp (np_2 \bullet \; np_1)^{\perp}$$

which leads to the De Morgan laws

$$(A \bullet B)^{\perp} \; \equiv \; B^{\perp} \wp\, A^{\perp} \quad and \quad (A \wp B)^{\perp} \; \equiv \; B^{\perp} \; \bullet \; A^{\perp}$$

But then what about negation? If we consider negation as a side exchange connective, we see that

$$A^{\perp} \bullet B^{\perp} \vdash \quad is\ equivalent\ to \quad \vdash B \wp A$$

which entails that

$$A^{\perp}, B^{\perp} \vdash \qquad is\ equivalent\ to \qquad \vdash B, A$$

In other words, in negation, the order of formulae is permuted, hence the rules

$$\frac{\Gamma \vdash A, \Delta}{A^{\perp}, \Gamma \vdash \Delta} \qquad \frac{A, \Gamma \vdash \Delta}{\Gamma \vdash A^{\perp}, \Delta}$$

Obviously, it is noticeable that given a sequent like

$$A_1 \vdash A_2, A_3, \ldots, A_n$$

the rules for negation allow us to have also, by transfer to the right,

$$\vdash A_1^{\perp}, A_2, A_3, \ldots, A_n \quad \textit{and} \quad \vdash A_2, A_3, \ldots, A_n, A_1^{\perp}$$

In order to avoid undesirable effects, it is natural to admit then the rule of *cyclic exchange*

$$\frac{\vdash \Gamma, A}{\vdash A, \Gamma}$$

which assumes we work in a one-sided calculus. Hence, finally, the formulation of the cyclic linear logic (Yetter, 1990; Retoré, 1996):

$$\frac{\vdash \Gamma, A, B, \Delta}{\vdash \Gamma, A \wp B, \Delta}[\wp] \qquad \frac{\vdash \Gamma, A \qquad \vdash B, \Gamma'}{\vdash \Gamma, A \bullet B, \Gamma'}[\bullet]$$

with the cyclic exchange rule, the axiom and the cut rule given hereafter.

$$\frac{\vdash \Gamma, A \qquad \vdash A^{\perp}, \Delta}{\vdash \Gamma, \Delta}$$

The cut-elimination theorem may be demonstrated for this calculus.

We may also retrieve the Lambek calculus from this logic. We use for that the following translation:

$$A \backslash B := A^{\perp} \wp B$$
$$B / A := B \wp A^{\perp}$$
$$A_1, A_2, \ldots, A_n \vdash B := \vdash A_n^{\perp}, A_{n-1}^{\perp}, \ldots, A_1^{\perp}, B$$

In order to get the Lambek calculus, strictly speaking, we nevertheless have to specify that, in the use of the rule $[\wp]$, at least one of the two lists Γ and Δ be not empty. If $\perp$ is introduced into the calculus and if we wish to keep the equivalent of $A \multimap \perp \equiv A^{\perp}$, then, rigorously, there must be two negations: a left one, ${}^{\perp}A \equiv A \backslash \perp$, and a right one, $A^{\perp} \equiv \perp / A$. This gives a version of non-commutative linear logic which owes much to Abrusci (Abrusci and Ruet, 2000) (see Section 6.6.2). Identifying ${}^{\perp}A$ and $A^{\perp}$ leads to the cyclic exchange rule.

In effect, Abrusci's non-commutative multiplicative linear logic (MNLL) has the two following axioms:

$$\vdash A^{\perp}, A \qquad \vdash A, {}^{\perp}A$$

but not

$$\nvdash A, A^{\perp} \quad \nvdash^{\perp} A, A$$

In the same way, the following cut rules are admissible:

$$\frac{\vdash \Gamma, A, \Gamma' \qquad \vdash A^{\perp}, \Delta}{\vdash \Gamma, \Delta, \Gamma'} \qquad \frac{\vdash \Gamma, A, \Gamma' \qquad \vdash \Delta, {}^{\perp}A}{\vdash \Gamma, \Delta, \Gamma'}$$

If ${}^{\perp}A$ and $A^{\perp}$ are identified, then they collapse into only one rule, the one of cyclic logic, provided that the cyclic exchange rule allows us to replace $\vdash A^{\perp}, \Delta$ with $\vdash \Delta, A^{\perp}$.

4.8 Linguistic Applications of the Additives

The additive connectives of LL have interesting applications in linguistics, concerning cases of underspecification. An expression is said to be underspecified with regard to some feature (syntactic, morphologic or semantic) if it may appear in contexts of expressions which demand distinct values of this feature. For instance, in French, the article *les* is underspecified for gender. In German, the plural name *Frauen* is underspecified for case: it may appear in a context which commands the accusative as well as in a context which commands the dative. *Männer* is different: it is only accusative, and *Kindern* only dative. The peculiarity of an underspecified expression is that it may appear in strange contexts (which may be said to be overspecified) which demand a superposition of values which are in principle incompatible. For instance, the German verbal coordination *findet und hilft* asks at the same time for an accusative (*findet*) and a dative (*hilft*), which makes ungrammatical sequences like

**Er findet und hilft Männer*
**Er findet und hilft Kindern*

when, on the contrary, the following is possible:

Er findet und hilft Frauen

The solution sometimes proposed, which is to assume that *Frauen* has a conjunctive type $n \wedge acc \wedge dat$, does not work since *dat* and *acc* are exclusive and therefore $acc \wedge dat$ is "false"! It is however easy to realize that each instance of "and" in this conjunction has a different meaning. If *Frauen* is altogether of syntactic category n and of case value acc "and" dat, this expression is not at the same time acc and dat, but is ready to accept a demand for an accusative ($acc^\perp$) as well as a demand for a dative ($dat^\perp$), and it is therefore of type $n \otimes (acc^\perp \oplus dat^\perp)$. Concerning the coordination *findet und hilft*, it will give a VP (vp) if it receives an n and if it can "choose" acc or dat, which amounts to having the type $sv \wp (n^\perp \wp (acc \,\&\, dat))$. We get the following deduction:

$$\dfrac{vp \vdash vp \qquad \dfrac{\dfrac{n^\perp, n \vdash \qquad \dfrac{\dfrac{\dfrac{}{acc, acc^\perp \vdash}}{acc, acc^\perp \oplus dat^\perp \vdash} \quad \dfrac{\dfrac{}{dat, dat^\perp \vdash}}{dat, acc^\perp \oplus dat^\perp \vdash}}{acc \,\&\, dat, acc^\perp \oplus dat^\perp \vdash}}{n^\perp \wp (acc \,\&\, dat), n, acc^\perp \oplus dat^\perp \vdash}}{n^\perp \wp (acc \,\&\, dat), n \otimes (acc^\perp \oplus dat^\perp) \vdash}}{vp \wp (n^\perp \wp (acc \,\&\, dat)), n \otimes (acc^\perp \oplus dat^\perp) \vdash vp}$$

4.9 Proof Nets

4.9.1 *A geometrization of logic*

Among all the things brought by linear logic, one of the most important is the concept of *proof nets*, made possible by its "consumption of resources" aspect, and opening the field of geometrization (or topologization) of logic. A proof net is in fact, as we will see, the *geometrical* essence of a proof. The concept comes from the elimination of redundancies in the calculus, something realized by better management of contexts. Let us consider, for instance, the two following proofs of the valid sequent in LL:

$$\vdash A^\perp \wp B^\perp, (A \otimes B) \otimes C, C^\perp$$

$$\dfrac{\dfrac{\dfrac{\dfrac{\vdash A, A^\perp \qquad \vdash B, B^\perp}{\vdash A \otimes B, A^\perp, B^\perp} \qquad \vdash C, C^\perp}{\vdash (A \otimes B) \otimes C, A^\perp, B^\perp, C^\perp}}{\vdash A^\perp, B^\perp, (A \otimes B) \otimes C, C^\perp}}{\vdash A^\perp \wp B^\perp, (A \otimes B) \otimes C, C^\perp} \qquad \dfrac{\dfrac{\dfrac{\dfrac{\vdash A, A^\perp \qquad \vdash B, B^\perp}{\vdash A \otimes B, A^\perp, B^\perp}}{\vdash A \otimes B, A^\perp \wp B^\perp} \qquad \vdash C, C^\perp}{\vdash (A \otimes B) \otimes C, A^\perp \wp B^\perp, C^\perp}}{\vdash A^\perp \wp B^\perp, (A \otimes B) \otimes C, C^\perp}$$

They only differ in the order in which rules are applied. Moreover, at each step of the deduction, some part of the bottom sequent has been copied even though it is inactive. For instance, in the following step,

$$\dfrac{\dfrac{\cdots\cdots}{\vdash A \otimes B, A^{\perp} \wp B^{\perp} \qquad \vdash C, C^{\perp}}}{\dfrac{(A \otimes B) \otimes C, A^{\perp} \wp B^{\perp}}{\cdots}}$$

while only the formula $(A \otimes B) \otimes C$ is active, the context $A^{\perp} \wp B^{\perp}$ is carried from the bottom to the top. It will become active only at the following step.

Let us therefore look at what we get if we content ourselves to keep at each step only the active formula (the one which is decomposed at this step) and if we connect by an edge the active formula to each of the two subformulae which appear after decomposing. We get the following graph:

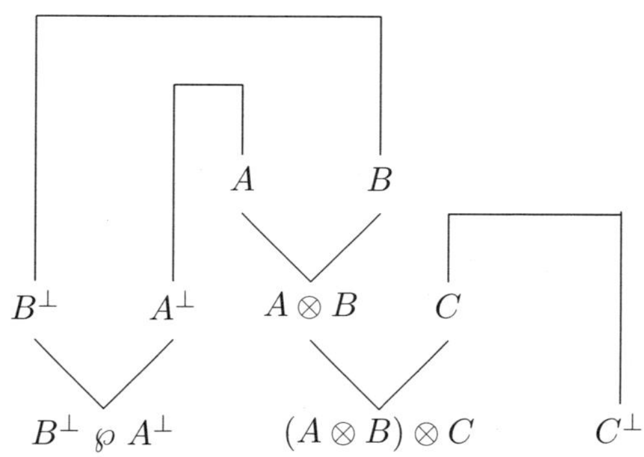

It is common to the two above demonstrations.

Here we deduce a graph from a demonstration, but it is interesting of course to build directly such a net, without passing through a deduction. For that aim, let us write for each formula of the one-sided sequent its subformulae tree, then link the positive and negative instances of the same atom. We get then a *proof structure*. Nevertheless, not every proof structure is a proof net. For instance, we could obtain a proof structure for the following sequent: $\vdash B^{\perp} \otimes A^{\perp}, A \otimes B$, as shown by the figure

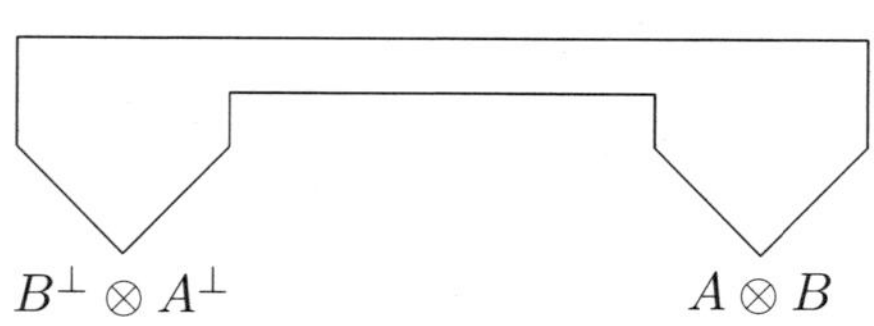

but this sequent is not valid.

However, the sequent $\vdash B^{\perp} \wp A^{\perp}, A \otimes B$ is provable. It is therefore necessary to identify *correction criteria* for proofs. Historically, the first criteria to be used were expressed in terms of *trips*. A trip in a net amounts to the move of a token according to precise instructions associated with each kind of link (depending on its being a tensor or a par). The long-trip condition expressed that a token had necessarily to travel among all the nodes of the net (two times, once by going up and once by going down). Other criteria were also introduced, like the one by Danos and Régnier (1995).

Let us define a set of links:
"Par" and "tensor" links:

(The root nodes of these links are called *conclusions*, the leaf nodes are called *premises*.)

"Axiom" link:

"Cut" link:

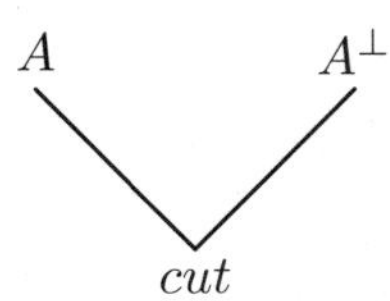

NB: Concerning axiom and cut links, the order of atoms or of formulae is immaterial. For the axiom link, it is possible to restrict it to the case of A being an atom, but this is of course not the case for the cut rule.

We may then define a *proof structure* as any such graph built only by means of these links such that each formula is the conclusion of exactly one link and the premise of at most one link.

Definition 4.1 (Correction criterion). A proof structure is a proof net if and only if the graph which results from the removal, for each $\wp$ link ("par" link) in the structure, of one of the two edges is connected and has no cycle (that is in fact a tree).

Example: The following proof structure is indeed a proof net since the two graphs in Fig. 4.9 are trees.

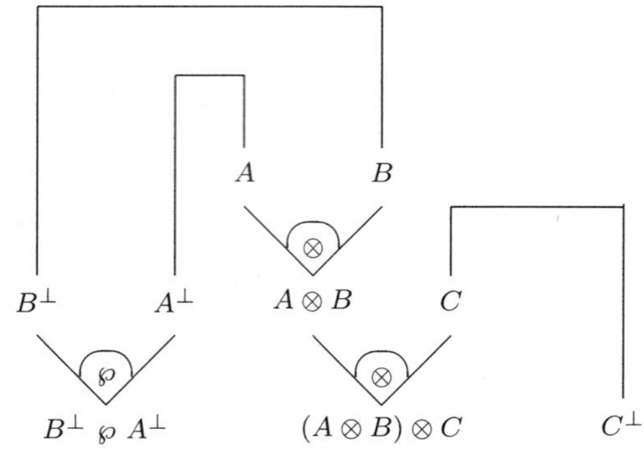

However, the following structure is not a net and, as we can see, does not correspond to any demonstration in LL.

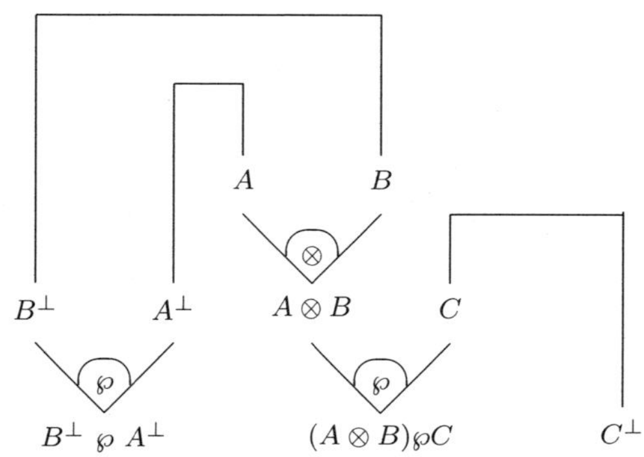

The removal of one edge of a $\wp$ reveals a disconnection (but also a cycle), as shown below.

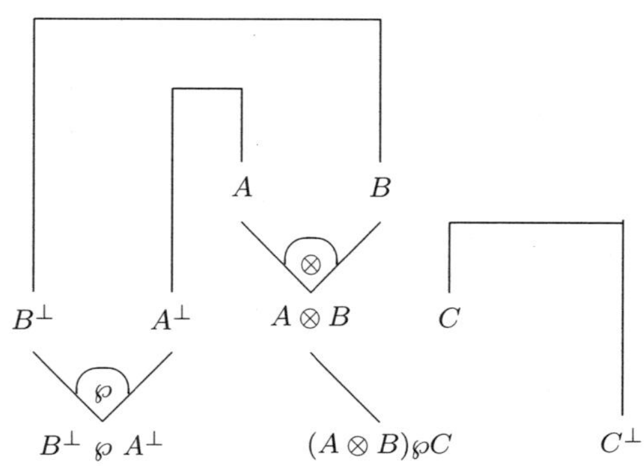

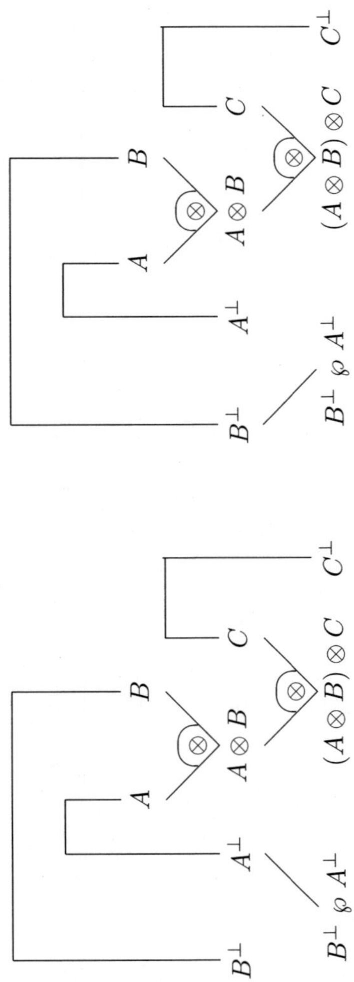

Fig. 4.9 Connected acyclic graphs

4.9.2 *Cut-elimination in proof nets*

It is interesting to notice how cut-elimination is performed in proof nets. Let us consider the proof net in Fig. 4.10, which contains a cut link between two conclusions. The elimination of the first cut generates two new cuts, but between simpler formulae (see Fig. 4.11). Eliminating cuts in the case of cuts with an axiom leads to the cut-free net in Fig. 4.12.

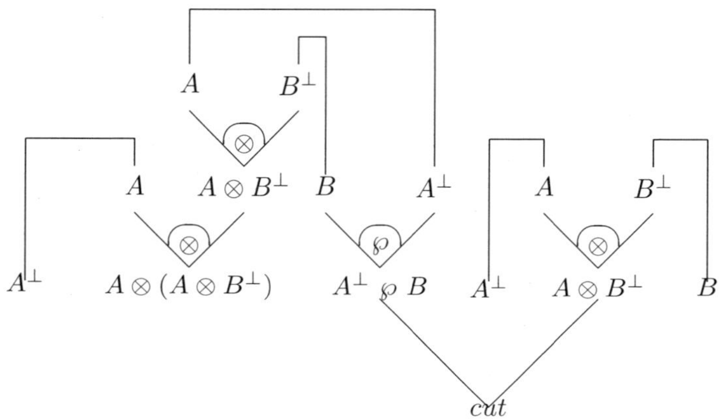

Fig. 4.10 A proof net with a cut link

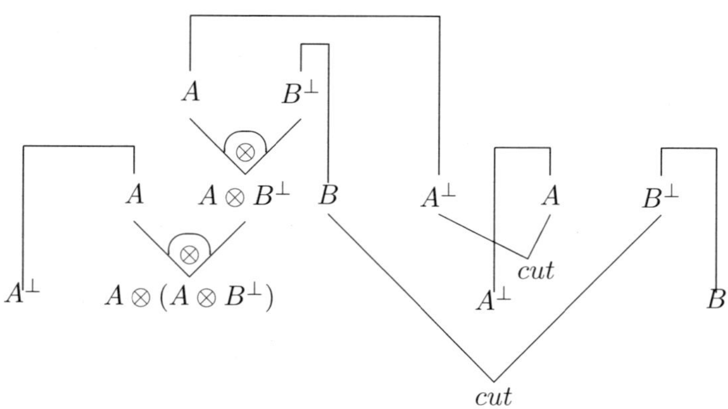

Fig. 4.11 Cut-elimination

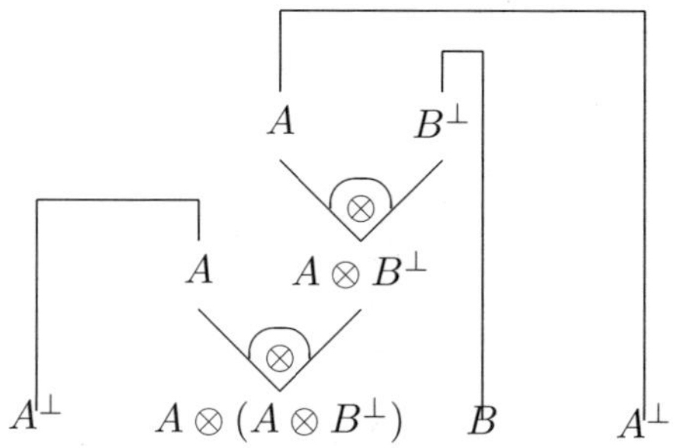

Fig. 4.12 Cut-free net

4.10 Proof Nets for the Lambek Calculus

The method of proof nets may be easily applied to the Lambek calculus. Let us take the example of the nominal phrase *the cat that Peter feeds*. The demonstration of its typing as an np amounts to constructing the proof of the following sequent:

$$np/n, n, (n\backslash n)/(s/np), np, (np\backslash s)/np \vdash np$$

which itself amounts to proving the following one-sided sequent in cyclic linear logic:

$$\vdash np \otimes (s^{\perp} \otimes np), np^{\perp}, (s \wp np^{\perp}) \otimes (n^{\perp} \otimes n), n^{\perp}, n \otimes np^{\perp}, np$$

It suffices then to prove that there is a proof net for this sequent. The only difference we have with regard to proof nets of the multiplicative linear logic is that here, we are not allowed to freely permute the premises of links at the same level: their left-right order is fixed and therefore a criterion of *planarity* is added to the criterion of acyclicity.

A proof structure is a net for cyclic linear logic if and only if

(1) the structure is such that removing one edge of each $\wp$ link always gives a connected acyclic graph,
(2) the structure of links is planar.[5]

[5]A graph is said to be planar if it is possible to find a representation of it in the plane such that any two edges never cross each other.

Let us check if it is the case for our example.

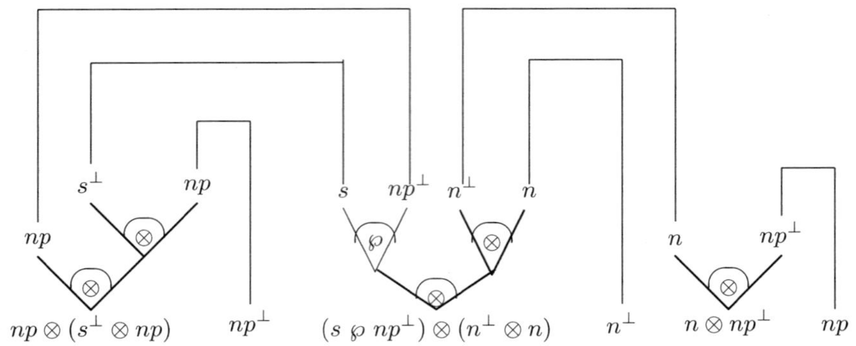

Links do not cross and the only $\wp$ link of the graph is such that if one of the two edges is suppressed, the graph remains connected and is acyclic.

4.11 Concluding Remarks

This chapter introduced linear logic, a currently active field of research. Linear logic mainly brought to logic the possibility of a more fine-grained analysis of proofs, leading finally to the geometric notion of proof net. We may already notice that, by means of these graphical structures, explicit formulae lose their importance in logic, since a proof net may exist independently of any decomposition of a given formula. The quality of being a proof rests on a mere topological property, and nodes may be replaced by simple addresses: this is the root of further developments which lead to ludics.

In principle, this chapter should contain more on the semantics of linear logic. Actually it appears that, since linear logic mainly concentrates on the proof process, there is no particular interest in searching for a semantics of *formulae*. An interpretation of this framework in terms of truth and falsity would be at odds, even if this has already been attempted. The most convenient semantics is a semantics of *proofs*, which may be developed in terms of *coherent spaces*. Another semantics is provided by *phase semantics*.

Chapter 5

Curry–Howard Correspondence

5.1 Introduction

We study in this chapter the computational content of proof systems, which was worked out mainly by Curry, but also by Howard (1980). Definitions of λ-terms, λ-equality and typed λ-calculus were already given in Section 2.4. Let us simply add the following definition.

Definition 5.1.

- We call *redex* any subterm of the form $(\lambda x.M\ N)$;
- a λ-term is said to be *normal* if it contains no redex;
- a λ-term is said to be *normalizable* if it may be reduced by β-reduction, in a finite number of steps, to a normal form;
- a calculus is said to be *normalizable* if all its terms are normalizable;
- a calculus is said to be *confluent* if for all normalizable terms, any reduction yields the same result.

It happens (from the Church–Rosser theorem[1]) that the λ-calculus with β-reduction is confluent. However, it is not normalizable, since it contains some λ-terms which are not.

Example: $\lambda x.(x\ x)$ is a λ-term, as well as $(\lambda x.(x\ x)\ \lambda x.(x\ x))$. We notice however that

$$(\lambda x.(x\ x)\ \lambda x.(x\ x)) \rightarrow (\lambda x.(x\ x)\ \lambda x.(x\ x))$$

and thus indefinitely!

There is therefore no sequence of β-reduction steps leading in that case to a redex-free form. This is not astonishing when looking at this λ-term: it

[1] A demonstration of which can be found in Girard *et al.* (1989).

represents the function which, given any object, applies it to itself, and we don't ignore that self-reference is at the source of problems in logic!

Of course, that does not arise if we add constraints to the well-formedness of the λ-terms, if for instance when building $(M\ N)$ we impose that M be a function $A \rightarrow B$ and N be an object in A. This amounts to *typing* the λ-terms, as already explained in Section 2.4.

5.2 A Correspondence Between Types and Formula

We may observe an extensive similarity between the types and the formulae handled by proofs in the systems we studied above. Between $\alpha \rightarrow \beta$ and $A \Rightarrow B$, there is only a difference of notation. The application rule, which allows us to build a λ-term $(M\ N)$ starting from λ-terms of respective types $\alpha \rightarrow \beta$ and α, is exactly the elimination rule of the implication. In the same way, the abstraction rule, which allows us to build a λ-term $\lambda x.M$ starting from a variable of type α and a λ-term of type β, exactly corresponds to the introduction rule of the implication. This is still not astonishing since building a *function* as a *process* which transforms a given value x into an "image" y amounts to taking an arbitrary value of x, then calculating its image and finally abstracting over this value. The value x from which we start is like a hypothesis, which is discharged when we want to express the function itself. We therefore come to the following parallelism, between two worlds (the logic and the calculus):

Logic	Calculus
formula	type
$[\Rightarrow\ Elim]$	application
$[\Rightarrow\ Intro]$	abstraction
active hypothesis	free variable
discharged hypothesis	λ-bound variable
proof	λ-term

To illustrate the last line of this table, let us take again one of the proofs made in the preceding chapter.

$$\dfrac{\dfrac{\dfrac{A \Rightarrow (A \Rightarrow B) \quad [A]^1}{A \Rightarrow B} \qquad [A]^1}{B}}{A \Rightarrow B}\ (1)$$

In that proof, let us associate the variable x with the hypothesis. Let us assume that f is a constant of type $A \Rightarrow (A \Rightarrow B)$, then, each step in the proof translates into an operation of λ-calculus, which allows us to label each formula with a λ-term, until the conclusion $A \Rightarrow B$ is reached, which will be therefore labelled by a λ-term which represents the whole proof. We get

$$\dfrac{\dfrac{\dfrac{f : A \Rightarrow (A \Rightarrow B) \quad [x : A]^1}{(f\ x) : A \Rightarrow B} \qquad [x : A]^1}{((f\ x)\ x) : B}}{\lambda x.((f\ x)\ x) : A \Rightarrow B}\ (1)$$

Notice that to the theorem $(A \Rightarrow (A \Rightarrow B)) \Rightarrow (A \Rightarrow B)$ corresponds a constant-free λ-term, without free variables. Such a term is called a *combinator*:

$$\lambda y.\lambda x.((y\ x)\ x)$$

From a linguistic viewpoint, this combinator may be applied to a transitive verb, for instance, *to shave*, in order to transform it into a reflexive verb (*to shave oneself*).

Labelling with λ-terms (which code proofs) may also be realized in "sequent calculus" versions. For natural deduction sequents, we have (limiting ourselves to implication and conjunction) the following labelling. Notice that, rigorously, λ-terms code proofs and not formulae. This is the reason why sequents must be labelled, and not formulae.

Axiom:

$$x : (\Gamma, A \vdash A)$$

Introduction rules:

$$\dfrac{u : (\Gamma \vdash A) \quad v : (\Gamma \vdash B)}{(u, v) : (\Gamma \vdash A \wedge B)}\ [\wedge\ I]$$

$$\dfrac{u : (\Gamma, x : A \vdash B)}{\lambda x.u : (\Gamma \vdash A \Rightarrow B)}\ [\Rightarrow\ I]$$

Elimination rules:

$$\dfrac{z : (\Gamma \vdash A \wedge B)}{\pi_1(z) : (\Gamma \vdash A)}\ [\wedge_1\ E] \qquad\qquad \dfrac{z : (\Gamma \vdash A \wedge B)}{\pi_2(z) : (\Gamma \vdash B)}\ [\wedge_2\ E]$$

where π_1 and π_2 correspond to *projections* in case λ-calculus is enriched in order to have pairing.

$$\frac{u : (\Gamma \vdash A \Rightarrow B) \quad v : (\Gamma \vdash A)}{(u\ v) : (\Gamma \vdash B)} [\Rightarrow\ E]$$

where "( )" corresponds to *pairing*.

It is however convenient, usually, to put the active labels (those which are used at the current step) on the active terms in the application of the rules, hence:

Axiom:

$$\Gamma, x : A \vdash x : A$$

Introduction rules:

$$\frac{\Gamma \vdash u : A \quad \Gamma \vdash v : B}{\Gamma \vdash (u, v) : A \wedge B} [\wedge\ I]$$

$$\frac{\Gamma, x : A \vdash u : B}{\Gamma \vdash \lambda x.u : A \Rightarrow B} [\Rightarrow\ I]$$

Elimination rules:

$$\frac{\Gamma \vdash z : A \wedge B}{\Gamma \vdash \pi_1(z) : A} [\wedge_1\ E] \qquad \frac{\Gamma \vdash z : A \wedge B}{\Gamma \vdash \pi_2(z) : B} [\wedge_2\ E]$$

$$\frac{\Gamma \vdash u : A \Rightarrow B \quad \Gamma \vdash v : A}{\Gamma \vdash (u\ v) : B} [\Rightarrow\ E]$$

Concerning now the intuitionistic sequent calculus (always limited to implication and conjunction), we have the following, which is immediately transferrable to the Lambek calculus:

Identity rules:
Axiom:

$$\Gamma, x : A \vdash x : A$$

Cut:

$$\frac{\Gamma \vdash \tau : A \quad \Gamma, x : A \vdash \gamma : C}{\Gamma \vdash [\tau/x]\gamma : C} [cut]$$

Logical rules:
Implication:

$$\frac{\Gamma \vdash u : A \quad \Gamma, (z\ u) : B \vdash \gamma : C}{\Gamma, z : A \Rightarrow B \vdash \gamma : C}\,[\Rightarrow L] \qquad \frac{\Gamma, x : A \vdash u : B}{\Gamma \vdash \lambda x.u : A \Rightarrow B}\,[\Rightarrow R]$$

Conjunction:

$$\frac{\Gamma, \pi_1(z) : A \vdash \gamma : C}{\Gamma, z : A \wedge B \vdash \gamma : C}\,[\wedge_1 L] \qquad \frac{\Gamma, \pi_2(z) : B \vdash \gamma : C}{\Gamma, z : A \wedge B \vdash \gamma : C}\,[\wedge_2 L]$$

$$\frac{\Gamma \vdash u : A \quad \Gamma \vdash v : B}{\Gamma \vdash (u, v) : A \wedge B}\,[\wedge R]$$

NB: It would also be possible to write the left rules in the following way:

Implication:

$$\frac{\Gamma \vdash u : A \quad \Gamma, x : B \vdash \gamma : C}{\Gamma, z : A \Rightarrow B \vdash [(z\ u)/x]\gamma : C}\,[\Rightarrow L]$$

Conjunction:

$$\frac{\Gamma, x : A \vdash \gamma : C}{\Gamma, z : A \wedge B \vdash [\pi_1(z)/x]\gamma : C}\,[\wedge_1 L] \quad \frac{\Gamma, x : B \vdash \gamma : C}{\Gamma, z : A \wedge B \vdash [\pi_2(z)/x]\gamma : C}\,[\wedge_2 L]$$

In the Lambek calculus, the above rules are obviously not valid in this form since the rules $[\wedge_1\ L]$ and $[\wedge_2\ L]$ entail weakening. The Lambek calculus rules (extended to the product, here denoted as $\bullet$) become (see examples of application in the next chapter)

$$\frac{}{x : A \vdash x : A}\,[Axiom]$$

$$\frac{\Theta \vdash v : B \qquad \Gamma, (u\ v) : A, \Delta \vdash \gamma : C}{\Gamma, u : A/B, \Theta, \Delta \vdash \gamma : C}\,[/L]$$

$$\frac{\Theta \vdash v : B \qquad \Gamma, (u\ v) : A, \Delta \vdash \gamma : C}{\Gamma, \Theta, u : B\backslash A, \Delta \vdash \gamma : C}\,[\backslash L]$$

$$\frac{\Gamma, x : B \vdash u : A}{\Gamma \vdash \lambda x.u : A/B}\,[/R] \qquad \frac{x : B, \Gamma \vdash u : A}{\Gamma \vdash \lambda x.u : B\backslash A}\,[\backslash R]$$

$$\frac{\Gamma \vdash u : A \qquad \Delta, x : A, \Delta' \vdash \gamma : B}{\Delta, \Gamma, \Delta \vdash [u/x]\gamma : B}\,[Cut]$$

$$\frac{\Gamma, \pi_1(z) : A, \pi_2(z) : B \vdash \gamma : C}{\Gamma, z : A \bullet B \vdash \gamma : C}\,[\bullet L] \qquad \frac{\Gamma \vdash u : A \quad \Gamma \vdash v : B}{\Gamma \vdash (u, v) : A \bullet B}\,[\bullet R]$$

5.3 An Example of a Combinator

Another example of a combinator is given by the so-called *commutator*, that is, the function which transforms any function of two arguments a and b, given in that order, into the same function where a and b have been permuted. This case is interesting: let us in effect suppose we work with associativity, then if the semantic representation of the verb is modified in the convenient way, the proof of the sequent

$$np,\ (np\backslash s)/np,\ np \vdash s$$

may be transformed into the proof of

$$np,\ np\backslash(s/np),\ np \vdash s$$

without modifying the semantic representation. To see this, let us take as an example the following small lexicon:

pierre	:	np	:	p
marie	:	np	:	m
regarde	:	$(np\backslash s)/np$	:	r

The proof without using associativity, labelled by the formulae, is

$$\frac{\dfrac{\dfrac{}{p : np \vdash p : np} \quad \dfrac{}{((r\ m)\ p) : s \vdash ((r\ m)\ p) : s}}{p : np,\ (r\ m) : np\backslash s \vdash ((r\ m)\ p) : s} \qquad \dfrac{}{m : np \vdash m : np}}{p : np,\ r : (np\backslash s)/np,\ m : np \vdash ((r\ m)\ p) : s}$$

Let us suppose there is a slightly different lexical entry for *regarde* (look at), in which the syntactic type would be $np\backslash(s/np)$ and the semantic counterpart would be a. We should also have a proof not using associativity for

$$p : np,\ a : np\backslash(s/np),\ m : np \vdash ((a\ p)\ m) : s$$

It remains to know the link between these two proofs. In fact the second one may be obtained from the first by the cut-elimination procedure applied to the proof obtained from a cut between the first one and a proof of

$$\alpha : np\backslash(s/np) \vdash \beta : (np\backslash s)/np$$

We have

$$\dfrac{\dfrac{\vdots}{a : np\backslash(s/np) \vdash \beta : (np\backslash s)/np} \qquad p : np,\ r : (np\backslash s)/np,\ m : np \vdash ((r\ m)\ p) : s}{p : np,\ a : np\backslash(s/np),\ m : np \vdash [\beta/r]((r\ m)\ p) : s}\,[cut]$$

The proof of the left premise is

$$\dfrac{\dfrac{\dfrac{\dfrac{}{y : np \vdash y : np} \qquad \dfrac{\dfrac{}{x : np \vdash x : np} \quad \dfrac{}{((a\ y)\ x) : s \vdash ((a\ y)\ x) : s}}{(a\ y) : s/np,\ x : np \vdash ((a\ y)\ x) : s}}{y : np,\ a : np\backslash(s/np),\ x : np \vdash ((a\ y)\ x) : s}}{a : np\backslash(s/np),\ x : np \vdash \lambda y.((a\ y)\ x) : np\backslash s}}{a : np\backslash(s/np) \vdash \lambda x.\lambda y.((a\ y)\ x) : (np\backslash s)/np}$$

Hence, $\beta = \lambda x.\lambda y.((a\ y)\ x)$ and the concluding sequent of the proof obtained with a cut is

$$p : np,\ a : np\backslash(s/np),\ m : np \vdash ((\lambda x.\lambda y.((a\ y)\ x)\ m)\ p) : s$$

where the right λ-term reduces to

$$p : np,\ a : np\backslash(s/np),\ m : np \vdash ((a\ p)\ m) : s$$

It results that if we want to have in both cases the same representation $((r\ m)\ p)$ for the sentence *Pierre regarde Marie* (Peter looks at Mary), it will be necessary to give a the value $\lambda x.\lambda y.((r\ y)\ x)$. In conclusion, changing a type in the lexicon always results in changing the associated λ-term, the new one being obtained by applying to the old one the combinator extracted from the proof of the validity of the type change.[2]

[2]Combinators have been intensively studied after the works by Curry and Feys (1958). They were actually invented much earlier by Schönfinkel (1924) who aimed at getting rid of variables in the study of functions. In fact, systems of combinators are designed to perform the same task as systems of λ-conversion. For instance, the commutativity law which normally expresses, for a function with two variables, as $\forall x, y\ A(x, y) = A(y, x)$ may be simply expressed as $\mathbf{C}A = A$, where $\mathbf{C}$ is the *commutator*, that is, the combinator

5.4 Concluding Remarks

This short chapter introduced the straightforward way of extracting a semantic recipe from a proof. It relies on the Curry–Howard correspondence and it gives a good solution to most of the problems addressed in the first chapter of this book. Nevertheless, as indicated before, the pure Lambek calculus is not powerful enough to give a complete account of observed phenomena in syntax. It is the reason why we shall explore other systems in the sequel, which are essentially based on applications of linear logic. The introduction of combinators can also be fruitful and it provides a basis for other approaches like Steedman's and Jacobson's. We shall say a word on them further.

---

such that $\mathbf{C}fxy = fyx$. Other examples of combinators are **B**, the *compositor*; **I**, the *identity*; **K**, the *eliminator*; and **S**, defined by the property $\mathbf{S}(f, g)(x) = f(x, g(x))$. Many authors have thought to use combinators in linguistic theory, to begin with Steedman (Ades and Steedman, 1982; Steedman, 1996, 2000) whose whole framework, combinatory categorial grammar (CCG) is based on them, but also Jacobson (1996, 2000) (for both see Chapter 6), Szabolcsi (1987) and in France in the 1980s Desclés (1990).

PART III

Proof Theory Applied to Linguistics

Chapter 6

Using the Lambek Calculus and Its Variants

6.1 Introduction

In this chapter, we make the most intensive use of the Lambek calculus in order to achieve the goal we had in mind since the first chapter, which is to render questions of binding in the most straightforward way, beginning by treating sentences with quantifiers (with neither quantifier raising nor *ad hoc* binding) and questions.

6.2 Using the Lambek Calculus

6.2.1 *Summary of the rules*

$$\frac{}{A \vdash A}\,[Axiom]$$

$$\frac{\Theta \vdash B \qquad \Gamma, A, \Delta \vdash C}{\Gamma, A/B, \Theta, \Delta \vdash C}\,[/L] \qquad \frac{\Theta \vdash B \qquad \Gamma, A, \Delta \vdash C}{\Gamma, \Theta, B\backslash A, \Delta \vdash C}\,[\backslash L]$$

$$\frac{\Gamma, B \vdash A}{\Gamma \vdash A/B}\,[/R] \qquad \frac{B, \Gamma \vdash A}{\Gamma \vdash B\backslash A}\,[\backslash R]$$

$$\frac{\Gamma \vdash A \qquad \Delta, A, \Delta' \vdash B}{\Delta, \Gamma, \Delta \vdash B}\,[Cut]$$

It is very easy in that system to prove the validity of a sequent corresponding to a sentence, by building a *cut-free* proof. Consider for instance the sentence

Example 6.1. The cat that Paul feeds hits the bottle

with the following assignment:

the	np/n
cat	n
that	$(n\backslash n)/(s/np)$
Paul	np
feeds	$(np\backslash s)/np$
hits	$(np\backslash s)/np$
bottle	n

We get the following deduction, where δ represents the right subproof:

$$\dfrac{\dfrac{\dfrac{\dfrac{}{np \vdash np}[Ax] \quad \dfrac{\dfrac{}{np \vdash np}[Ax] \quad \dfrac{}{s \vdash s}[Ax]}{np, np\backslash s \vdash s}[\backslash L]}{np, (np\backslash s)/np, np \vdash s}[/L]}{np, (np\backslash s)/np \vdash s/np}[/R] \qquad \dfrac{\delta}{np/n, n, n\backslash n, (np\backslash s)/np, np/n, n \vdash s}}{np/n, n, (n\backslash n)/(s/np), np, (np\backslash s)/np, (np\backslash s)/np, np/n, n \vdash s}[/L]$$

with δ consisting of

$$\dfrac{\dfrac{}{n \vdash n}[Ax] \quad \dfrac{\dfrac{}{n \vdash n}[Ax] \quad \dfrac{\dfrac{}{n \vdash n}[Ax] \quad \dfrac{\dfrac{}{np \vdash np}[Ax] \quad \dfrac{\dfrac{}{np \vdash np}[Ax] \quad \dfrac{}{s \vdash s}[Ax]}{np, np\backslash s \vdash s}[\backslash L]}{np, (np\backslash s)/np, np \vdash s}[/L]}{np, (np\backslash s)/np, np/n, n \vdash s}[/L]}{np/n, n, (np\backslash s)/np, np/n, n \vdash s}[/L]}{np/n, n, n\backslash n, (np\backslash s)/np, np/n, n \vdash s}[\backslash L]$$

6.2.2 *Natural deduction presentation*

Rules for the Lambek calculus (restricted to / and \, without •) may be expressed in the natural deduction format as

$$\dfrac{A/B \qquad B}{A}[e/] \qquad\qquad \dfrac{B \qquad B\backslash A}{A}[e\backslash]$$

Comment: With two premises A/B and B (resp. B and $B\backslash A$) in this order, we may deduce a category A (that is exactly the functional application rules, FA).

$$\dfrac{\begin{matrix}\Gamma & [B]^i \\ \vdots & \\ A & \end{matrix}}{A/B}\,[i\,/]^i \qquad\qquad \dfrac{\begin{matrix}[B]^i & \Gamma \\ & \vdots \\ & A \end{matrix}}{B\backslash A}\,[i\,\backslash]^i$$

Comment: If, from a sequence of types Γ to which we add on the right (resp. on the left) a category B, we may deduce A, then from Γ alone, we deduce A/B (resp. $B\backslash A$). The index i is used to indicate which hypothesis is discharged during the use of the i-rule.

6.2.3 *Examples*

Let us now examine some examples.

Example 6.2. Mary likes a Japanese writer

This sentence gives rise to the following sequence of syntactic types:

Mary : np
likes : $(np\backslash s)/np$
a : $((s/np)\backslash s)/n$
Japanese : n/n
writer : n

Of course the analysis of *a Japanese writer* is easily done by means of a repetitive application of the FA rules, or more logically speaking, the *elimination* rules:

$$\dfrac{a : ((s/np)\backslash s)/n \qquad \dfrac{Japanese : n/n \quad writer : n}{Japanese\ writer : n}\,[e/]}{a\ Japanese\ writer : (s/np)\backslash s}\,[e/]$$

At this stage, there is no direct connection between the verbal type $((np\backslash s)/np)$ and the type of the nominal phrase, which is the high-order type $(s/np)\backslash s$. We intuitively know that things will fall into place because we know from the type-raising rules that the np expected by the verb may also be seen as a $(s/np)\backslash s$, but instead of using this "extra" knowledge, we may stay inside the calculus itself and resolve the problem *by making use of a hypothesis*; therefore let us introduce a hypothesis np just on the right of the verb (with an index 1):

Mary : np likes : $(np\backslash s)/np$ $[np]^1$ a Japanese writer : $(s/np)\backslash s$

This leads to the following successive subproofs:

$$Mary : np \qquad \dfrac{likes : (np\backslash s)/np \quad [np]^1}{likes\ \epsilon : np\backslash s} \qquad a\ Japanese\ writer : (s/np)\backslash s$$

$$\dfrac{Mary : np \qquad \dfrac{likes : (np\backslash s)/np \quad [np]^1}{likes\ \epsilon : np\backslash s}}{Mary\ likes\ \epsilon : s} \qquad a\ Japanese\ writer : (s/np)\backslash s$$

At this stage, the *np* hypothesis may be discharged, and we get

$$\dfrac{\dfrac{Mary : np \qquad \dfrac{likes : (np\backslash s)/np \quad [np]^1}{likes\ \epsilon : np\backslash s}}{Mary\ likes\ \epsilon : s}}{Mary\ likes : s/np}\,[i\ /]^1 \qquad a\ Japanese\ writer : (s/np)\backslash s$$

with now the left-hand part which can be applied to the right-hand side by elimination rule from the left ($[e\backslash]$):

$$\dfrac{\dfrac{\dfrac{Mary : np \qquad \dfrac{likes : (np\backslash s)/np \quad [np]^1}{likes\ \epsilon : np\backslash s}}{Mary\ likes\ \epsilon : s}}{Mary\ likes : s/np}\,[i\ /]^1 \qquad a\ Japanese\ writer : (s/np)\backslash s}{Mary\ likes\ a\ Japanese\ writer : s}\,[e\backslash]$$

Informally we have used concatenation between strings in order to represent which expression each obtained type covers during the deduction. ϵ (the empty string) was used to phonetically mark a hypothesis.

It is interesting to represent this proof tree more as a syntactic tree by turning it upside down, thus giving

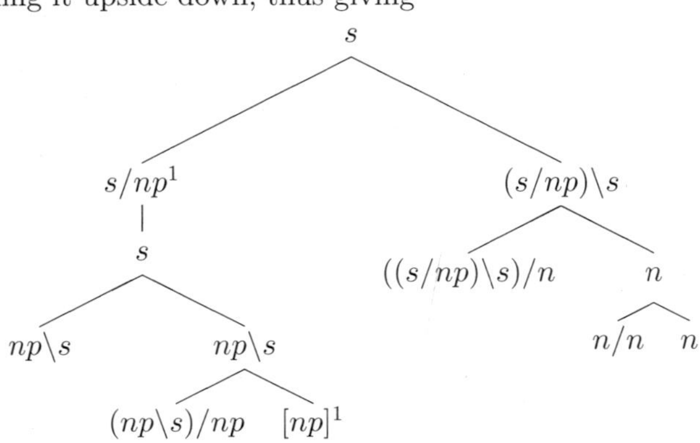

It is further illuminating to compare it with a syntactic tree more in the generative tradition, with quantifier raising:

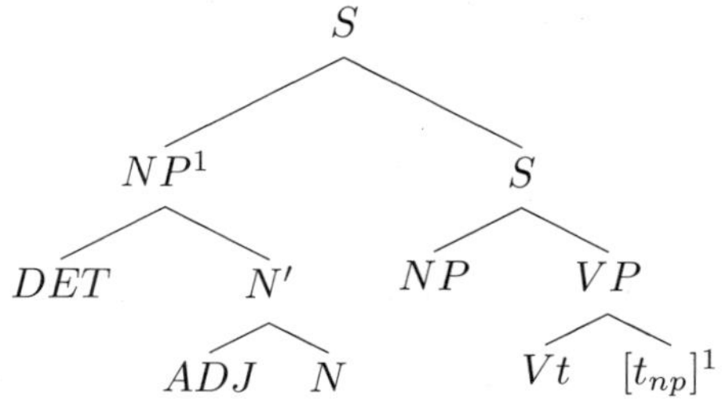

They are of course quite similar: in the first tree, the hypothesis has the role played by the trace in the second one. But in the first tree, the quantified nominal phrase does not move properly speaking: it is the hole ready to accept it which moves, from the complement position of the verb to the level of the whole sentence, before being absorbed by the expectation of the quantified nominal phrase.

Example 6.3. Which book do you think that Mary read?

This question gives rise to the following sequence of types:

which : $(s/(s/np))/n$	book : n	do : s/s	you : $s/(np\backslash s)$
think : $(np\backslash s)/np$	that : np/s	Mary : np	read : $(np\backslash s)/np$

That Mary read is analyzed as follows:

$$\dfrac{\dfrac{that : np/s \qquad \dfrac{Mary : np \qquad \dfrac{read : (np\backslash s)/np \quad (hyp) : [np]^1}{read\ \epsilon : np\backslash s}}{Mary\ read\ \epsilon : s}}{that\ Mary\ read\ \epsilon : np}}{that\ Mary\ read : np/np}\,[i\ /]^1$$

In a similar way, we get

$$do\ you\ think\ that\ Mary\ read : s/np$$

And by easy steps of /-elimination, we get the correct s-typing of the whole sentence. The inverted[1] proof tree is given by

---

[1] An "inverted" proof tree means the rules are in the reverse order, with the conclusion on top and premises at the bottom, in order to show the resemblance with a syntactic tree.

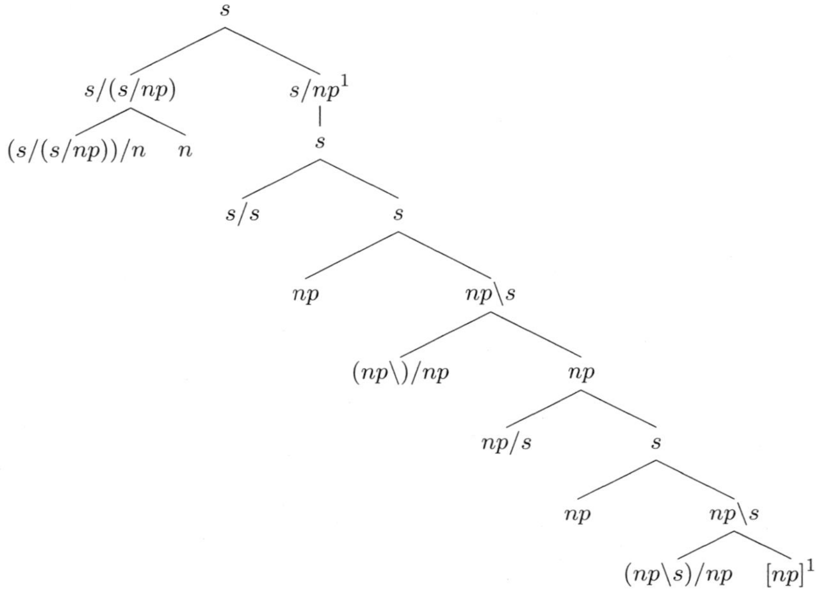

This proof tree may be compared with the derivation for the same sentence in a Heim and Kratzer-like grammar:

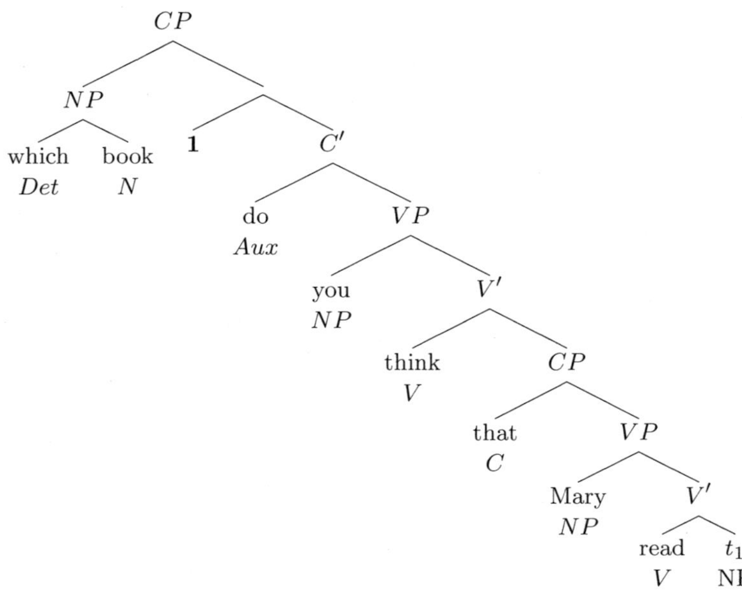

and with the intuition we had that the mechanism of displacement leaving a trace behind is exactly similar to hypothetical reasoning. Instead of introducing a binder **1**, it seems to be more appropriate to have a unary branch expressing the step where the hypothesis is discharged:

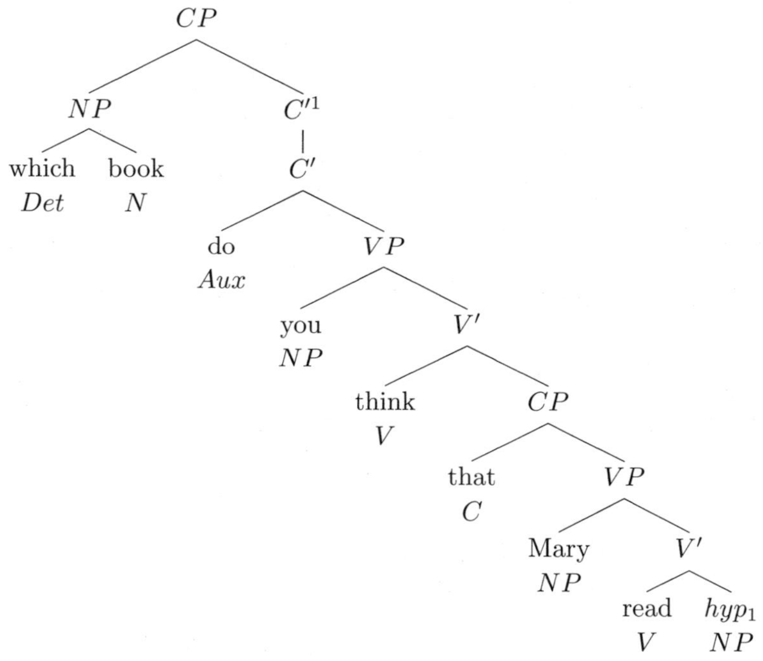

6.2.4 *Compositional semantics*

In Montague semantics, like in Heim and Kratzer's, we saw that λ-abstraction must be used at some steps: this is the case in the S_3 and S_{14} rules in Montague semantics, and in predicate abstraction in Heim and Kratzer's. Here, applying the Curry–Howard correspondence, we get Fig. 6.1.

Let us suppose that a term f is associated with A/B or $B\backslash A$ in the elimination rules, and that a term τ is associated with B, then A will be associated with the *application* of f to τ.

Let us suppose otherwise that a *variable* is associated with each hypothesis, for instance x with $[B]^i$, and that the conclusion of the proof, in the introduction rules, is associated with the term f. Then by discharging the hypothesis, we get the λ-term $\lambda x.f$. So, the rules stated above may be completed in the following manner:

$$\frac{A/B : f \qquad B : \tau}{A : f(\tau)}[e/] \qquad \frac{B : \tau \qquad B\backslash A : f}{A : f(\tau)}[e\backslash]$$

$$\frac{\begin{matrix}\Gamma \qquad [B : x]^i \\ \vdots \\ A : f\end{matrix}}{A/B : \lambda x.f}[i\ /]^i \qquad \frac{\begin{matrix}[B : x]^i \qquad \Gamma \\ \vdots \\ A : f\end{matrix}}{B\backslash A : \lambda x.f}[i\ \backslash]^i$$

Figure 6.2 shows this correspondence in the sequent calculus format, where $\gamma[x \leftarrow \mu]$ denotes the substitution of μ for x everywhere x occurs in the

$$\frac{A/B \qquad B}{A}[e/] \qquad \frac{B \qquad B\backslash A}{A}[e\backslash]$$

$$\frac{\begin{matrix}\Gamma \qquad [B]^i \\ \vdots \\ A\end{matrix}}{A/B}[i\ /]^i \qquad \frac{\begin{matrix}[B]^i \qquad \Gamma \\ \vdots \\ A\end{matrix}}{B\backslash A}[i\ \backslash]^i$$

Fig. 6.1 Lambek rules in natural deduction

$$\frac{}{A : x \vdash A : x}[Axiom]$$

$$\frac{\Theta \vdash B : \alpha \qquad \Gamma, A : x, \Delta \vdash C : \gamma}{\Gamma, A/B : f, \Theta, \Delta \vdash C : \gamma[x \leftarrow f(\alpha)]}[/L]$$

$$\frac{\Theta \vdash B : \alpha \qquad \Gamma, A : x, \Delta \vdash C : \gamma}{\Gamma, \Theta, B\backslash A : f, \Delta \vdash C : \gamma[x \leftarrow f(\alpha)]}[\backslash L]$$

$$\frac{\Gamma, B : x \vdash A : f}{\Gamma \vdash A/B : \lambda x.f}[/R] \qquad \frac{B : x, \Gamma \vdash A : f}{\Gamma \vdash B\backslash A : \lambda x.f}[\backslash R]$$

$$\frac{\Gamma \vdash A : \alpha \qquad \Delta, A : x, \Delta' \vdash B : \beta}{\Delta, \Gamma, \Delta \vdash B : \beta[x \leftarrow \alpha]}[Cut]$$

Fig. 6.2 Lambek rules in the sequent calculus

$$\cfrac{np : mary \qquad \cfrac{(np\backslash s)/np : \lambda u \lambda v.like(v,u) \quad [x : np]^1}{np\backslash s : \lambda v.like(v,x)}}{\cfrac{\cfrac{s : like(mary,x)}{s/np : \lambda x.like(mary,x)}[i\ /]^1 \qquad (s/np)\backslash s : \lambda P.\exists z(JW(z) \wedge P(z))}{s : \exists z(JW(z) \wedge like(mary,z))}[e\backslash]}$$

Fig. 6.3 Mary likes a Japanese writer

expression γ, and where, in the $[/\ L]$ and $[\backslash\ L]$ rules, the variable x is not already used in the terms associated with the other types occurring in the sequent. (Same restriction concerning the right rules.)

As it may be noticed, the cut rule has a functional interpretation as *substitution* in a term β.

Using the Curry–Howard correspondence between proofs and λ-terms, we may compute the semantic representations of the two previous sentences (here without the string labels): see Figs. 6.3 and 6.4.

6.2.5 *Limitations of the Lambek calculus*

The Lambek calculus thus provides a very elegant and efficient way for analyzing sentences and deducing their semantic representation. Unfortunately, it suffers from some limitations. They are direct consequences of its main properties as they have been studied now for half a century. In particular, in 1963, Chomsky already predicted that categorial grammars were not more powerful than context-free grammars, that is, the kind of language they can describe is exactly the same class of languages that context-free grammars can generate. In fact, Chomsky probably thought only of what we have called earlier AB-grammars, and in fact that result was proven by Gaifman and Shamir (Bar-Hillel *et al.*, 1960). But for Lambek grammars, such a proof waited to be found for 30 years. It was only in 1993 that Mati Pentus (1993) showed that in fact the weak generative power of Lambek grammars was the same as the context-free ones. The proof was quite intricate and used refined techniques of proof theory (like the interpolation theorem, which was proven to be true for the Lambek calculus by Dirk Roorda (1991) a couple of years before). What is the power of this theorem? It states that for any language L which is recognized by a Lambek grammar, there is a context-free grammar G

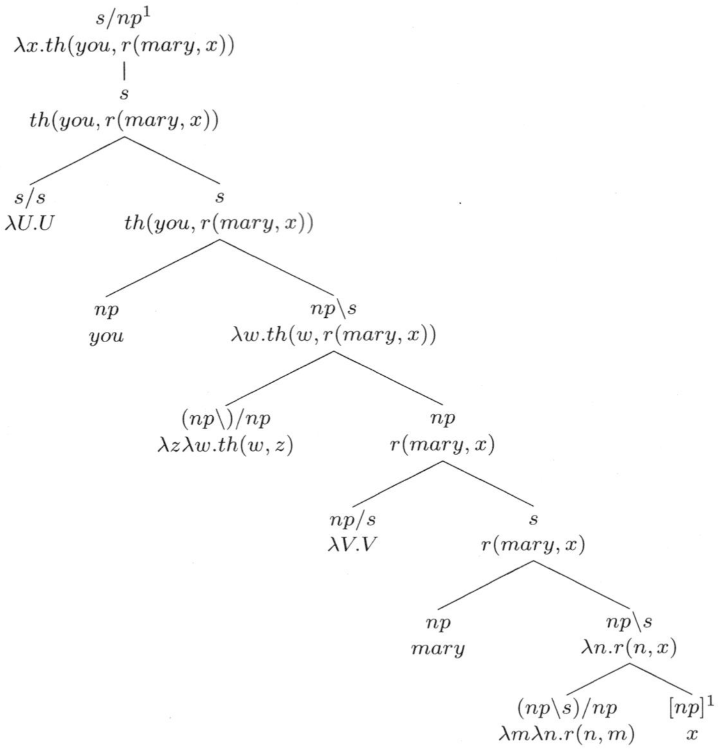

Fig. 6.4 *You think that Mary read?* (Here in the inverted proof tree format)

which generates it ($L = L(G)$) and reciprocally. We say that a language L is recognized by a Lambek grammar L_k if, and only if, there is an assignment of types A and a distinguished type s such that for every sentence σ in this language, made of the words $\sigma_1, \sigma_2, \ldots, \sigma_i, \ldots, \sigma_m$, the sequence[2]

$$A(\sigma_1), A(\sigma_2), \ldots, A(\sigma_i), \ldots, A(\sigma_m) \vdash s$$

is provable. Of course when saying that, we don't pay attention to the proof tree obtained to prove this sequent. Neither do we attend to the derivation

[2]This definition is valid when the assignment A is *rigid*, that is to say A is an application from words to types. If not, that is, if A is an application from words to $\wp(Typ) - \{\emptyset\}$, we must say that there is a type τ_i inside each $A(\sigma_i)$ such that the sequence of τ_i reduces to s.

tree in the context-free grammar. We therefore say that *Lambek grammars are weakly equivalent to context-free grammars.*

The point concerning *strong* equivalence is more delicate. Nevertheless, we understand that if Lambek grammars have just the context-free power, they are not sufficient for analyzing human languages, since it has been shown that those languages belong to a slightly larger class which has been dubbed mildly context-sensitive languages.[3]

A strong limitation of the (pure) Lambek calculus is that, if appropriate for analyzing peripheral extractions (as we did with a sentence like *Which book do you think that Mary read?*), it has no means to analyze medial extraction, like in *Which book do you think that Mary read on the beach?* or *The novel that Paul gave to his friend Bill.* It could be possible to have some *ad hoc* grammar to do that job, but that would necessitate some unlikely verbs like *read on the beach* or *give to [someone's] friend Bill*! If we wish to get a convenient analysis of these facts, we must have a way to point to a medial hole and then to fill the gap. In Lambek formalism, things are easy for a right or a left extraction: for a right (resp. left) extraction, we know the hypothesis was made on the extreme right (resp. left) of the sentence, that gives an X/np type (or $np\backslash X$) by using the introduction rule on the right (resp. on the left). That amounts to saying that in the Lambek calculus, rules come in symmetrical pairs: $[/\ L]$ is symmetrical to $[/\ R]$ and $[\backslash\ L]$ to $[\backslash\ R]$, but we cannot formulate a pair of an introduction rule and an elimination rule for a connective which would express the lack of something "somewhere inside" the string of words, so that these two rules would be symmetrical, one with regard to the other.[4]

---

[3]See note on p. 81.

[4]Steedman's combinatory categorial grammar (CCG) does not encounter such limitations. This is due to the fact that CCG includes special rules which extend the generative power of categorial grammar. For instance, the so-called *crossing functional composition rules* (Steedman and Baldridge, 2007) — here given in the Lambek notation

$$X/Y : f\ Z\backslash Y : g \vdash Z\backslash X : \lambda z.f(g(z))$$
$$Y/Z : g\ Y\backslash X : f \vdash X/Z : \lambda z.f(g(z))$$

which are of course not theorems of the Lambek calculus. The inclusion of these rules into the Lambek calculus causes its collapse into its permutational variant **LP**. Steedman first solved this problem by making the use of such rules depend on some categories, which is really a case of an *ad hoc* issue. More recently, Baldridge (Baldridge and Kruijff, 2003) proposed to distinguish several slashes according to distinct flavours: this amounts to using different composition modes, making CCG close to Moortgat's multimodal framework (see Chapter 7).

It may be thought that such limitations come from the fact that we try to manage altogether the syntactic types (the "narrow syntax" in Chomsky's sense) *and* the word order, a phenomenon which seems to be more related to another component of the grammar (phonological interpretation, for instance). There will therefore be two ways of trying to overcome these difficulties with Lambek grammars:

- Enrich the Lambek calculus with new operators (*structural modalities*) which would give it more flexibility in dealing with word order questions: it is the solution taken by Morrill (1994), Hepple (1990), Moortgat (1997), etc.
- Get rid of the word order phenomena at the purely syntactic level, and go back to ideas dating back to Curry (1961), according to whom the architectonic of grammar is made of two levels: the phenotypic one and the genotypic one, with the second only concerned with pure types (without order) and only the first concerned with this order. It is the solution taken by many formalisms, like λ-grammars (Muskens, 2003), abstract categorial grammar (de Groote, 2001a), convergent grammars (Pollard and Hana, 2003; Pollard, 2007), linear grammars (Anoun and Lecomte, 2006), etc.

6.3 Flexible Types

6.3.1 *Flexible Montague grammar*

A relevant application of the rules of (a variant of) the Lambek calculus concerning the problems we came up with in the Montagovian framework, and then in Heim and Kratzer's, was made by Hendriks, who proposed to enrich Montague's framework with *flexible types.* The calculus is based on the commutative variant of the Lambek calculus (**LP**) (or "non-directional" Lambek calculus), which was studied by Van Benthem (1986) in the 1980s. This calculus only contains the identity axiom, cut rule, \ (written $\rightarrow$) -elimination rule, and $\rightarrow$ -introduction rule. λ-terms are assigned to the sequents (see Chap. 5). It is a practical calculus but is nevertheless too rich, in particular because, without directionality, it has nothing to indicate the right order of the arguments in an elementary sentence like *Peter loves Mary.* Taking this drawback into account, Hendricks (1993) proposed a calculus based on derived rules excluding the commutativity rule,[5] the Geach rule[6] and the composition law.[7] He only keeps

- value raising: $\overrightarrow{a} \to b \vdash \overrightarrow{a} \to (b \to c) \to c$, with its corresponding λ-term (or combinator) $\lambda\overrightarrow{x}.\lambda y.y(z(\overrightarrow{x}))$,
- argument raising: $a \to (\overrightarrow{c} \to b) \vdash ((a \to b) \to b) \to (\overrightarrow{c} \to b)$, with its corresponding λ-term $\lambda v.\overrightarrow{x}_c.v(\lambda z_a.y(z)(\overrightarrow{x}))$,
- argument lowering: $\overrightarrow{a} \to (((b \to t) \to t) \to d) \vdash \overrightarrow{a} \to (b \to d)$, with its λ-term $\lambda x.\lambda y.z(\overrightarrow{x})(\lambda v.v(y))$,
- cut: if $a \vdash b$ with λ-term $\tau_b(x_a)$ and $b \vdash c$ with λ-term $\tau_c(y_b)$, then $a \vdash c$ with λ-term $\tau_c[x_b := \tau_b]$.

It is easy in this system to show that we get scope ambiguities and *de re/de dicto* ambiguities (left to the reader!).

6.3.2 *Variable-free semantics*

Pauline Jacobson (1996, 2000) uses combinators (see Section 5.3) to describe binding phenomena, thus avoiding variables. If an expression ϕ is of type X, then the same expression where a nominal expression is replaced by an unbound pronoun is of type $e \to X$ (like *he lost*, represented by $\lambda x.[[lost]](x)$) (if n nominal expressions are replaced by unbound pronouns, then the type is $e \to \cdots \to e \to X$ (n *occurrences of* e)). It results that an unbound pronoun itself is of type $e \to e$, and represented by the identity function $\lambda x.x$. But how to combine the verb *lost*, of semantics $[[lost]]$ of type $e \to t$, with the pronoun *he* of type $e \to e$? That requires that *lost* be transformed into something waiting for an $e \to e$ argument, instead of an e argument. This of course has a flavour of the Geach rule (or division rule), according to which every expression of type $a \to b$ has also the type $(c \to a) \to (c \to b)$. The λ-term associated with the proof of the corresponding sequent provides the combinator $\mathbf{g}$, so that, finally,

$$\mathbf{g}([[lost]])([[he]]) = \lambda f.\lambda x.[[lost]](f(x))(\lambda y.y) = [[lost]]^{8}$$

Dealing then with sentences like

[5] The commutativity rule is the rule according to which it is possible to exchange the arguments in a formula $a \to (b \to c)$.

[6] The Geach rule is the rule which transforms a type $b \to c$ into $(a \to b) \to (a \to c)$.

[7] As we know, composition is $a \to b, b \to c \vdash a \to c$.

[8] This way has much to do with *continuation semantics*, as will be seen in Chapter 10.

Example 6.4. Every student$_1$ loves his$_1$ mother

Jacobson analyzes the semantics of *his mother* as the function:

$$\lambda x.[[the^\frown mother^\frown of(x)]] \tag{6.1}$$

which is an expression of type $e \to e$. Then to compose *love* with *his mother* amounts to combining an expression of type $e \to e \to t$ with an expression of type $e \to e$, trying to bind the variable to which the function is applied with the subject of the verb. In order to do that, she uses another combinator, which, properly speaking, is not deduced from a linear framework, like the Lambek calculus, but from the intuitionistic (non-linear) one. It can be shown in effect in simple type theory that every expression of type $a \to (e \to b)$ is also of type $(e \to a) \to (e \to b)$. The proof of the associated sequent provides a combinator $\mathbf{z}$, so that if f is a function of type $a \to (e \to b)$, then $\mathbf{z}(f)$ is of type $(e \to a) \to (e \to b)$, where $\mathbf{z}(f) = \lambda Y^{e \to a}.\lambda x^a.f(Y(x))(x)$. Finally, *loves his mother* is analyzed as[9,10]

$$\mathbf{z}([[love]])(the\ mother\ of) = \lambda x.([[love]]\ [[the^\frown mother^\frown of(x)]], x) \tag{6.2}$$

[9]On the syntactic side, meanings are combined by categorial rules (application, composition and type raising). There is nevertheless a risk of confusion between the verbal entry *lost*, of syntactic type $NP\backslash S$, and the sentence with a pronoun *he lost*, since they both share the same semantic representation, of type $e \to t$! To avoid that, Jacobson distinguishes between $NP\backslash S$ and S^{NP}, the latter being the syntactic type of the sentence with a pronoun. She then admits that for every expression α of type A/B, there is an expression β with the same phonology, of type A^C/B^C, thus passing up the superscript from the argument to the resulting category. That amounts in fact to still adopting a framework with several (at least two) composition modes.

[10]Other approaches of type change exist, but which are not based on combinators. Let us remember that a combinator comes from the demonstration of a logical law. It is different from a simple *operator* or *functor* which transforms a term of a certain type a into a term of a certain type b, where b is not deducible from a. For instance Partee (2004) observes that an NP may not only have e and $(e \to t) \to t$ types, but also in some contexts an $e \to t$ type (see also for this observation Dobrovie-Saurin and Beyssade, 2004). This appears in a sentence like *Mary considers John an authority*, which is parallel to *Mary considers John competent* and therefore requires that *an authority* be of semantic type $e \to t$, whereas there is of course neither a proof deducing $e \to t$ from e nor from $(e \to t) \to t$! In this case, Partee introduces *ad hoc* functors: *ident*, *iota*, *nom* and *pred*, defined by $ident: j \mapsto \lambda x.[x = j]$, $iota: P \mapsto \iota x.[P(x)]$, $nom: P \mapsto^{\cap} P$ and $pred: x \mapsto^{\cup} x$.

6.4 Non-associative Lambek Calculus

A few years later, in his "Mathematics of Sentence Stucture" (1961), Lambek suggested suppression of the associativity rule from his calculus. This led to a system which is not only sensitive to the quantity and the order of resources, but also to their structuring. In the associative calculus **L**, we had

$$A, B, C \vdash A \bullet B \bullet C$$

This is not in the non-associative calculus **NL**. In the latter we have on one side

$$(A, (B, C)) \vdash A \bullet (B \bullet C)$$

and on the other side

$$((A, B), C) \vdash (A \bullet B) \bullet C$$

which means that from now on, the comma is not only a separator expressing a product (when applying the left rule of $\bullet$), but a functor which expresses some structure, here in fact a *tree structure*. In its sequential form, the calculus may be formulated like this:

$$\frac{}{A \vdash A}\,[Axiom]$$

$$\frac{\Theta \vdash B \qquad \Gamma[A] \vdash C}{\Gamma[(A/B, \Theta)] \vdash C}\,[/L] \qquad \frac{\Theta \vdash B \qquad \Gamma[A] \vdash C}{\Gamma[(\Theta, B\backslash A)] \vdash C}\,[\backslash L]$$

$$\frac{(\Gamma, B) \vdash A}{\Gamma \vdash A/B}\,[/R] \qquad \frac{(B, \Gamma) \vdash A}{\Gamma \vdash B\backslash A}\,[\backslash R]$$

$$\frac{\Gamma[(A, B)] \vdash C}{\Gamma[(A \bullet B)] \vdash C}\,[\bullet L] \qquad \frac{\Gamma \vdash A \quad \Delta \vdash B}{(\Gamma, \Delta) \vdash (A \bullet B)}\,[\bullet R]$$

$$\frac{\Gamma \vdash A \qquad \Delta[A] \vdash B}{\Delta[\Gamma] \vdash B}\,[Cut]$$

where the notation $\Gamma[\phi]$ means "a tree structure of formulae of which ϕ is a sub-structure". When, in a rule like cut, we pass from $\Delta[A]$ to $\Delta[\Gamma]$, this must be understood as the substitution of the structure Γ for A at the precise location of the structure Δ where A was located. From now on, in such a calculus, we shall consider the left-hand side of a sequent as a kind of *structured database*.

Like with the pure calculus **L**, it is possible to give **NL** a categorical formulation, that is, a formulation which takes the deduction relations between formulae as primitive objects. If we compare this with Section 2.1.2, we see that (b) and (b′) have disappeared, and (c), (c′), (d) and (d′) may be summarized into

$$xy \rightarrow z \ if \ and \ only \ if \ x \rightarrow z/y \ and \ if \ and \ only \ if \ y \rightarrow x\backslash z$$

and (e) is preserved. Moortgat calls this system a "pure residuation logic". It is made of

(REFL)	$A \rightarrow A$
(TRANS)	$if\ A \rightarrow B\ and\ B \rightarrow C\ then\ A \rightarrow C$
(RES)	$A \rightarrow C/B\ iff\ A \bullet B \rightarrow C\ iff\ B \rightarrow A\backslash C$

It is the minimal system we can have for a product and its associated division. Because the product is supposed to be non-commutative, we have of course two associated divisions. It is worth noting that the idea of a "non-associative" product often disturbs mathematicians. As noted by Girard, the lack of associativity goes against the kind of categoric semantics which is expected for logical systems in general. For him and other mathematicians, the risk is to fall into improper structures with regard to mathematical requirements, since, actually, only linguistics seems to have the need for such weird mathematical structures!

On the linguistic side, the interest of **NL** lies in its sensitivity to constituency, which we know is a fundamental feature of the syntax of natural languages. We shall note that in **NL** a simple sentence like *Peter likes Mary* no longer has two cut-free demonstrations (one associated with (*Peter likes*) *Mary*, the other with *Peter* (*likes Mary*)), but only one, which corresponds to the phrase structure that linguists accept, which is the second one (provided that the transitive verb has the type $(np\backslash s)/np$ rather than $np\backslash(s/np)$).

The lack of associativity entails unfortunate consequences, like the lack of the composition rule $A/B\ B/C \rightarrow A/C$ which was fruitful for describing displacement phenomena. Thus the nominal phrase *the book that Peter bought* may no longer be analyzed!

If we try to derive this phrase, the structure of which is

$$(np, (n, ((n\backslash n)/(s/np), (np, (np\backslash s)/np)))) \vdash np$$

we are led to

$$\dfrac{\dfrac{((np,(np\backslash s)/np),np)\vdash s}{(np,(np\backslash s)/np)\vdash s/np} \quad \dfrac{\dots}{(np,(n,n\backslash n))\vdash np}}{(np,(n,((n\backslash n)/(s/np),(np,(np\backslash s)/np))))\vdash np}$$

but since there is no associativity here, there is no means to go from

$$((np,(np\backslash s)/np),np)\vdash s$$

to

$$(np,((np\backslash s)/np,np))\vdash s$$

which would be the only means to go further in the deduction.

It would be necessary to add the rule (readable from the bottom to the top as well as from the top to the bottom)

$$\frac{\Gamma[(A,(B,C))]\vdash D}{\Gamma[((A,B),C)]\vdash D}\,[Assoc]$$

but we would be back to **L**! Nevertheless, one advantage of the calculus **NL** is its polynomiality (see de Groote, 1999), contrarily to **L**, which is NP-complete as was proven by Pentus (2003).[11] There is also a completeness theorem, established by Došen (1992), according to which the deduction relation $A \to B$ is provable in **NL** if and only if, for all valuation v on a ternary frame, $v(A) \subseteq v(B)$. This suggests that we survey some results about the completeness of Lambek systems.

6.5 Semantics of the Lambek Calculus

6.5.1 *Monoidal semantics*

There exists a quite obvious denotational semantics for the Lambek calculus **L**. Given that this calculus only concerns string concatenation, it may be well understood that its semantics may be expressed in *monoid* terms. Let

[11] This must not confuse the reader: **L** is equivalent to context-free grammar in the sense that for each **L**-grammar, that is, each assignment A of types to words in a lexicon, there is a context-free grammar that generates the language recognized by the **L**-grammar. Even if we know that context-free languages may be parsed in polynomial time — for instance by means of the Earley algorithm — that does not entail that the calculus **L** is polynomial. That only entails that every **L**-grammar is, when converted into a context-free grammar. But it is not said that this conversion is polynomial. In fact it is not.

us recall here that a monoid is an algebraic structure obtained from an associative operation (that we shall denote "$\frown$") on a set, with a neutral element (ϵ).

Definition 6.1 (L-structure). An **L**-structure for a set of primitive types At will be defined by a monoid M and an application V (valuation) from At to $\wp(M)$.

Definition 6.2 (Extension). Let Typ be the set of types for a Lambek grammar G. Given a valuation V on $\wp(M)$, we may define an extension $V^\star$ of V to Typ in the following way:

- *if* $p \in At$, *then* $V^\star(p) = V(p)$;
- $V^\star(A \bullet B) = \{z \in M; \exists x \in V^\star(A),\ \exists y \in V^\star(B); z = x^\frown y\}$;
- $V^\star(A/B) = \{x \in M; \forall y\ y \in V^\star(B) \Rightarrow x^\frown y \in V^\star(A)\}$;
- $V^\star(B\backslash A) = \{x \in M; \forall y\ y \in V^\star(B) \Rightarrow y^\frown x \in V^\star(A)\}$.

Theorem 6.1. ***L*** *is complete with regard to this monoidal interpretation. In other words,* $\forall \phi, \psi \in Typ : \phi \vdash \psi$ *is provable in* ***L*** *if and only if for every* ***L****-structure* (M, V), $V^\star(\phi) \subset V^\star(\psi)$.

Example: It is easy to show in effect that $V^\star((A/B) \bullet (B/C))) \subset V^\star(A/C)$.

Let $z \in V^\star((A/B) \bullet (B/C)))$. By definition, $\exists x \in V^\star(A/B), \exists y \in V^\star(B/C)$ such that $z = x^\frown y$. Let u be such that $u \in V^\star(C)$. We have $z^\frown u = (x^\frown y)^\frown u = x^\frown(y^\frown u)$. Since $y \in V^\star(B/C)$, $y^\frown u \in V^\star(B)$ and since $x \in V^\star(A/B)$, $x^\frown(y^\frown u) \in V^\star(A)$, therefore $z^\frown u \in V^\star(A)$, therefore $z \in V^\star(A/C)$.

This interpretation works for **L** but not for the non-associative calculus **NL**. This leads to a new semantics, *à la* Kripke.

6.5.2 *Relational semantics*

A semantics *à la* Kripke supposes that we interpret formulae (and therefore types) as sets of possible worlds. A syntactic type will interpret more precisely into a set of accessible resources. Like in modal logic, we begin by defining the notion of a *frame*.

Definition 6.3. A ternary frame is provided by a set W of accessible resources, and a ternary relation R such that $Rzxy$ is read as "the resource z obtains from composition of resources x and y".

A valuation V on a frame (W, R) is a mapping from At to $\wp(W)$ which maps all atomic types on a set of accessible resources.

Such a valuation extends to Typ by means of the following definition:

- $V^\star(p) = V(p)$ for all atomic type p,
- $V^\star(A \bullet B) = \{z \in W; \exists x \in V^\star(A), \exists y \in V^\star(B), Rzxy\}$,
- $V^\star(A/B) = \{y \in W; \forall x\ x \in V^\star(B) \Rightarrow \exists z \in V^\star(A), Rzyx\}$,
- $V^\star(B\backslash A) = \{y \in W; \forall x\ x \in V^\star(B) \Rightarrow \exists z \in V^\star(A), Rzxy\}$.

If no restriction is put on R, we get an interpretation for **NL**, which corresponds to Moortgat's residuation calculus. Other calculi are obtained by imposing particular structural conditions (*frame conditions*).

Associativity (L): $\forall x, y, z, u \in W$:
$\exists t, Rxyt \wedge Rtzu \Leftrightarrow \exists v, Rvyz \wedge Rxvu$

Commutativity (LP): $\forall x, y, z \in W$:
$Rxyz \Leftrightarrow Rxzy$

In a general way, completeness is shown for each of these calculi, relatively to the class of frames which characterizes them. It is easy to prove at first

$$\forall \phi, \psi \in Typ, \vdash_S \phi \rightarrow \psi \Rightarrow for\ all\ (W, R, V),\ V^\star(\phi) \subset V^\star(\psi)$$

where S may be **L**, **NL** or **LP** according to whether R has no constraint, obeys associativity or obeys associativity + commutativity. The demonstration is made by induction on the structure of the formula.

Basis case: $\phi = \psi = p$ where p is an atom.

Induction step: See the example above. For instance, let us suppose ψ is obtained by applying at the last step of the proof the rule $A \bullet C \rightarrow B \Rightarrow A \rightarrow B/C$. That is, $\psi = B/C$, $\phi = A$. The induction hypothesis is that $V^\star(A \bullet C) \subset V^\star(B)$. We have to prove that $V^\star(A) \subset V^\star(B/C)$. For that, let us take $y \in V^\star(A)$. By definition of $V^\star(A \bullet C)$, for all $z \in V^\star(C)$, there exists a $u \in V^\star(A \bullet C)$ such that $R(u, y, z)$. By the induction hypothesis, in this case, $u \in V^\star(B)$. Therefore, for all $y \in V^\star(A)$ and for all $z \in V^\star(C)$, there exists $u \in V^\star(B)$ such that $R(u, y, z)$, and this makes it valid to say that $y \in V^\star(B/C)$.

The converse entailment is more difficult; we have to prove

$$\forall \phi, \psi \in Typ, if\ for\ all\ (W, R, V),\ V^\star(\phi) \subset V^\star(\psi)\ then\ \vdash_S \phi \rightarrow \psi$$

It is usual for this kind of proof to prove the contrapositive; in other words, if $\nvdash_S \phi \rightarrow \psi$, it is then possible to find a model (W, R, V) such that $V^\star(\phi) \not\subset V^\star(\psi)$. Such a model is often provided by the so-called "canonical model", for which

$$\begin{array}{rcl} W & = & Typ \\ R(A, B, C) & \text{iff} & \vdash_S C \bullet B \rightarrow A \end{array}$$

It allows us to define the following valuation:

$$V(p) = \text{the set of formulae } \phi \text{ such that } \vdash_S \phi \rightarrow p$$

We may then prove (*truth lemma*) Theorem 6.2.

Theorem 6.2. *For every formula ϕ of S, $A \in V^\star(\phi)$ iff $\vdash_S A \rightarrow \phi$.*

Proof: Let us show the entailment in the first direction. If $\vdash_S A \rightarrow \phi$ then $A \in V^\star(\phi)$. The proof is still made by induction on the length of the deduction.

Basis case: If ϕ is an atom p, then by definition of $V^\star$, $A \in V^\star(\phi)$.

Induction step: Let us suppose, for instance, that $\phi = B/C$, then if $\vdash_S A \rightarrow B/C$, we have (at the previous step of the deduction) $\vdash_S A \bullet C \rightarrow B$. In other words, by definition of R, we have $R(B, C, A)$ and by induction hypothesis: $A \bullet C \in V^\star(B)$. Now, let $X \in V^\star(C)$. By definition we have $\vdash_S X \rightarrow C$, therefore $\vdash_S A \bullet X \rightarrow A \bullet C$, therefore $R(A \bullet C, X, A)$. That is for all $X \in V^\star(C)$, there exists a formula $Z \in V^\star(B)$, $A \bullet C$, such that $R(Z, X, A)$ which proves that $A \in V^\star(B/C)$.

The proof is analogous in other cases.

Let us show now the other direction, that is, if $A \in V^\star(\phi)$ then $\vdash_S A \rightarrow \phi$. Let us review all the possible cases. Let us suppose for instance that $\phi = B/C$. Let $A \in V^\star(B/C)$. By definition of the valuation $V^\star$, for all $Z \in V^\star(C)$ and in particular for $Z = C$, there exists $Y \in V^\star(B)$ such that $R(Y, C, A)$, that is, such that $\vdash_S A \bullet C \rightarrow Y$. By induction hypothesis, $\vdash_S Y \rightarrow B$, therefore by transitivity, $\vdash_S A \bullet C \rightarrow B$, therefore $\vdash_S A \rightarrow B/C$.

The completeness theorem follows by noticing that if $\nvdash_S \phi \rightarrow \psi$, then $\phi \notin V^\star(\psi)$ for the canonical model, which means that there exists a formula A $(A = \phi)$ such that $A \in V^\star(\phi)$ but $A \notin V^\star(\psi)$, in other terms $V^\star(\phi) \not\subset V^\star(\psi)$.

6.6 An Extension of the Lambek Calculus: The Lambek–Grishin Calculus

6.6.1 *Adding new connectives*

The most straightforward extension of the Lambek calculus was given by Lambek (1993) himself and, almost at the same time, by Grishin (1983). Both extensions consist in giving the Lambek calculus the symmetries it has not. As said earlier in Section 4.7, the Lambek calculus is not complete in the sense that it lacks some logical operators. If structural rules are added, of course the tensor product ($\otimes$) collapses into $\wedge$. But what collapses into $\vee$? The answer is: nothing! Hence the idea to add to this calculus the product's dual, in the form of a kind of disjunction $\oplus$, *which should not be confused with the $\oplus$ of linear logic.* To avoid risks of confusion, we shall here denote this operator $\ominus$. It may be nevertheless argued that linear logic was already such an extension, with a connective ("$\wp$") which plays the same role as this new $\ominus$. In Lambek's and Grishin's views, this operator comes from its possible definition in the monoidal semantics. They put

$$v(A \ominus B) = \{c; \forall a \forall b(c = ab \Rightarrow a \in A \vee b \in B)\} \tag{6.3}$$

where the defining formula is exactly the dual of that of $v(A \otimes B)$.

If it is assumed that the monoid is provided with a neutral element 1, thus allowing a unary connective $\mathbf{1} = \{1\}$ to be added to the Lambek calculus, and if we define $\mathbf{0}$ to be the complementary set the following proposition can be easily shown.

Proposition 6.1. $A \ominus \mathbf{0} = A$

Proof:

$$\begin{aligned} A \ominus \mathbf{0} &= \{c; \forall a \forall b(c = ab \Rightarrow a \in A \vee b \in \mathbf{0})\} \\ &= \{c; \forall a \forall b(c = ab \Rightarrow a \in A \vee b \neq 1)\} \\ &= \{c; \forall a \forall b(c = ab \wedge b = 1 \Rightarrow a \in A)\} \\ &= \{c; \forall a \forall b(c = a \Rightarrow a \in A)\} = A \end{aligned}$$

The following proposition can also be proven.

Proposition 6.2. $(A \ominus B) \ominus C = A \ominus (B \ominus C)$

Residuals for this new connective are also added to the calculus. By definition, they must be such that

$$C \subset A \ominus B \Leftrightarrow C \oslash B \subset A \Leftrightarrow A \obslash C \subset B \tag{6.4}$$

We may simply define

$$C \oslash B = \{x; \exists b(xb \in C \wedge b \notin B\} \tag{6.5}$$

$$A \obslash C = \{x; \exists a(ax \in C \wedge a \notin A\} \tag{6.6}$$

This logic with $\otimes, /, \backslash, \ominus, \oslash, \obslash$ is called the weak form of *bilinear logic*. Its rules may be presented in the following format, where it is assumed that sequents for weak bilinear logic are always of the form $\Gamma \vdash \Delta$, with $\Gamma = A_1, \ldots, A_m$ and $\Delta = B_1, B_2, \ldots, B_n$, with commas interpreted on the left-hand side as $\otimes$ and on the right-hand side as $\ominus$:

$$A_1 \otimes \cdots \otimes A_m \rightarrow B_1 \ominus \cdots \ominus B_n \tag{6.7}$$

6.6.2 *Rules*

Identity and cut:

$$A \vdash A$$

$$\frac{\Lambda \vdash A \quad \Gamma, A, \Delta \vdash \Theta}{\Gamma, \Lambda, \Delta \vdash \Theta}[cut_1] \qquad \frac{\Lambda \vdash \Phi, A, \Psi \quad A \vdash \Theta}{\Lambda \vdash \Phi, \Theta, \Psi}[cut_2]$$

Logical rules:

$$\frac{\Gamma, \Delta \vdash \Theta}{\Gamma, \mathbf{1}, \Delta \vdash \Theta} \qquad \vdash \mathbf{1}$$

$$\frac{\Gamma, A, B, \Delta \vdash \Theta}{\Gamma, A \otimes B, \Delta \vdash \Theta} \qquad A, B \vdash A \otimes B$$

$$B/A \quad A \vdash B \qquad \frac{\Gamma, A \vdash B}{\Gamma \vdash B/A}$$

$$\Lambda \vdash \Phi, \top, \Psi$$

$$A \wedge B \vdash A \qquad \frac{\Lambda \vdash \Phi, A, \Psi \quad \Lambda \vdash \Phi, B, \Psi}{\Lambda \vdash \Phi, A \wedge B, \Psi}$$

with all the rules we may obtain by inverting the arrows, changing $\otimes$ into $\ominus$ and $\wedge$ into $\vee$. Moreover, the two cut rules may be absorbed into the introduction rules in order to give a sequent calculus enjoying the cut-elimination property. The resulting rules are the following:

$$\frac{\Gamma \vdash A \qquad \Delta \vdash B}{\Gamma, \Delta \vdash A \otimes B} \qquad \frac{A \vdash \Gamma \qquad B \vdash \Delta}{A \ominus B \vdash \Gamma}$$

This calculus is not yet complete in the sense that it has no negation (the slashes and co-slashes are not defined by using the product and co-product with negation). It must therefore be compared with Abrusci's non-commutative classical linear logic (Abrusci, 1991; Abrusci and Ruet, 2000) which involves negation and the usual laws for defining implications. Of course, since no structural rule in this system exists (not even the cyclic exchange rule), it has two negations, one on the left and one on the right ($^{\perp}A$ and $A^{\perp}$). Moreover, $\oslash$ and $\obslash$ are not given, but they could be easily defined by the equalities $A \oslash B = A \ominus B^{\perp}$ and $B \obslash A =^{\perp} B \ominus A$.

More generally, in many aspects, weak bilinear logic may be compared with linear logic. We must nevertheless notice that the two systems do not aim at the same kind of semantics. Whereas the so-called "bilinear" logic aims at a semantics for formulae (it has not only a monoidal semantics but also a Kripke semantics, as shown by Moortgat (2009)), linear logic is provided with a semantics for proofs (*coherent spaces* semantics). The bilinear calculus seems to be a good option for dealing with linguistic phenomena in the traditional way, as shown below, whereas linear logic seems to call for another kind of treatment (like the one we shall envisage in Chapter 12), based on proofs.

6.6.3 *The Grishin postulates*

The weak version of bilinear logic adopted by Moortgat (2009) does not offer us the expressivity needed to address the usual questions which arise from linguistics,[12] like binding. We need in fact more connections between the operators of the $\otimes$ family and those of the $\ominus$ family. Grishin pointed out that some relations easily proven in the models provided by monoids could not be proven in the calculus. It was therefore possible to add the corresponding postulates to the rules of logic, thus keeping soundness with regard to such models. For instance, the following proposition can be shown.

Proposition 6.3.

$$(A \ominus B)/C \to A \ominus (B/C) \qquad\qquad C\backslash(B \ominus A) \to (C\backslash B) \ominus A$$

---

[12] The version adopted by Moortgat lacks associativity, like it is the case for the calculus **NL**. It is one of the reasons why postulates must be added. If we took Abrusci and Ruet's non-commutative linear logic, some of these postulates (not all) would be derived by associativity, but in this case, the structures built by the calculus would have no *constituents*, in the linguistic sense.

Proof (*example*): Let $x \in (A \ominus B)/C$ and $c \in C$, then $xc \in A \ominus B$, therefore if $xc = ab$, we get $a \in A$ or $b \in B$. Let us assume $x = ay$ and $a \notin A$, then $xc = ab = (ay)c = a(yc) \in A \ominus B$. Since $a \notin A$, it follows that $yc \in B$. Therefore for every $c \in C$, $yc \in B$, we deduce that $y \in B/C$. Therefore if $x = ay$ and $a \notin A$, we get $y \in B/C$, therefore $x \in A \ominus (B/C)$.

These postulates are called *interaction postulates*. Moortgat observes that they come in two groups of four. In his terminology, the two above are G1′ and G3′. In the same group, there are also

$$(A \ominus B)/C \rightarrow (A/C) \ominus B \qquad\qquad C\backslash(B \ominus A) \rightarrow B \ominus C\backslash A$$

But another group is

$$(A \oslash B) \otimes C \rightarrow A \oslash (B \otimes C) \quad (G1) \quad C \otimes (B \oslash A) \rightarrow (C \otimes B) \oslash A \quad (G3)$$

$$C \otimes (A \oslash B) \rightarrow A \oslash (C \otimes B) \quad (G2) \quad (B \oslash A) \otimes C \rightarrow (B \otimes C) \oslash A \quad (G4)$$

Adding such postulates, it becomes easy to derive sentences with *in situ* binding, like[13]

Example 6.5. Alice suspects someone is cheating

where the interpretation is *de re*, that is,

$$\exists x.\mathbf{suspects}(\mathbf{Alice}, \mathbf{cheating}(x)) \tag{6.8}$$

by means of the following assignment: someone $(s \oslash s) \oslash np$. It must be noticed that the usual Lambek type is derivable: it can be shown in effect that

$$(s \oslash s) \oslash np \rightarrow s/(np\backslash s) \tag{6.9}$$

A fragment of the proof is given below:

$$\dfrac{\dfrac{\dfrac{np \otimes (((np\backslash s)/s) \otimes (np \otimes (np\backslash s))) \rightarrow s \qquad s \rightarrow (s \oslash s) \ominus s}{np \otimes (((np\backslash s)/s) \otimes (np \otimes (np\backslash s))) \rightarrow ((s \oslash s) \ominus s)}\ trans}{((s \oslash s) \oslash (np \otimes (((np\backslash s)/s) \otimes (np \otimes (np\backslash s))))) \rightarrow s}\ drp}{np \otimes (((np\backslash s)/s) \otimes (((s \oslash s) \oslash np) \otimes (np\backslash s))) \rightarrow s}\ G2$$

(where "drp" means dual residuation principle).

In Chapter 10, we shall see the semantics which fits well this syntactic calculus, as it has been given by Moortgat and Bernardi (2009) and

[13]The example comes from Moortgat (2009).

Moortgat (2010). In short, a *continuation semantics* will be defined on this calculus, that is, a semantics which uses μ and $\tilde{\mu}$ terms, and the corresponding extension of the Curry–Howard correspondence will provide semantic recipes exactly like in the case of the usual Lambek calculus, except that they will contain these μ- and $\tilde{\mu}$-abstractors. By means of translations (the so-called "CPS-transforms"), these recipes will be converted into λ-terms. But because such classical calculi are not deterministic, several translations will be possible: they will result from the adopted strategy, either *call-by-name* or *call-by-value*.

Chapter 7

Grammatical Reasoning

7.1 Motivations

The exponentials of LL have already given an idea of the expected role of the notion of *structural modality*: they have to locally reintroduce the structural rules which have been removed globally. For instance, some linguistic resources (words, expressions) may be permuted, others allow some flexibility of the phrase structure, even some contraction phenomena.

In the case of parasitic gapping, like in *the book*$_1$ *that John shelves* e_1 *without reading* e_1, let us suppose we reach the conclusion that *John shelves* e_1 *without reading* e_1 is of type $s/(np \bullet np)$, which seems natural because this sentence lacks two nominal phrases to be rightly saturated. Then it would be necessary, assuming the type of *that* is $(n\backslash n)/(s/np)$, to show that

$$s/(np \bullet np) \vdash s/np$$

which would lead us to prove that

$$s/(np \bullet np), np \vdash s$$

and therefore that

$$np \vdash np \bullet np$$

which is only possible by using the left contraction rule.

In a similar way, we saw that by restricting ourselves to the non-associative calculus **NL**, some phenomena (like phrase extraction) can no longer be analyzed.

In order to solve these problems, we will try, like in the case of exponentials, to reintroduce locally what we lost globally, hence the introduction of structural modalities which will affect the formulae, requiring a relaxation of (order or structure) constraints.

7.2 Modal Preliminary

7.2.1 *Necessity and possibility*

Modal logic dates at least from Aristotle's *De Interpretatione*, Chapter 9, where he sets the problem of *contingent futures*. We cannot take a statement concerning the future as being "true" or "false", since if it is true (resp. false), it is true (resp. false) at any point of time, that is, it is true (resp. false) from the beginning of Time, and there is therefore no contingence; our future is totally pre-determined. In order to avoid such a consequence, Aristotle introduced the difference between "true" (resp. "false") at some point and *necessarily* true (resp. false), that is, true or false at any point of time.

Nowadays, *necessity* is seen as a modality and is represented by $\Box$. With it, we may easily discover a second modality, *possibility*, simply by using negation: $\Diamond P \equiv \neg\Box\neg P$.

The same considerations have been made about temporality. Suppose **H** means: "it has been always the case that", then $\neg\mathbf{H}\neg P$ is interpreted as "it has not been always the case that not P", hence a dual modality **P** which is interpreted as "it has been the case at least once".

The same can be made with regard to a second temporal modality: **G** = "it will be always the case that", which has, as its dual, **F** = "it will be the case at least once". We don't expand much on modal logic since a huge literature exists on that topic; let us only notice the following results.

7.2.2 *Axiomatization*

Every propositional modal logic takes as axioms at least all the axioms of the propositional calculus, and the following one:

$$(K) \quad \vdash \Box(\phi \Rightarrow \psi) \Rightarrow (\Box\phi \Rightarrow \Box\psi)$$

In other terms, say, in the *alethic* interpretation, if ϕ necessarily implies ψ, then if necessarily ϕ, necessarily ψ.

Every modal logic also admits, with the *modus ponens* rule, the so-called *necessitation* rule:

$$\frac{\vdash \phi}{\vdash \Box\phi}$$

which is interpreted as "if ϕ is proven, then it is proven that ϕ is necessary" (for instance axioms are necessary truths, like theorems, but of course contingent facts are not).

Moreover, other axioms may be introduced, which allow us to interpret arbitrary sequences of modalities. What would, for instance, $\Box\Box P$ or $\Box\Diamond\Box\Diamond P$ mean? Is "necessity of necessity" different from "necessity"?

Hence the following axioms:

$$\mathbf{T}: \quad \vdash \Box\phi \Rightarrow \phi \tag{7.1}$$

$$\mathbf{S}_4: \quad \vdash \Box\phi \Rightarrow \Box\Box\phi \tag{7.2}$$

$$\mathbf{S}_5: \quad \vdash \Diamond\Box\phi \Rightarrow \phi \tag{7.3}$$

which allow us to distinguish between several modal logics:

- **K** which only contains (except axioms of **CP**) the axiom K,
- **T** = **K** + the axiom T,
- S_4 = **T** + the axiom S_4,
- $S_5 = S_4$ + the axiom S_5.

It is possible to introduce other axioms and then to define a lot of modal logics (the interpretations of which are questionable...). Note for instance that we can introduce an axiom (D) such that

$$(D) \quad \vdash \Box\phi \Rightarrow \Diamond\phi$$

which allows us to define a logic **KD** which would be convenient for the formalization of some kinds of *deontic* logic: $\Box$ would be interpreted as *mandatory* and $\Diamond$ as *permitted*. Of course, if P is mandatory, it must be permitted, but P is not necessarily the case. If it is obligatory to wait for the red light to cross the street, of course you should be allowed to cross this street in this case, but it may also happen (above all, in Latin countries) that you cross the street when the light is green.

A disadvantage of modal logics is that they don't easily allow a natural deduction presentation. Of course we can admit a rule like

$$\frac{\Gamma \vdash \Box A}{\Gamma \vdash A}\,[\Box\ E]$$

but the rule we would like to give to translate *necessitation* would be

$$\frac{\Gamma \vdash A}{\Gamma \vdash \Box A}\,[\Box\ I]$$

which does not work since the system we should then get would collapse into propositional logic: there would be no difference between P and $\Box P$! Prawitz suggested restricting the use of $[\Box\ I]$ only to the cases where all the formulae in Γ are boxed. In this way, by denoting with $\Box\Gamma$ a list of boxed formulae, we should get

$$\frac{\Box\Gamma \vdash A}{\Box\Gamma \vdash \Box A}[\Box\ I]$$

which gives the S_4 logic, which would then be, from the proof-theoretical viewpoint, the only "good" modal logic. We may however notice that *this rule is exactly the introduction rule for the exponential* "!" in intuitionistic linear logic. We shall also notice that these two rules easily transform into their corresponding rules in the sequent calculus.

$$\frac{\Gamma, A \vdash B}{\Gamma, \Box A \vdash B}[\Box\ L] \qquad \frac{\Box\Gamma \vdash A}{\Box\Gamma \vdash \Box A}[\Box\ R]$$

We may then notice what follows by duality (always in the case of S_4). Of course a correct treatment of duality requires we adopt a symmetric sequent calculus, the rules of which for modalities are

$$\frac{\Gamma, A \vdash \Delta}{\Gamma, \Box A \vdash \Delta}[\Box\ L] \qquad \frac{\Box\Gamma \vdash A, \Diamond\Delta}{\Box\Gamma \vdash \Box A, \Diamond\Delta}[\Box\ R]$$

Let us consider the modality $\Diamond$. We get by duality from the previous rules

$$\frac{\Gamma \vdash A, \Delta}{\Gamma \vdash \Diamond A, \Delta}[\Diamond\ R] \qquad \frac{\Box\Gamma, A \vdash \Diamond\Delta}{\Box\Gamma, \Diamond A \vdash \Diamond\Delta}[\Diamond\ L]$$

By comparing with linear logic, we observe that "!" and "?" are "$\Box$" and "$\Diamond$" in S_4. Nevertheless, another aspect emerges with linear logic: modalities are associated with structural rules that non-modalized types cannot trigger. This approach is generalized in what follows.

7.3 Residuation and Modalities

As indicated by its name, a structural modality is used to manage some structure of resources. We already know that in **NL**, resources are structured as trees, while in **L**, they form lists. If we consider associativity, we may imagine intermediary structures (between lists and trees) such that some parts at least of the whole structure may support transformations. These structures may be trees but affected by some mark indicating what type of structural change may be applied. We shall then have structures of the form $\Gamma[(\Delta)^{\Diamond}]$, which means a tree structure inside which there is a substructure Δ allowing the change. In the same way that binary constructors (like simple commas, or columns) translate into binary products at the level of formulae ($\bullet$, $\otimes$, etc.), a mark like $()^{\Diamond}$ will translate into some unary operator at the formulae level, or in other words, into a structural modality ($\Diamond^{\downarrow}$). The rules concerning the introduction of these modalities quite obviously are[1]

$$\frac{\Gamma \vdash A}{(\Gamma)^{\Diamond} \vdash \Diamond^{\downarrow} A}\,[\Diamond^{\downarrow}\ R] \qquad\qquad \frac{\Gamma[(A)^{\Diamond}] \vdash B}{\Gamma[\Diamond^{\downarrow} A] \vdash B}\,[\Diamond^{\downarrow}\ L]$$

- If a structure Γ reduces to the type A, then the same structure allowing structural change will reduce to the modalized type $\Diamond^{\downarrow} A$.
- If, thanks to a substructure reducing to A, marked for structural change, we get B, then it is possible to replace the substructure with the modal type $\Diamond^{\downarrow} A$.

A sequent $(\Gamma)^{\Diamond} \vdash X$ occurs in conclusion of a rule ($[\Diamond^{\downarrow}\ R]$). We may ask then what happens if such a sequent must be used as a premis. Of course, if $(\Gamma)^{\Diamond}$ allows us to get A, the removal of the structure $()^{\Diamond}$ allows us to get a type which lacks this structure. In the same way, we may expect that in order to prove a sequent of the form $\Gamma[(X)^{\Diamond}] \vdash B$, we have to replace X with a type which precisely lacks the structure $()^{\Diamond}$. We may compare this with the rules of sequent calculus for **L**: the same reasonings are applied to a unary connective as to a binary one. In particular, we get a sequent of the form $\Gamma[X \bullet B] \vdash C$ if we replace X with A/B and if we may prove the sequent $\Gamma[A] \vdash C$. In the case of a binary product, the lack of a resource amounts to a $/$ or a $\backslash$ (it is said that they are the residuates of the product, or the associated *divisions*). In the case of a unary connective (modality), the lack translates into another modality (which will be said to be dual), denoted by $\Box^{\downarrow}$, and we shall get the rules

$$\frac{(\Gamma)^{\Diamond} \vdash A}{\Gamma \vdash \Box^{\downarrow} A}\,[\Box^{\downarrow}\ R] \qquad\qquad \frac{\Gamma[A] \vdash B}{\Gamma[(\Box^{\downarrow} A)^{\Diamond}] \vdash B}\,[\Box^{\downarrow}\ L]$$

A comparison with the S_4 logic shows that, obviously, it is not the same system since "structural" and therefore "extra-logical" considerations come into play. Nevertheless, we may show that it is easy to simulate the S_4 logic by means of the translation

$$(\Box A)^t = \Diamond^{\downarrow}\Box^{\downarrow}(A)^t$$

[1]Structural modalities are denoted by $\Diamond^{\downarrow}$ and $\Box^{\downarrow}$ to avoid confusion with ordinary modalities, since they do not have the same rules.

Duality is naturally expressed by

$$\frac{\Diamond^{\downarrow} A \vdash B}{\overline{\overline{A \vdash \Box^{\downarrow} B}}}$$

It is in effect possible to formulate the following proof:

$$\cfrac{\cfrac{\cfrac{}{A \vdash A}}{(A)^{\Diamond} \vdash \Diamond^{\downarrow} A}[\Diamond^{\downarrow} R] \quad \Diamond^{\downarrow} A \vdash B}{\cfrac{(A)^{\Diamond} \vdash B}{A \vdash \Box^{\downarrow} B}[\Box^{\downarrow} R]}[cut]$$

a rule by means of which it is possible to deduce the two theorems

$$\Diamond^{\downarrow}\Box^{\downarrow} A \vdash A \quad A \vdash \Box^{\downarrow} \Diamond^{\downarrow} A$$

with the proofs

$$\frac{\Box^{\downarrow} A \vdash \Box^{\downarrow} A}{\Diamond^{\downarrow}\Box^{\downarrow} A \vdash A} \quad \frac{\Diamond^{\downarrow} A \vdash \Diamond^{\downarrow} A}{A \vdash \Box^{\downarrow} \Diamond^{\downarrow} A}$$

We could also expect to have

$$\frac{A \vdash \Diamond^{\downarrow} B}{\overline{\overline{\Box^{\downarrow} A \vdash B}}}$$

Actually, we do not have this perfect symmetry (unfortunately!). To be convinced, let us define the structural modality $\Diamond^{\downarrow} A$ by

$$\Diamond^{\downarrow} A = A \bullet q$$

where q is a type constant (or a propositional constant, like $\top$). Then $\Box^{\downarrow} A$ is A/q. We should have the previous rule, since $A \vdash B/q$ is equivalent to $A \bullet q \vdash B$, but it is false that from $A \vdash B \bullet q$ we may deduce $A/q \vdash B$ since $(B \bullet q)/q$ is not in general equal to B (cf. np/np is not the neutral type, as witnessed by the subsequence *the book of which I know* _ in *the book of which I know the title*). Notice that in terms of deduction relations, the above rule (to be read in both directions) amounts to a residuation rule, like in the case of the binary connectives of the calculus **NL**. We have

$$(MOD)\ \Diamond^{\downarrow} A \rightarrow B \textit{ iff } A \rightarrow \Box^{\downarrow} B \tag{7.4}$$

We may also notice the formal similarity of this modality with the *temporal* modality **F**, which can be interpreted as *it is possible in the future that*, the dual of which, **G**, is interpreted as *it was necessary in the past that.* In effect, if the possibility in the future of p depends on the realization (today) of q, then if p is realized, it is necessarily because q occurred in the past. However, if the fact that p occurs today allows that q may be realized in the future, the fact that p was necessarily the case in the past does not imply that q is the case today.

Let us consider now the example of the calculus **NL** enriched by a modality for associativity $\Diamond^{\downarrow}$ with the following structural rules.

$$\frac{\Gamma[((\Diamond^{\downarrow}A,B),C)]\vdash D}{\Gamma[(\Diamond^{\downarrow}A,(B,C))]\vdash D} \quad \frac{\Gamma[((A,\Diamond^{\downarrow}B),C)]\vdash D}{\Gamma[(A,(\Diamond^{\downarrow}B,C))]\vdash D} \quad \frac{\Gamma[((A,B),\Diamond^{\downarrow}C)]\vdash D}{\Gamma[(A,(B,\Diamond^{\downarrow}C))]\vdash D}$$

In categoric terms, these three structural rules may be expressed by

$$(ASS_1) \qquad (\Diamond^{\downarrow}A\bullet(B\bullet C))\leftrightarrow((\Diamond^{\downarrow}A\bullet B)\bullet C) \tag{7.5}$$

$$(ASS_2) \qquad (A\bullet(\Diamond^{\downarrow}B\bullet C))\leftrightarrow((A\bullet\Diamond^{\downarrow}B)\bullet C) \tag{7.6}$$

$$(ASS_3) \qquad (A\bullet(B\bullet\Diamond^{\downarrow}C))\leftrightarrow((A\bullet B)\bullet\Diamond^{\downarrow}C) \tag{7.7}$$

We shall call them *structural postulates.*

7.4 Linguistic Applications

The previous section suggests the following strategy: since on one hand, every modal type prefixed by $\Diamond^{\downarrow}$ triggers the associativity we need to derive a phrase like *the book which John wrote* and, on the other hand, every type of the form $\Diamond^{\downarrow}\Box^{\downarrow}A$ reduces to A, let us give types of this form to the assumptions required in the deduction. Let us therefore admit in the lexicon the following entry for the relative pronoun *which*:

$$which \quad : \quad (n\backslash n)/(s/\Diamond^{\downarrow}\Box^{\downarrow}np)$$

Then the deduction which previously failed may now be continued:

$$\dfrac{\dfrac{\dfrac{\dfrac{\dfrac{\cdots}{\Diamond^{\downarrow}\Box^{\downarrow}np\vdash np} \quad \dfrac{\cdots}{(np,(np\backslash s))\vdash s}}{(np,((np\backslash s)/np,\Diamond^{\downarrow}\Box^{\downarrow}np))\vdash s}}{((np,(np\backslash s)/np),\Diamond^{\downarrow}\Box^{\downarrow}np)\vdash s}}{(np,(np\backslash s)/np)\vdash s/\Diamond^{\downarrow}\Box^{\downarrow}np} \quad \dfrac{\cdots}{(np,(n,n\backslash n))\vdash np}}{(np,(n,((n\backslash n)/(s/\Diamond^{\downarrow}\Box^{\downarrow}np),(np,(np\backslash s)/np))))\vdash np}$$

The prefix $\Diamond^{\downarrow}\Box^{\downarrow}$ allows restructuring and then the cancellation rule for modalities makes it possible to go back to the non-modalized type which is required as an argument of the verb.

Structural modalities thus solve some of the under-generation problems in the Lambek calculus (even in its associative version) like its lack of expressivity for non-peripheral extractions. While it is possible to analyze, with coherent types (that is the phrase types which linguistically make sense) *the book which I read* — it is not possible to analyze, with the same types, *the book which I read — yesterday.*

The solution amounts to moving the assumption associated with the gap ("_") towards the periphery, but this move is possible only by restructuring the phrase structure of the sentence, which is only possible by means of a modalized type. Let us keep then the same lexicon as previously (notably with regard to *which*) but let us change the structural postulates:

$$(ASS) \quad (A \bullet B) \bullet \Diamond^{\downarrow} C \rightarrow A \bullet (B \bullet \Diamond^{\downarrow} C) \tag{7.8}$$

$$(DIS) \quad (A \bullet B) \bullet \Diamond^{\downarrow} C \rightarrow (A \bullet \Diamond^{\downarrow} C) \bullet B \tag{7.9}$$

Notice that these arrows are equivalent, in the sequent format, to the structural rules:

$$\frac{\Gamma[(A,(B,\Diamond^{\downarrow}C))] \vdash D}{\Gamma[((A,B),\Diamond^{\downarrow}C)] \vdash D}[Ass] \qquad \frac{\Gamma[((A,\Diamond^{\downarrow}C),B)] \vdash D}{\Gamma[((A,B),\Diamond^{\downarrow}C)] \vdash D}[Dis]$$

Then for the sequent

$$(np/n,(n,((n\backslash n)/(s/\Diamond^{\downarrow}\Box^{\downarrow}np),((np,(np\backslash s)/np),(np\backslash s)\backslash(np\backslash s))))) \vdash np$$

comes the deduction in Fig. 7.1, where *Annul* is the sequent translation of the cancellation rule for modalities. In this deduction, the type of *which*

$$\dfrac{\dfrac{\dfrac{\dfrac{\dfrac{\dfrac{\cdots}{((np,((np\backslash s)/np,np)),(np\backslash s)\backslash(np\backslash s)) \vdash s}}{((np,((np\backslash s)/np,\Diamond^{\downarrow}\Box^{\downarrow}np)),(np\backslash s)\backslash(np\backslash s)) \vdash s}[Annul]}{(((np,(np\backslash s)/np),\Diamond^{\downarrow}\Box^{\downarrow}np),(np\backslash s)\backslash(np\backslash s)) \vdash s}[Ass]}{(((np,(np\backslash s)/np),(np\backslash s)\backslash(np\backslash s)),\Diamond^{\downarrow}\Box^{\downarrow}np) \vdash s}[Dis]}{((np,(np\backslash s)/np),(np\backslash s)\backslash(np\backslash s)) \vdash s/\Diamond^{\downarrow}\Box^{\downarrow}np}[/R] \qquad \dfrac{\cdots}{(np/n,(n,n\backslash n)) \vdash np}}{(np/n,(n,((n\backslash n)/(s/\Diamond^{\downarrow}\Box^{\downarrow}np),((np,(np\backslash s)/np),(np\backslash s)\backslash(np\backslash s))))) \vdash np}[/L]$$

Fig. 7.1 Medial extraction with structural modalities

enforces the presence of an assumption $\Diamond^{\downarrow}\Box^{\downarrow}np$, at first in a peripheral position, which may then move until it is transformed into an np assumption in order to be absorbed by the transitive verb. If the deduction is read top-down, we may observe that the np assumption, needed by the transitive verb, is modalized to move towards the periphery. When it is moved, it can be discharged, making a functional type $s/\Diamond\Box np$ (the type of sentence which lacks a $\Diamond^{\downarrow}\Box^{\downarrow}np$ on its right) appear.

7.5 Back to Quantification

From the beginning of this book, we have been concerned with the question of quantifier scopes, particularly when the quantified expressions occur in the object position, like in

Example 7.1. Mr Pickwick lectures a pupil

for which we wish an existential reading: $\exists x$ x is a pupil & Mr Pickwick lectures x. We also plan to take into account sentences like

Example 7.2. Mary believes that Bill met someone

Example 7.3. Mary believes that someone met Bill

with all their ambiguities.

A sentence like Example 7.1 may be derived if we give a determiner two possible types: $(s/(np\backslash s))/n$ and $\Diamond^{\downarrow}\Box^{\downarrow}((s/(\Diamond^{\downarrow}\Box^{\downarrow}np\backslash s))/n)$. The first will be used for the subject position and the second one for the object position. We keep of course the structural postulates (Ass) and (Dis). We get

$$\dfrac{\dfrac{\dfrac{\dfrac{\dfrac{\dfrac{\dfrac{\cdots}{(np,((np\backslash s)/np,np))\vdash s}}{(np,((np\backslash s)/np,\Diamond^{\downarrow}\Box^{\downarrow}np))\vdash s}[Annul]}{((np,(np\backslash s)/np),\Diamond^{\downarrow}\Box^{\downarrow}np)\vdash s}[Ass]}{(np,(np\backslash s)/np)\vdash s/\Diamond^{\downarrow}\Box^{\downarrow}np} \qquad \dfrac{}{s\vdash s}}{((np,(np\backslash s)/np),(s/\Diamond^{\downarrow}\Box^{\downarrow}np)\backslash s)\vdash s}[\backslash L]}{((np,(np\backslash s)/np),\Diamond^{\downarrow}\Box^{\downarrow}(s/\Diamond^{\downarrow}\Box^{\downarrow}np)\backslash s)\vdash s}[Annul]}{(np,((np\backslash s)/np,\Diamond^{\downarrow}\Box^{\downarrow}(s/\Diamond^{\downarrow}\Box^{\downarrow}np)\backslash s))\vdash s}[Ass]$$

We may check that because the Curry–Howard translation only applies to logical rules (and not to the structural ones, which remain semantically neutral), the semantic representation that we get is indeed the expected one. We may also check (exercise) that the scope ambiguities of a sentence like

Example 7.4. Every student must meet a teacher are also obtained by means of this ambiguous typing.

The result is in fact not so remote from that obtained in Montague grammar, which amounted to having a particular syntactic construction for the objectual case, in that the restructuring allowed by modalities precisely amounts to performing the syntactic transformation of an implicit phrase structure. In one case as well as in the other the problem was that in a sentence like *Mr Pickwick lectures a pupil*, the quantified expression must be seen as a functor, the argument of which is associated with a phrase, while *Mr Pickwick lectures* is not a phrase. In the Montague approach, a kind of move of the object is performed such that after this move, the object c-commands the incomplete sentence *Mr Pickwick lectures*, while in modal categorial grammars, the structure is modified so that the incomplete sentence may be seen as a phrase.

A more difficult case is provided by a sentence like *Mary believes that someone met Bill* with the *de re* reading, that is, the reading according to which *there exists x such that Mary believes that x met Bill.* The problem here comes from the fact that the quantifier scope is not on the left or on the right of the quantifier, but wraps it, something known as *quantifying in*: the variable introduced as an argument (to the verb *met* here) is internal to the phrase and is linked by an external quantifier. Previous modalized types do not allow this kind of analysis. If we study the simpler type we can imagine for a quantified nominal phrase $s/(np\backslash s)$; it is clear that by giving this type to a quantified np, we oblige the scope of the expression to be of type $np\backslash s$, but to prove that it has this type amounts to introducing an np assumption and then proving that we have an s. In other words, the sequence of types completed with an np trace is a sentence. If we have proven that, then it will remain to prove that the rest of the sequence of types reduces to s. We may then say that an expression of type $s/(np\backslash s)$ *links* a variable of type np inside a *domain* of type s, by *producing* a type s. Moortgat proposed a ternary connective $q(A, B, C)$ which behaves exactly this way.

It may be proven easily that, in **L**, the types $C/(A\backslash B)$ obey the following derived rule:

$$\frac{A, \Delta \vdash B \quad \Gamma, C \vdash D}{\Gamma, C/(A\backslash B), \Delta \vdash D}$$

Moortgat then proposes to have a more general type, with the following rule:

$$\frac{\Delta[A] \vdash B \quad \Gamma[C] \vdash D}{\Gamma[\Delta[q(A, B, C)]] \vdash D}$$

Let us give *someone* the type $q(np, s, s)$. The derivation will be

$$\frac{\dfrac{\dots}{(np, ((np\backslash s)/np, (np/s, (np, ((np\backslash s)/np, np))))) \vdash \mathtt{s}_1} \quad \dfrac{}{\mathtt{s}_2 \vdash s}}{(np, ((np\backslash s)/np, (np/s, (q(np, \mathtt{s}_1, \mathtt{s}_2), ((np\backslash s)/np, np))))) \vdash s}$$

We may of course ask whether this left introduction rule is coherent with the other rules. In fact, Moortgat (1996a) shows that $q(A, B, C)$ may be expressed by means of several structural modalities. We enter then the domain of multimodal categorial grammars (Moortgat, 1996b).

7.6 Kripke Semantics

We may take again the relational semantic of **NL** and enrich it by adding to the ternary relation R^3 a binary relation R^2 to interpret a unary modality.

Definition 7.1. We shall call a frame for a one-mode categorial calculus a 3-tuple (W, R^3, R^2), where W is a set of available resources, R^3 a ternary relation on W and R^2 a binary relation on W.

A valuation V now extends to modalized types by setting

- $V^\star(\Box^\downarrow A) = \{z \in W; \forall y\ R^2yz \Rightarrow y \in V^\star(A)\}$
- $V^\star(\Diamond^\downarrow A) = \{y \in W; \exists z \in V^\star(A)\ R^2yz\}$

We get the same completeness results, with similar demonstrations, as for **L**, **NL** and **LP**.

The intuitive interpretation is the following. $\Box^\downarrow A$ is interpreted as the set of available resources which give, when they are transferred by the relation R^2, a resource of type A (in other words, they are indeed objects which lack a structural feature to become of type A), and $\Diamond^\downarrow A$ as the set of available resources which come by this transfer from at least an A-type resource (that is, they have the structural feature in question).

7.7 Concluding Remarks and Observations

This chapter showed how to overcome the limitations of the pure Lambek calculus by adding structural modalities. The presentation we gave of it is along the lines of Moortgat's multimodal categorial grammar (MMCG) (which departs in some respects from Morrill's and Hepple's views). This view is often referred to as "grammatical reasoning": this puts the emphasis on a particular conception of grammar as a logical system. According to this view, the production of language would not be essentially different from other cognitive capacities, like reasoning. This of course may be questioned. There are many arguments in favour of a neat distinction between a linguistic device and the logical modules used in natural reasoning (see Pinker, 1994). Nevertheless, there are also arguments in favour of mathematical structures having their roots in language. If, like it is claimed by Hauser *et al.* (2002), elementary arithmetics originates from *recursivity*, seen as the essential property of human language, we may also assume that other basically mathematical properties like *residuation* come from other essential properties of language. The fundamental unsaturation of certain linguistic units and the fact that they ask for some structuration of resources are likely to be such properties.

Drawbacks of MMCG come from other considerations. It has for instance been shown that the Lambek calculus with only one structural modality (Carpenter, 1996) becomes Turing-complete. This amounts to saying that modal grammars can generate as many languages as Turing machines. They are therefore not atuned to human language specifically. Structural modalities may appear as extra-devices with regard to grammar, lacking an independent motivation.[2]

---

[2]With regard to the connection of multimodal categorial grammar with the minimalist framework, it must be noted that an in-depth work was done by [Vermaat (1999)], a former student of Moortgat's. She showed in her master's thesis how to encode minimalist grammars. Essentially, selectors become argument categories, licensors become indexed $\Box^{\downarrow}$'s and licensees $\Diamond^{\downarrow}$'s. For instance, a lexical entry like $= \texttt{v}\ + \texttt{wh}\ \texttt{c}$ becomes $\Box^{\downarrow}_{wh}c/v$.

Of course, some of the disadvantages of MMCG are avoided in the new paradigm of grammar proposed by Moortgat and Bernardi, based on the Lambek–Grishin calculus (see previous chapter), where structural modalities may be gotten rid of. Nevertheless, the Lambek–Grishin calculus has been shown to be NP-complete (Bransen, 2010).

Chapter 8

A Type-Theoretical Version of Minimalist Grammars

8.1 Inserting Chains

In Section 3.4.6, we underlined the strong resemblance between minimalist grammars and categorial grammars. This link is obvious concerning the operation Merge since, as we have seen, selectors may be understood as argument categories. Instead of ordering the words during the construction of the syntactic tree, like it is the case in categorial grammar, we may assume that we simply aim at resolving the second task, leaving the first one to another module or to the effects of displacement. In this case, argument categories under the right slash are seen as *complements* while those under the left slash are *specifiers*. There remains the operation Move. What corresponds to Move in Lambek grammars is, as we saw in the previous chapter, *hypothetical reasoning*. When a DP is moved, that amounts to making a hypothesis corresponding to its original site (before Move) and then to discharging it when a higher-order type logically obtained from it, for instance $s/(dp\backslash s)$, may absorb the phrase containing a hole. There is no such logical reasoning in minimalist grammars. Instead, we have *chains* (Brody, 1995), consisting in a leading XP followed by one or more empty expressions, which are inserted in specific positions, those which are already marked by "hypotheses"!

For instance, in the derivation of *What book do you think that Mary read?*, instead of moving twice *what book*, we may mark three positions (respectively corresponding to the position of the complement of *read*, the position of the specifier, and then the position of the specifier of the CP *do*) by "hypotheses" and then replace these hypotheses with the respective components of the chain:

$$< what\ book\ :\ dp[-\mathtt{wh}], \epsilon\ :\ dp, \epsilon\ :\ dp >$$

This solution leads us to work with only one operation, namely Merge, like the following example shows us.

We assume

(1) A selector $= \mathtt{x}$ may be resolved by merging the tree which has $= \mathtt{x}$ as the first feature of its head, with a provisional $\mathtt{x}$, with an empty phonetical feature (ϵ), that we call a *hypothesis*. The *rank* of a hypothesis in a tree is the level at which it is introduced.
(2) Every licensor $+\mathtt{f}$ is resolved by merging the tree which has $+\mathtt{f}$ as the first feature of its head, also with a *hypothesis*, that is a provisional $-\mathtt{f}$, with an empty phonetical feature (ϵ). The hypothesis is strong ($[-\mathtt{F}]$) if the licensor is strong; it is weak ($[-\mathtt{f}]$) in the contrary case.
(3) A phrase Φ which is requested to *overtly* move has at least one feature $-\mathtt{f}$ corresponding to a strong feature in the tree in construction. It is represented by a so-called *chain* with as many components as there are formal features in Φ up to its last licensee.
(4) A phrase Φ which is requested to *covertly* move has at least one feature $-\mathtt{f}$ in the tree in construction, and no strong feature.
(5) In case of *overt* movement, only the component of a chain corresponding to the highest hypothesis associated with a strong feature is not phonetically empty, while all the others are.
(6) In case of *covert* movement, only the component of a chain corresponding to the lowest hypothesis is not phonetically empty, while all the others are.
(7) Chains are inserted into the tree in construction as soon as all the hypotheses they need have been introduced.

Example: Let us derive the tree corresponding to *What book does Mary read?* (without DO-inversion).

We build at first a tree by simple Merge operations:

(1) Merge with a hypothesis d:

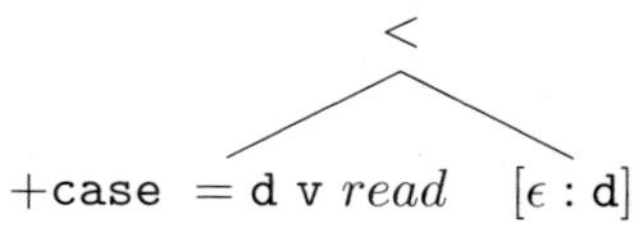

(2) Merge with a hypothesis −case:

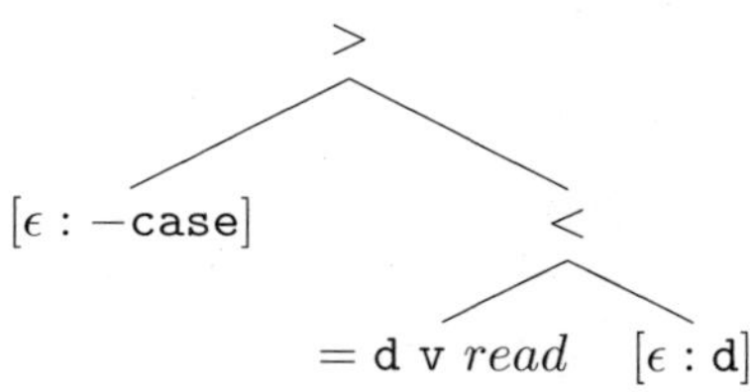

(3) Merge with a new hypothesis d:

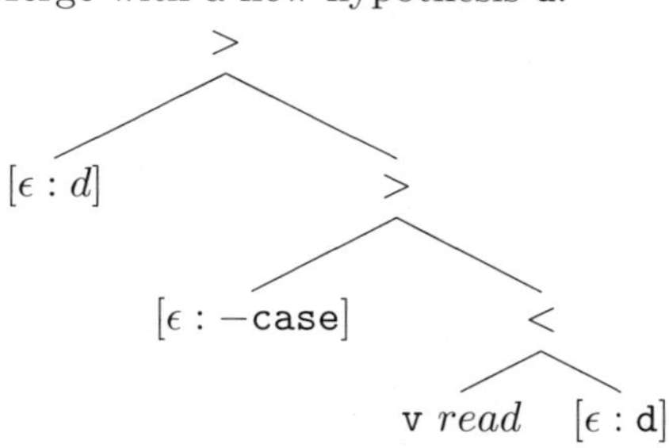

(4) Merge with =v +CASE t /*does*/:

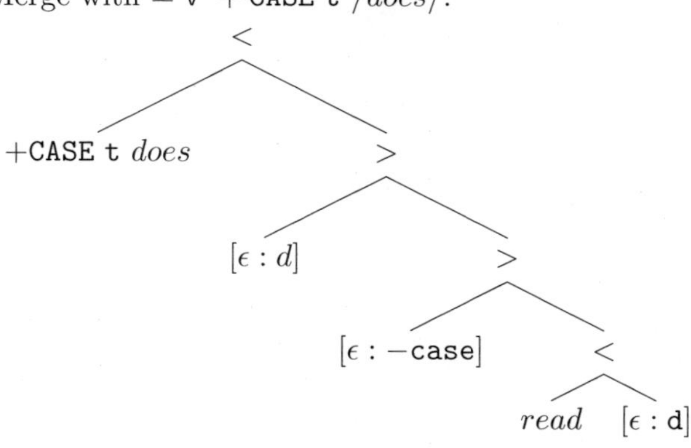

(5) Merge with a hypothesis [−CASE]:

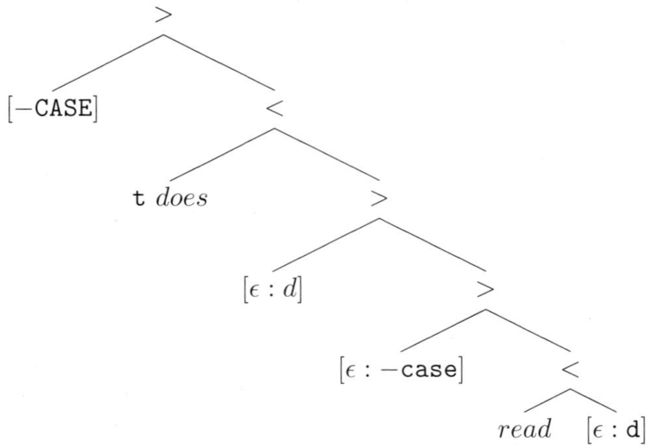

(6) Merge with = T + WH c:

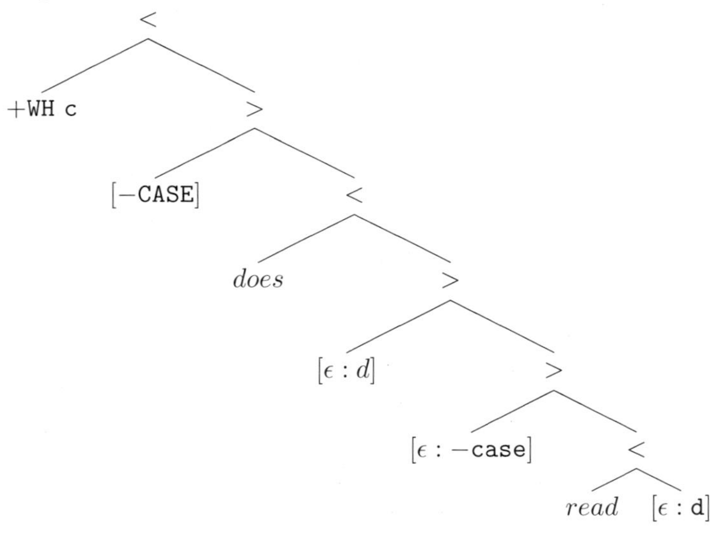

(7) Merge with a hypothesis [$-\mathtt{WH}$]:

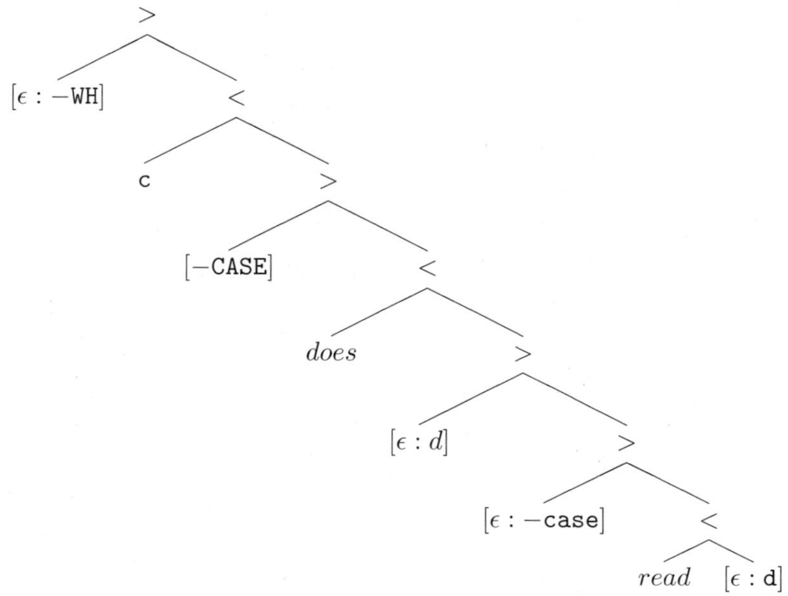

In this grammar and in this sentence, there are two expressions which are requested "to move" because they have licensees in their list of features. One is

$$\mathtt{d}\ -\mathtt{case}\ -\mathtt{wh}\ \textit{what book}$$

The other is

$$\mathtt{d}\ -\mathtt{case}\ Mary$$

Moreover, in the tree, $+\mathtt{WH}$ is strong. The first one is therefore represented by the chain

$$<\textit{what book} : -\mathtt{wh}, \epsilon : -\mathtt{case}, \epsilon : \mathtt{d}>$$

and the second one by

$$<Mary : -\mathtt{case}, \epsilon : \mathtt{d}>$$

We say that the derivation converges if and only if chains may be inserted according to the following convention (shortest move condition or SMC):

> If there are two chains to insert and if both contain the same feature $-\mathtt{f}$, the one which is inserted before the other (by means of its previous component) in the tree continues to insert oneself at the lowest $[-\mathtt{f}]$ active hypothesis.

For instance, concerning our example, we have the following matches:

First, insertion of *what book*:

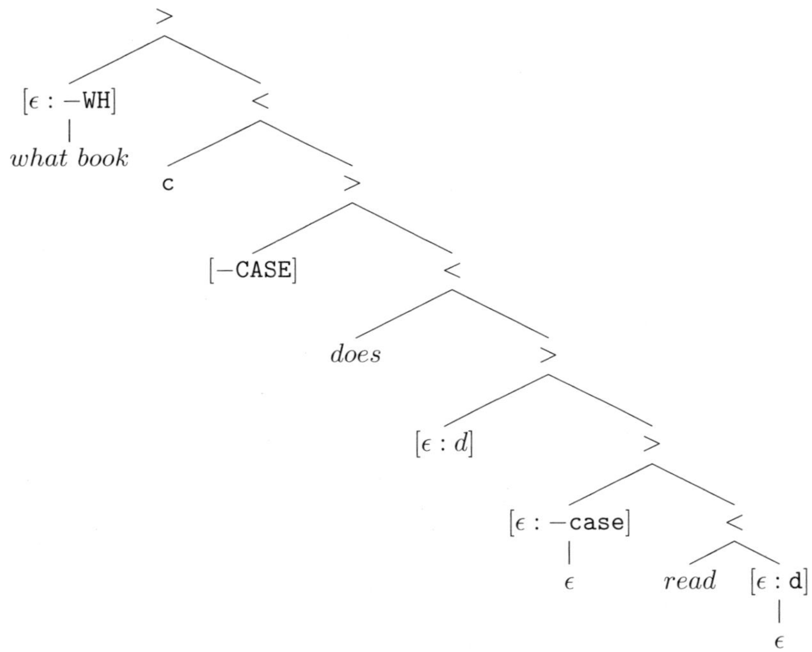

Then, insertion of *Mary*:

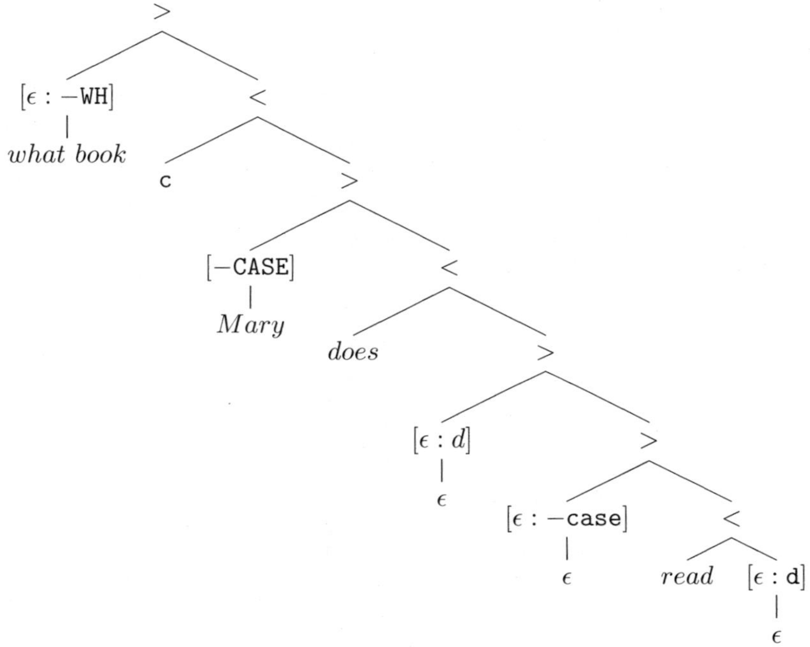

Hopefully, it is now clear that the operation we have performed when inserting chains actually looks like the *elimination of the product* in the natural deduction presentation of the Lambek grammar with product, which is something that we recall here:

$$\frac{A \bullet B \qquad \begin{array}{c}[A] \quad [B] \\ \vdots \\ C\end{array}}{C}$$

The difference is that in the Lambek calculus with product, the product is non-commutative, and that results in the fact that the two hypotheses in the rule above must be adjacent and in this order. The constraint is removed if the product is commutative. In this case, we shall denote by $\otimes$ the commutative product (see the tensor of linear logic in Chapter 4) and the rule becomes

$$\frac{A \otimes B \quad \begin{matrix} [A] \quad [B] \\ \vdots \\ C \end{matrix}}{C}$$

where A and B may be interchanged and where they can be separated by other formulae.

It may be noticed that an n-chain (that is a chain of length n) with $n > 2$ may be inserted in several steps.

$$\frac{D \otimes (A \otimes B) \quad \dfrac{[D] \quad \dfrac{[A \otimes B] \quad \begin{matrix} [A] \quad [B] \\ \vdots \\ C \end{matrix}}{C}}{E}}{E}$$

For instance, for a 3-chain like a chain $< \mathtt{-WH}, \mathtt{-CASE}, \mathtt{d} >$,

$$\frac{\mathtt{-WH} \otimes (\mathtt{d} \otimes \mathtt{-CASE}) \quad \dfrac{[\mathtt{-WH}] \quad \dfrac{[\mathtt{d} \otimes \mathtt{-CASE}] \quad \begin{matrix} [\mathtt{d}] \quad [\mathtt{-CASE}] \\ \vdots \\ C \end{matrix}}{C}}{E}}{E}$$

This may be generalized for an arbitrary n:

$$\frac{A_n \otimes (\ldots \otimes (A_2 \otimes A_1)\ldots) \quad \dfrac{[A_n] \quad \begin{matrix} \dfrac{[A_k \otimes (\ldots \otimes (A_2 \otimes A_1)\ldots)] \quad \dfrac{[A_k] \quad \begin{matrix} \dfrac{[A_2 \otimes A_1] \quad \begin{matrix} [A_2] \quad [A_1] \\ \vdots \\ C_1 \end{matrix}}{C_1} \\ \vdots \\ C_{k-1} \end{matrix}}{C_k}}{C_k} \\ \vdots \\ C_{n-1} \end{matrix}}{C_n}}{C_n}$$

Every tensor product discharges only two hypotheses: one is the most recent hypothesis, and the other is the tensor product of all the earlier ones, a tensor product which has been introduced at the previous ⊗-elimination step in order to eliminate the previous hypotheses and so on. Only the last tensor product is not a new hypothesis but corresponds to a lexical or syntactic expression. All the other ones are intermediary hypotheses, consisting in growing tensor products. The SMC should be adapted to this general case. Actually, in case of a choice, the hypothesis which is discharged at each elimination step of a product is the one which *minimizes* the distance between the two hypotheses which are discharged at the same time.

We shall see in more detail how to get a derivation for a sentence by means of the two following rules (we use only \ as a constructor of functional types; we could have denoted it by $\multimap$):

- *-elimination*:

$$\frac{\mathtt{x} \quad \mathtt{x}\backslash Y}{Y}$$

- ⊗*-elimination*:

$$\frac{A \otimes B \qquad \begin{matrix}[A] \quad [B] \\ \vdots \\ C\end{matrix}}{C}$$

where small letters denote primitive types and uppercase letters may denote any complex type.

We assume that word order is obtained by the way types are labelled and by some general statement[1] (in the spirit of *principles and parameters*) like *all the heads must be in the first position of the phrase*; more precisely, specifiers come first, then the head and then complements. At each derivation step, a binary structure is therefore built, indicating in what direction the head has to be found. This forces us to slightly adapt the two previous

[1]Such a statement actually corresponds to the value of a parameter.

rules, distinguishing, like in MGs, the case of a lexical head from the case of a non-lexical one.

- $\backslash$-*elimination*:

$$\frac{\mathtt{x} : X \quad \mathtt{x}\backslash Y : \Sigma}{Y : \{^{<}\Sigma, X\}}\ merge\ [\mathtt{x}\backslash Y : \Sigma \in Lex]$$

$$\frac{\mathtt{x} : X \quad \mathtt{x}\backslash Y : \Sigma}{Y : \{^{<}X, \Sigma\}}\ merge\ [\mathtt{x}\backslash Y : \Sigma \notin Lex]$$

- $\otimes$-*elimination*:

$$\frac{A \otimes B : (X, Y) \qquad \begin{array}{c}[A : x] \quad [B : y] \\ \vdots \\ C : \Sigma\end{array}}{C : \Sigma[x/X, y/Y]}$$

So-called "strong" features are indicated by a mark * put on primitive types, and which must be propagated so that only a hypothesis also marked by it may respond to the demand for this type. This star is also transmitted to the hypothesis which labels the starred hypothesized type. Afterwards, each product type corresponding to a moving phrase will have a corresponding starred component. When the $\otimes$-elimination rule is used, a sequence of terms (variables or constants) substitutes a sequence of variables. More precisely, let us define *terms* as *variables, constants* or *pairs of terms* and let us assume variables are individual variables $(x, x', x'', \ldots)$ or vector variables $(\overrightarrow{x} \ldots)$. Assuming the factorization of an n-product into a 2-product by means of associativity, a pair of terms substitutes a pair of variables at each $\otimes$-elimination step. Only a starred individual variable which corresponds to the highest starred hypothesis will receive a non-empty phonetic feature.

The *height* of a hypothesis is its rank on a sequence of active hypotheses (not yet discharged) counted from right to left.

$$\frac{A^* \otimes B : (X, \overrightarrow{\epsilon}) \qquad \begin{array}{c}[A^* : x^*] \quad [B : \overrightarrow{y}] \\ \vdots \\ C : \Sigma\end{array}}{C : \Sigma[x^*/X, \overrightarrow{y}/\overrightarrow{\epsilon}]}\ X \neq \epsilon \qquad \frac{A \otimes B^* : (\epsilon, \overrightarrow{Y}) \qquad \begin{array}{c}[A : x] \quad [B^* : \overrightarrow{y}] \\ \vdots \\ C : \Sigma\end{array}}{C : \Sigma[x/\epsilon, \overrightarrow{y}/\overrightarrow{Y}]}\ \overrightarrow{Y} \neq \overrightarrow{\epsilon}$$

In the right-hand option, in case B is a product type, the star * on B indicates that at least one factor of B is starred and only the component of $\overrightarrow{Y}$ which corresponds to the highest starred hypothesis in the sequence $\overrightarrow{y}$ is different from ϵ.

In case there is no strong feature, all the hypotheses are discharged with the substitution of the empty string until the last one, which receives the plain phonetic feature:

$$\frac{A \otimes B : (\epsilon, \overrightarrow{Y}) \qquad \begin{array}{c}[A : x] \quad [B : \overrightarrow{y}] \\ \vdots \\ C : \Sigma\end{array}}{C : \Sigma[x/\epsilon, \overrightarrow{y}/\overrightarrow{Y}]} \textit{ only the last component of } \overrightarrow{Y} \neq \epsilon$$

The steps of the derivation are then

(1) First step: *what book*:

$$\frac{\mathtt{n} : \textit{book} \qquad \mathtt{n}\backslash(\mathtt{wh} \otimes (\mathtt{case} \otimes \mathtt{d})) : \textit{what}}{\mathtt{wh} \otimes (\mathtt{case} \otimes \mathtt{d}) : \{^{<}\textit{what}, \textit{book}\}} \textit{merge}$$

(2) Second step: introducing a $\mathtt{d}$ hypothesis:

$$\frac{[\mathtt{d} : z] \qquad \mathtt{d}\backslash(\mathtt{case}^*\backslash(\mathtt{d}\backslash\mathtt{v})) : \textit{read}}{\mathtt{case}^*\backslash(\mathtt{d}\backslash\mathtt{v}) : \{^{<}\textit{read}, z\}}$$

(3) Third step: introducing a $\mathtt{case}$ hypothesis:

$$\frac{[\mathtt{case}^* : z'^*] \qquad \mathtt{case}^*\backslash(\mathtt{d}\backslash\mathtt{v}) : \{^{<}\textit{read}, z\}}{(\mathtt{d}\backslash\mathtt{v}) : \{^{>}z'^*, \{^{<}\textit{read}, z\}\}}$$

(4) Fourth step: eliminating two hypotheses:

$$\frac{[\mathtt{case}^* \otimes \mathtt{d} : (u', u)]^* \quad (\mathtt{d}\backslash\mathtt{v}) : \{^{>}z'^*, \{^{<}\textit{read}, z\}\}}{(\mathtt{d}\backslash\mathtt{v}) : \{^{>}u'^*, \{^{<}\textit{read}, u\}\}}$$

(5) Fifth step: introducing a new $\mathtt{d}$ hypothesis:

$$\frac{[\mathtt{d} : y] \qquad (\mathtt{d}\backslash\mathtt{v}) : \{^{>}u'^*, \{^{<}\textit{read}, u\}\}}{\mathtt{v} : \{^{>}y, \{^{>}u'^*, \{^{<}\textit{read}, u\}\}\}}$$

(6) Sixth step: introducing *does*:

$$\frac{\mathtt{v} : \{^{>}y, \{^{>}u'^*, \{^{<}\textit{read}, u\}\}\} \quad \mathtt{v}\backslash(\mathtt{case}^*\backslash\mathtt{t}) : \textit{does}}{\mathtt{case}^*\backslash\mathtt{t} : \{^{<}\textit{does}, \{^{>}y, \{^{>}u'^*, \{^{<}\textit{read}, u\}\}\}\}}$$

(7) Seventh step: introducing a new case hypothesis:

$$\frac{[\texttt{case}^* : y'^*] \qquad \texttt{case}^*\backslash\texttt{t} : \{^< does, \{^> y, \{^> u'^*, \{^< read, u\}\}\}\}}{\texttt{t} : \{^> y'^*, \{^< does, \{^> y, \{^> u'^*, \{^< read, u\}\}\}\}\}}$$

(8) Eighth step: eliminating the two hypotheses by a lexical entry:

$$\frac{\texttt{case}^* \otimes \texttt{d} : (Mary, \epsilon) \quad \texttt{t} : \{^> y'^*, \{^< does, \{^> y, \{^> u'^*, \{^< read, u\}\}\}\}\}}{\texttt{t} : \{^> Mary, \{^< does, \{^> \epsilon, \{^> u'^*, \{^< read, u\}\}\}\}\}}$$

(9) Ninth step: introducing c:

$$\frac{\texttt{t} : \{^> Mary, \{^< does, \{^> \epsilon, \{^> u'^*, \{^< read, u\}\}\}\}\} \quad \texttt{t}\backslash(\texttt{wh}^*\backslash\texttt{c}) : \epsilon}{\texttt{wh}^*\backslash\texttt{c} : \{^< \epsilon, \{^> Mary, \{^< does, \{^> \epsilon, \{^> u'^*, \{^< read, u\}\}\}\}\}\}}$$

(10) Tenth step: introducing a wh hypothesis:

$$\frac{[\texttt{wh}^* : z''^*] \quad \texttt{wh}^*\backslash\texttt{c} : \{^< \epsilon, \{^> Mary, \{^< does, \{^> \epsilon, \{^> u'^*, \{^< read, u\}\}\}\}\}\}}{\texttt{c} : \{^> z''^*, \{^< \epsilon, \{^> Mary, \{^< does, \{^> \epsilon, \{^> u'^*, \{^< read, u\}\}\}\}\}\}\}}$$

(11) Eleventh step: eliminating all the hypotheses:

$$\frac{\texttt{wh}^* \otimes (\texttt{case} \otimes \texttt{d}) : (\{^< what, book\}, \epsilon, \epsilon) \quad \texttt{c} : \{^> z''^*, \{^< \epsilon, ..., \{^> u'^*, \{^< read, u\}\}\}\}}{\texttt{c} : \{^> \{^< what, book\}, \{^< \epsilon, \{^> Mary, \{^< does, \{^> \epsilon, \{^> \epsilon, \{^< read, \epsilon\}\}\}\}\}\}\}}$$

Let us suppose now English is an *in situ* language, that is, a language where the interrogative phrase does not move, and let us also assume that case is a weak feature. That would result in

(10) Tenth step:

$$\frac{[\texttt{wh} : z''] \quad \texttt{wh}\backslash\texttt{c} : \{^< \epsilon, \{^> Mary, \{^< does, \{^> \epsilon, \{^> u'^*, \{^< read, u\}\}\}\}\}\}}{\texttt{c} : \{^> z'', \{^< \epsilon, \{^> Mary, \{^< does, \{^> \epsilon, \{^> u', \{^< read, u\}\}\}\}\}\}\}}$$

(11) Eleventh step: eliminating all the hypotheses:

$$\frac{\texttt{wh} \otimes (\texttt{case} \otimes \texttt{d}) : (\epsilon, (\epsilon, \{^< what, book\})) \quad \texttt{c} : \{^> z''^*, \{^< \epsilon, ..., \{^> u'^*, \{^< read, u\}\}\}\}}{\texttt{c} : \{^> \epsilon, \{^< \epsilon, \{^> Mary, \{^< does, \{^> \epsilon, \{^> \epsilon, \{^< read, \{^< what, book\}\}\}\}\}\}\}\}}$$

In case wh is weak but case is strong, we get

$$\frac{\texttt{wh} \otimes (\texttt{case}^* \otimes \texttt{d})^* : (\epsilon, (\{^< what, book\}, \epsilon)) \quad \texttt{c} : \{^> z''^*, \{^< \epsilon, ..., \{^> u'^*, \{^< read, u\}\}\}\}}{\texttt{c} : \{^> \epsilon, \{^< \epsilon, \{^> Mary, \{^< does, \{^> \epsilon, \{^> \{^< what, book\}, \{^< read, \epsilon\}\}\}\}\}\}\}}$$

Comments: At the fourth step, hypotheses are eliminated immediately after a `case` hypothesis has been introduced: this is due to the *shortest move constraint*, according to which *no two hypotheses of the same type must be active at the same time*. In order to prevent this configuration, `d` is eliminated together with `case` by introducing a complex product hypothesis of type $\texttt{case} \otimes \texttt{d}$.

Obviously, other derivations could have been obtained, choosing for instance to discharge the first `d` and `case` by means of the proper noun *Mary*, but then, the `wh` feature could have been checked only together with the second `d` and `case`, thus resulting in

$$\{^{>}\epsilon, \{^{<}\epsilon, \{^{>}\{^{<}what, book\}, \{^{<}does, \{^{>}\epsilon, \{^{>}\{^{<}Mary\}, \{^{<}read, \epsilon\}\}\}\}\}\}\}$$

that is, $\{^{<}what, book\}$ in the subject position and $Mary$ in the object position.

8.2 Head Movement

We saw that in MGs, *head movement* was in fact a variant of Merge, a variant where the head of the selectee comes to adjoin the head of the head syntactic phrase (the one which possesses the selector), on the left or on the right, depending on the form of the selector: $= \texttt{C}$ or $\texttt{C} =$. Selectors which trigger such an operation were denoted by uppercase letters.

In our type-logical frame, therefore, nothing *logically speaking* will distinguish head movement from Merge. It is only on the "interpretative" side, that is, in the projection of the derivation on the phonetic component, that changes may occur. Like "*" in the previous section, which was a mere diacritic (without a logical meaning) indicating where phonetic features would have to be inserted, a "^{h}" diacritic will be introduced here in order to select the *head adjunct* variant of Merge. This requires we define a function *head* which, given a phrase, returns its head.

$$\begin{aligned} head(\texttt{x}) &= \texttt{x} \\ head(\{^{<}X, Y\}) &= head(X) \\ head(\{^{>}X, Y\}) &= head(Y) \end{aligned}$$

$$\frac{\texttt{x} : X \quad \texttt{x}^h \backslash Y : \Sigma}{Y : \{^{<}\{head(X), \Sigma\}, X[head(X)/\epsilon]\}}\ H\ merge$$

The previous derivation may now be adapted to the problem of DO-inversion by assuming $\texttt{c}$ actually selects a $\texttt{t}^h$.

(9) Ninth step: introducing $\texttt{c}$:

$$\frac{\texttt{t} : \{^{>}Mary, \{^{<}does, \{^{>}\epsilon, \{^{>}u'^{*}, \{^{<}read, u\}\}\}\}\} \quad \texttt{t}^h\backslash(\texttt{wh}^*\backslash\texttt{c}) : \epsilon}{\texttt{wh}^*\backslash\texttt{c} : \{^{<}\{does, \epsilon\}, \{^{>}Mary, \{^{<}\epsilon, \{^{>}\epsilon, \{^{>}u'^{*}, \{^{<}read, u\}\}\}\}\}\}}$$

(10) Tenth step: introducing a $\texttt{wh}$ hypothesis:

$$\frac{[\texttt{wh}^* : z''^{*}] \quad \texttt{wh}^*\backslash\texttt{c} : \{^{<}\{does, \epsilon\}, \{^{>}Mary, \{^{<}\epsilon, \{^{>}\epsilon, \{^{>}u'^{*}, \{^{<}read, u\}\}\}\}\}\}}{\texttt{c} : \{^{>}z''^{*}, \{^{<}\{does, \epsilon\}, \{^{>}Mary, \{^{<}\epsilon, \{^{>}\epsilon, \{^{>}u'^{*}, \{^{<}read, u\}\}\}\}\}\}\}}$$

(11) Eleventh step: eliminating all the hypotheses:

$$\frac{\texttt{wh}^* \otimes (\texttt{case} \otimes \texttt{d}) : (\{^{<}what, book\}, \epsilon, \epsilon) \quad \texttt{c} : \{^{>}z''^{*}, \{does, \epsilon\}, ...\epsilon, \{^{<}read, \epsilon\}\}\}}{\texttt{c} : \{^{>}\{^{<}what, book\}, \{^{<}\{does, \epsilon\}, \{^{>}Mary, \{^{<}\epsilon, \{^{>}\epsilon, \{^{>}\epsilon, \{^{<}read, \epsilon\}\}\}\}\}\}\}}$$

8.3 Adjoining and Scrambling

The type-logical framework we develop here for minimalist grammars is particularly suitable for dealing with *adjunction* and *scrambling.* Let us recall what they are.

- *Adjunction* consists in adding adjuncts (like adjectives, adverbs, relatives, adverbial phrases and so on), it is an operation which is *optional* and may be *reiterated.* Moreover it is *type-preserving*: adding an adjunct to a noun for instance does not change the syntactic category of the phrase: it is still a noun.
- Like adjunction is a kind of Merge (but "asymmetrical"), *scrambling* is a kind of Move, which leads to the adjunction of a moved XP to a host YP. It may be reiterated too, like adjunction, and it is also type-preserving.

The type-preserving characteristic of an expression is easily reflected in categorial grammars by the assignment of a type $X\backslash X$ or X/X to that expression. There is therefore no need here to introduce some new device (like some specific rule) for expressing those operations. The way we design types is enough to express them. Nevertheless, there is always Lasnik's objection, according to which *adverbials are not heads*, but we may answer that we have separated here the notion of *semantic* head from that of *syntactic* head. If the former is still expressed by a functional treatment, the latter purely comes from the definition of Merge and is only marked in the structure obtained after derivation by the labels "<" and ">". Concerning adjoining, we shall simply eliminate any trace of headedness, thus defining a variant of the Merge rule which only applies to the case of Merge with a syntactic object of type $\mathtt{x}\backslash\mathtt{x}$ (and for each object of such a type):

- *adjoining*:

$$\frac{\mathtt{x} : X \quad \mathtt{x}\backslash\mathtt{x} : \Sigma}{\mathtt{x} : \{^a\Sigma, X\}}\ \textit{adjoining}$$

Let us study for instance the case of adverbs. We may assume at first glance the following assignments:

$$\begin{array}{ll} \textit{wall} & := \mathtt{n} \\ \textit{the} & := \mathtt{n}\backslash(\mathtt{d} \otimes \mathtt{k}) \\ \textit{Peter} & := \mathtt{d} \otimes \mathtt{k} \\ \textit{paint} & := \mathtt{d}\backslash(\mathtt{k}\backslash(\mathtt{d}\backslash\mathtt{v})) \\ (\textit{infl}) & := \mathtt{v}\backslash(\mathtt{k}\backslash\mathtt{t}) \\ \textit{meticulously} & := \mathtt{v}\backslash\mathtt{v} \end{array}$$

The derivation comes to be (in a tree format) like in Fig. 8.1.

Of course, adverbs may be inserted at various levels of the structure: we have here two solutions.

(1) Either we enrich the typing (using primitives types like $\mathtt{cpr}$, $\mathtt{spr}$, with entries corresponding to $\mathtt{v}\backslash\mathtt{cpr}$ and $\mathtt{cpr}\backslash(\mathtt{d}\backslash\mathtt{spr})$) (see Frey and Gärtner, 2002),[2] thus obtaining as many levels of insertion that we wish ($\mathtt{cpr}$ for core predicates and $\mathtt{spr}$ for sentential predicates),
(2) or we allow types with order more than 1, like $(\mathtt{d}\backslash\mathtt{v})\backslash(\mathtt{d}\backslash\mathtt{v})$.

Solutions are similar for scrambling. Let us take the German example: *dass die hemden Hans bügelt* (that Hans irons the shirts). The lexicon contains

$$\begin{array}{ll} \textit{die} & := \mathtt{n}\backslash(\mathtt{d} \otimes \mathtt{k} \otimes (\mathtt{cs}\backslash\mathtt{cs})) \\ \textit{dass} & := \mathtt{cs}\backslash\mathtt{c} \end{array}$$

[2]See also Gärtner and Michaelis (2008).

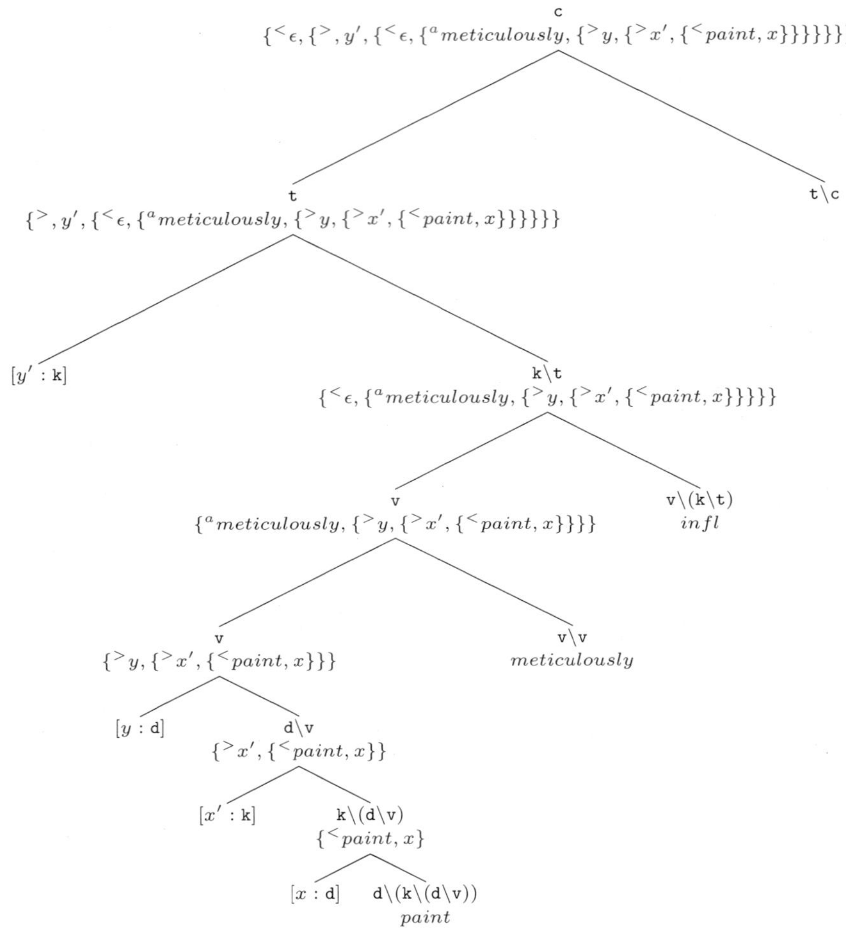

Fig. 8.1 Derivation of *Peter meticulously paints the wall*

where cs means "core sentence" and must necessarily be introduced if we want the entry for *dass* to be correctly inserted. We may assume for simplicity that cs is introduced after t. There will be a derivation incorporating *dass* and the complex noun phrase *die hemden* of type $\mathtt{d} \otimes \mathtt{k} \otimes (\mathtt{cs}\backslash\mathtt{cs})$ if hypotheses of order 1, like [cs\cs], are allowed. The derivation is given in Fig. 8.2, in several pieces.

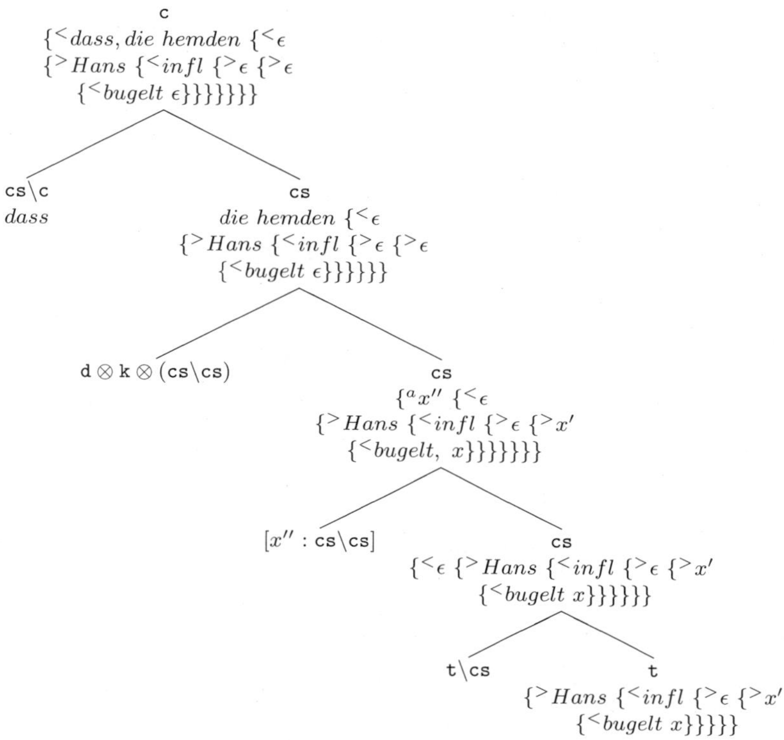

Fig. 8.2 *dass die hemden Hans bügelt*

8.4 Semantics Without Cooper Storage

8.4.1 *Cooper storage and hypothetical reasoning*

As it may seem obvious, there is absolutely no difference between the Cooper-storage analysis and the type-logical one, based on hypothetical reasoning. The first phase of Move (what we called above the Merge internal to Move) is simply the stage where an assumption is made. Such an assumption is kept in the proof tree until we decide to discharge it: there is no need for a particular "store". The "retrieval" stage is simply the step where the moved constituent (of high-order type) is applied to the subpart of the proof, the last step of which is a discharging rule.

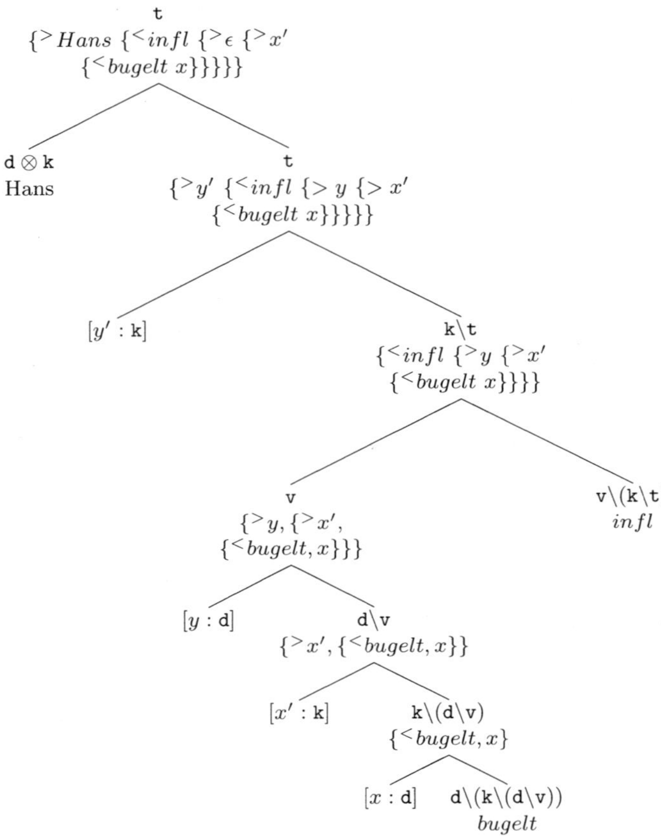

Fig. 8.2 (*Continued*)

What seems nevertheless odd in the type-logical approach is the discrepancy we get between the phonetic and the semantic sides. According to the Curry–Howard correspondence (Chapter 5), *abstraction* corresponds to the introduction of implication, and what corresponds to the elimination of the conjunction is the substitution of two variables by the two projections of a product. It seems here that, when performing a Move operation, translated as it is into the $\otimes$-elimination rule, *the phonetic and the semantic interpretations diverge.*

- On the phonetic side, as we have seen, we use *substitution*, as shown in the schema:

$$\dfrac{A \otimes B : (X, Y) \qquad \begin{matrix} [A : x] \quad [B : y] \\ \vdots \\ C : \Sigma \end{matrix}}{C : \Sigma[x/X, y/Y]}$$

- On the semantic side, we use *abstraction*, followed by *application*:

$$\dfrac{\phi : (T \to U) \to V \qquad \dfrac{\begin{matrix} [\xi : T] \\ \vdots \\ \gamma : U \end{matrix}}{\lambda\xi.\gamma : T \to U}}{\phi(\lambda\xi.\gamma) : V}$$

In the semantic rule, T is supposed to be the semantic type of B, ξ is the corresponding semantic variable and $(T \to U) \to V$ is supposed to be the semantic type of A, the semantic recipe of which is ϕ. V is the semantic type corresponding to C. Making the assumption ξ amounts to Cooper storage. The last rule is then the retrieval stage.

8.4.2 *One or two derivations?*

We are confronted with a choice: either performing *one and only one derivation*, having two interpretations (one phonetic, the other semantic, and the rules of each one being the simple images by a homomorphism of the rules applied in the derivation), or performing *two derivations* in parallel (one for the phonetic issue and the other for the semantic one, which use neither the same rules nor rules which logically correspond to each other). The second solution is taken in this chapter. We may sum it up by the following rules and correspondences:

- we assume the $\otimes$-elimination rule occurs as soon as the two necessary assumptions for applying it have been made. This is in accordance with the shortest move constraint, and does not alter the proof we can make at this stage. This leads us to practically use a ternary rule:

$$\frac{\vdash U : A \otimes B \quad x : A \vdash x : A \quad y : B, \Delta \vdash \{^{<}z, y\} : A\backslash C}{\Delta \vdash \{^{>}p(U), \{^{<}z, q(U)\}\} : C} \otimes e^3$$

or

$$\frac{U : A \otimes B \vdash U : A \otimes B \quad x : A \vdash x : A \quad y : B, \Delta \vdash \{^{<}z, y\} : A\backslash C}{U : A \otimes B, \Delta \vdash \{^{>}p(U), \{^{<}z, q(U)\}\} : C} \otimes e^3$$

depending on whether U is a constant or a variable (where p and q are projections of the product). If U is a constant, it is a pair made of a non-null string and ϵ.

- On the semantic side, we assume A corresponds to a high-order type $(T \to U) \to V$, and B to the type T of an assumption. There is no tensor and therefore no $\otimes$-elimination rule. Instead, only an assumption of type T is discharged, by the $\to$-introduction rule:

$$\frac{\Gamma \vdash \phi : (T \to U) \to V \quad \Delta \cup \{[x : T]\} \vdash \gamma : U}{\Gamma \cup \Delta \vdash \phi(\lambda x.\gamma) : V} [\to\ i] + [\to\ e]$$

or

$$\frac{\Gamma \vdash z : T \quad \Delta \cup \{[x : T]\} \vdash \gamma : U}{\Gamma \cup \Delta \vdash \lambda x.\gamma(z) : V} [\to\ i] + [\to\ e]$$

depending on whether we have a constant (ϕ) or a variable (z) as a displaced term.

We give in parallel two derivations for the question *Who does Mary like?*[3]:

- Phonetic step 1:

$$\frac{(u,v) : \mathtt{k} \otimes \mathtt{d} \vdash (u,v) : \mathtt{k} \otimes \mathtt{d} \quad y : \mathtt{k} \vdash y : \mathtt{k} \quad \dfrac{\vdash \mathit{like} : (\mathtt{k}\backslash(\mathtt{d}\backslash\mathtt{v}))/\mathtt{d} \quad x : \mathtt{d} \vdash x : \mathtt{d}}{x : \mathtt{d} \vdash \{^{<}\mathit{like}, x\} : \mathtt{k}\backslash(\mathtt{d}\backslash\mathtt{v})}}{(u,v) : \mathtt{k} \otimes \mathtt{d} \vdash \{^{>}u, \{^{<}\mathit{like}, v\}\} : \mathtt{d}\backslash\mathtt{v}}$$

[3]Where we assume both derivations held in tandem, associating each semantic step with a corresponding phonetic step, respecting the correspondence between rules, like that existing between the two previous rules given above.

- Semantic step 1:

$$\dfrac{y : e \vdash y : e \qquad \dfrac{\vdash like : e \to (e \to t) \quad x : e \vdash x : e}{x : e \vdash like(x) : e \to t}}{y : e \vdash (\lambda x.like(x))(y) : e \to t}$$

- Phonetic step 2:

$$\dfrac{\vdash does : (\mathtt{k}\backslash\mathtt{t})/\mathtt{v} \qquad \dfrac{z : \mathtt{d} \vdash z : \mathtt{d} \quad (u,v) : \mathtt{k} \otimes \mathtt{d} \vdash \{^{>}u, \{^{<}like, v\}\} : \mathtt{d}\backslash\mathtt{v}}{z : \mathtt{d}, (u,v) : \mathtt{k} \otimes \mathtt{d} \vdash \{^{>}z, \{^{>}u, \{^{<}like, v\}\}\} : \mathtt{v}}}{z : \mathtt{d}, (u,v) : \mathtt{k} \otimes \mathtt{d} \vdash \{^{<}does, \{^{>}z, \{^{>}u, \{^{<}like, v\}\}\}\} : \mathtt{k}\backslash\mathtt{t} \ (1)}$$

- Semantic step 2:

$$\dfrac{\vdash prt : t \to t \qquad \dfrac{z : e \quad y : e \vdash (\lambda x.like(x))(y) : e \to t}{z : e, y : e \vdash like(y)(z) : t}}{z : e, y : e \vdash prt(like(y)(z)) : t}$$

- Phonetic step 3:

$$\dfrac{\vdash (Mary, \epsilon) : \mathtt{k} \otimes \mathtt{d} \quad z' : \mathtt{k} \vdash z' : \mathtt{k} \quad z : \mathtt{d}, (u,v) : \mathtt{k} \otimes \mathtt{d} \vdash \ (1)}{(u,v) : \mathtt{k} \otimes \mathtt{d} \vdash \{^{>}Mary, \{^{<}does, \{^{>}\epsilon, \{^{>}u, \{^{<}like, v\}\}\}\}\} : \mathtt{t} \ (2)}$$

- Semantic step 3:

$$\dfrac{\vdash \lambda P.P(Mary) : (e \to t) \to t \quad z : e, y : e \vdash prt(like(y)(z)) : t}{y : e \vdash \lambda P.P(Mary)(\lambda z.prt(like(y)(z))) : t}$$

- Phonetic step 4:

$$\dfrac{\vdash who : \mathtt{wh} \otimes (\mathtt{k} \otimes \mathtt{d}) \quad w : \mathtt{wh} \vdash w : \mathtt{wh} \qquad \dfrac{\epsilon : (\mathtt{wh}\backslash\mathtt{c})/\mathtt{t} \quad (u,v) : \mathtt{k} \otimes \mathtt{d} \vdash \ (2)}{(u,v) : \mathtt{k} \otimes \mathtt{d} \vdash \ (3)}}{\vdash \{^{>}who, \{^{<}\epsilon, \{^{>}Mary, \{^{<}does, \{^{>}\epsilon, \{^{>}u, \{^{<}like, v\}\}\}\}\}\}\} : \mathtt{c}}$$

where (3) is

$$\{^{<}\epsilon, \{^{>}Mary, \{^{<}does, \{^{>}\epsilon, \{^{>}u, \{^{<}like, v\}\}\}\}\}\} : \mathtt{wh}\backslash\mathtt{c}$$

- Semantic step 4:

$$\dfrac{who : (e \to t) \to t \qquad \dfrac{Id : t \to t \quad y : e \vdash prt(like(y)(Mary)) : t}{y : e \vdash prt(like(y)(Mary)) : t}}{\vdash who(\lambda y.prt(like(y)(Mary))) : t}$$

$$\begin{array}{lll} \text{with} \quad \text{who} \quad := \quad \lambda Q.?z.Q(z) \quad : & & \\ \text{who}(\lambda y.prt(like(y)(Mary))) & \rightarrow & \lambda Q.?z.Q(z)(\lambda y.prt(like(y)(Mary))) \\ & \rightarrow & ?z.prt(like(z)(Mary))) \end{array}$$

These two parallel derivations may be equally well represented as the following trees which make the differences more apparent:

First phase:

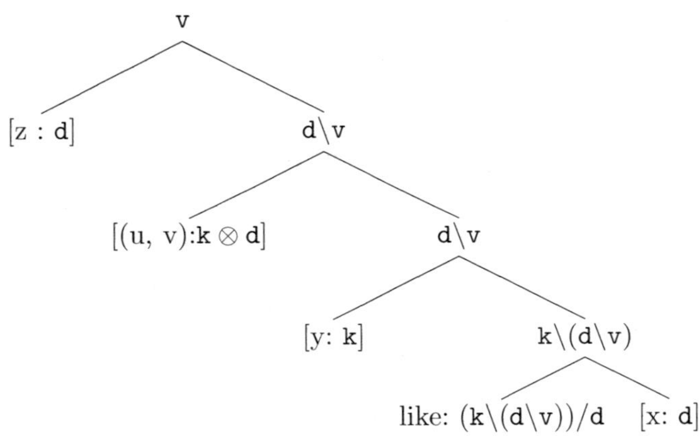

to which corresponds the following semantic tree:

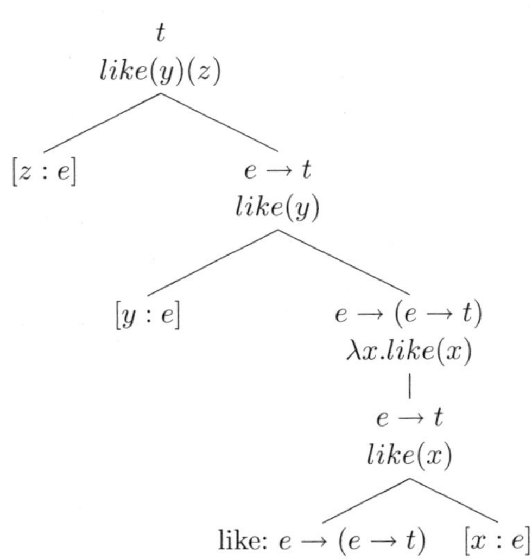

The insertion of case, associated with the variable y in the phonetic tree, is also associated with y in the semantic tree, but because this feature does not mark the position where the constituent stops moving, the semantic effect is vacuous: it is a mere *change of variable.*

The tree transformation which gives the semantic tree from the phonetic one is straightforward:

(1) *erase* the branch terminating at the last assumption which has been made (here k) (that results in a unary node);
(2) *replace* the branch terminating at the tensor product which discharges the two assumptions with a branch which either terminates at the last assumption we have made or at the constant object of high-order type semantically associated with the product;
(3) *compute* the values of the nodes appropriately.

Of course, from the viewpoint of generative grammars, convergence of the derivation is convergence of the two derivations, that is, the ability of computing the nodes (phase 3 of the transformation).

If we go on comparing the two trees, we get

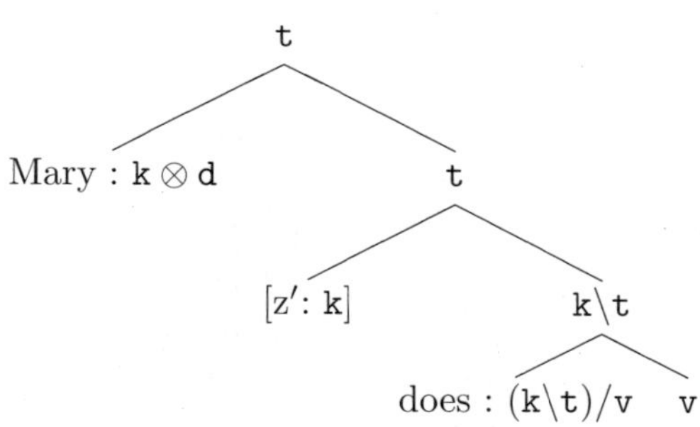

which yields by the previous transformation

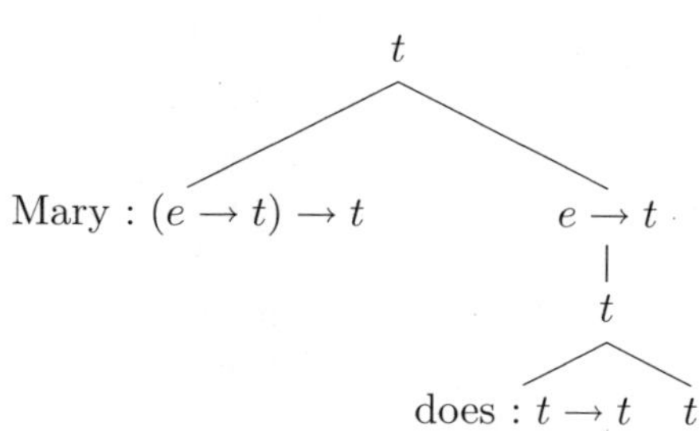

In order to compute the nodes, we need the following correspondence table:

Syntactic type	Semantic type
d	e
n	$e \to t$
k	$\top$
wh	$\top$
v	t
t	t
c	t

with the homomorphism of types

$$Hom(A\backslash B) = Hom(A) \to Hom(B)$$

We must add that every *tensor product* is associated with a term of high-order type $(e \to t) \to t$, so that for instance if we have an interrogative determiner like *which* of syntactic type $\mathtt{n}\backslash(\mathtt{wh} \otimes \mathtt{k} \otimes \mathtt{d})$, it will semantically result in

- a function from $e \to t$ to the type of $\mathtt{wh} \otimes \mathtt{k} \otimes \mathtt{d}$, which is $(e \to t) \to t$,
- and therefore an object of type $(e \to t) \to ((e \to t) \to t)$.

$\top$ may be interpreted as *true* if the arrow is seen as the implication (we know that in propositional logic, $\top \Rightarrow p \equiv p$), or, in set terms, as a singleton like $\{\mathbf{1}\}$, since there is a one-to-one correspondence between any set D and the set of functions from $\{\mathbf{1}\}$ to D. Those correspondences allow us to compute for instance

$$\begin{aligned} Hom(\mathtt{t}\backslash(\mathtt{wh}\backslash\mathtt{c})) &= Hom(\mathtt{t}) \to Hom(\mathtt{wh}\backslash\mathtt{c}) \\ &= t \to (\top \to t) \\ &= t \to t \end{aligned}$$

We thus grasp the fact that *formal features are semantically vacuous.* If they have nevertheless an impact on the meaning, it is only because they trigger operations which have such an impact, like when for instance they bring a constituent in a position where it may get its expected scope.

8.5 Concluding Remarks: Some Tracks to Explore

The idea of simulating Move by means of the tensor elimination rule came from remarks by Retoré and first appeared in Lecomte and Retoré (1999) (see also Lecomte and Retoré, 2001). A semantics based on this approach was first attempted in Lecomte (2005) and completed in Lecomte (2008). More results are given in Amblard's (2007) thesis, and a completely formalized view is given in Amblard *et al.* (in press). In particular, it is shown that it is possible to give a faithful approach to the treatment of the VP along the lines of the theory of external arguments developed by Kratzer (1994). The latter rests on discursive representation structures adapted to this framework, with the use of a continuation semantics (see Chapter 10).

Chapter 9

Grammars in Deductive Forms

9.1 Introduction

Besides Moortgat's multimodal framework and our first proposition of a type-theoretical version of minimalist grammars, there are other approaches of the use of proof theory in grammar, mainly elaborating on the convenience of the connection between proofs and λ-terms for extracting a semantic representation. As seen in the previous chapter, these approaches hesitate between the solution of only one derivation having two interpretations (one phonetic and the other semantic) and the solution of two distinct derivations, but related by some kind of correspondence between rules. Other approaches (Jacobson, 1996) do not use proof theory properly speaking but *combinators* (see Section 5.3).

We analyze here Pollard's proposal of *convergent grammars* (Pollard, 2007; Pollard and Hana, 2003), which belongs to the kind of grammars assuming two parallel derivations. Then, we propose a very close framework, developed by Anoun in her PhD thesis, which belongs to the first kind (see also Anoun and Lecomte, 2006).

9.2 Convergent Grammars

9.2.1 *CVG types and rules: Semantics*

Carl Pollard's convergent grammar is a system which comes from previous reflections by Pollard on *head-driven phrase structure grammar*, a framework that he invented and developed with Ivan Sag during the 1980s and 1990s. Convergent grammars have a special kind of sequent, of the general form $\Gamma \vdash a : A \dashv \Delta$, where Γ is called the context (which is made of variable/type pairs), and Δ is the Cooper store, or *co-context* (where it is said that, here, "co" must be understood as a mnemonic not only for "Cooper", but also for "commitment"). The CVG types are defined as follows:

- there are basic types;
- if A and B are types, then $A \to B$ is a type;
- if A, B and C are types, then A_B^C is a type, with A interpreted as the binding type, B as the continuation type and C as the result type.

On the latter clause, we must comment by saying that A_B^C is simply Moortgat's $q(A, B, C)$ (Section 7.5), and that, in a non-directed calculus, it simply amounts to $(A \to B) \to C$[1]: it is assumed there is a hole (or a hypothesis) of type A inside B, and by applying an object of this complex "operator" type to a B lacking this A, we get a C. As said above, this constructor is supposed to belong only to the semantic tier.

The Cooper stores contain operators waiting for their scope, and are therefore paired each with the variable they will eventually bind. We may have, for instance, $\mathbf{everyone}_x$ in a store. Generally, the stored terms will be denoted by a_x where a is an operator (like **everyone** above) and x the variable it is devoted to bind. Proceeding like this, it happens that there is no need for an explicit λ-abstractor. This λ would be necessary if there were an introduction rule in the logic, but because instead we have this mechanism of Cooper storage, such a rule will be dispensed with. The rules are then the following:

Identity:

$$\frac{}{x : A \vdash x : A \dashv}\ Id$$

where x is a fresh variable when the hypothesis is introduced. The variables introduced into the derivation by using this axiom are said to be *uncommitted.*

Non-logical axioms:

$$\frac{}{\vdash c : A \dashv}\ Lex$$

Such axioms come from the lexicon.

[1]Assuming the use of Pollard's arrow instead of the linear implication.

Elimination rule:

$$\frac{\Gamma \vdash f : A \rightarrow B \dashv \Delta \qquad \Gamma' \vdash a : A \dashv \Delta'}{\Gamma, \Gamma' \vdash f(a) : B \dashv \Delta, \Delta'} [\rightarrow\ e]$$

Schema C (Commitment):

$$\frac{\Gamma \vdash a : A_B^C \dashv \Delta}{\Gamma \vdash x : A \dashv a_x : A_B^C, \Delta} \, Comm.$$

where x is a fresh variable. Here, x is introduced as a *committed* variable, and the operator is *stored.*

Schema R (Responsibility):

$$\frac{\Gamma \vdash b : B \dashv a_x : A_B^C, \Delta}{\Gamma \vdash a_x(b) : C \dashv \Delta} \, Res.$$

By this schema R, the content of the store is *retrieved* and then applied to the present object of type B, thus giving something of type C, which is exactly what is expected from an object of type A_B^C.

Schema O (Outside assistance):

$$\frac{\Gamma \vdash a : A_B^C \dashv \Delta \qquad x : A, \Gamma' \vdash b : B \dashv \Delta'}{\Gamma, \Gamma' \vdash a_x(b) : C \dashv \Delta, \Delta'} \, O$$

This rule is necessary for dealing with cases of "overt" movements, like for wh-operators: they have a A_B^C type (for instance an interrogative phrase like *who* could have the type $(\mathbf{e} \rightarrow \mathbf{t}) \rightarrow \mathbf{t}^{wh}$ where $\mathbf{t}^{wh}$ would be the type of a question), and in the course of a derivation they have to meet a B where an $x : A$ has been hypothesized, resulting in binding the (uncommitted) variable x by means of the operator a.

9.2.2 *CVG types and rules: Syntax*

The architecture of convergent grammars is then such that there are in fact *two* proof theories in parallel, and therefore rules for their interfacing, instead of simply one proof theory like in categorial formalisms. Syntax is provided by a particular typing and some elementary rules:

- basic categories are S and NP;
- if A and B are categories, then $A \multimap_F B$ is a category, where F belongs to a set of grammatical function names;

- terms are therefore constants or functional terms like $[f\ a^C]$ and $[^S a\ f]$ (to read as, respectively, a is a complement of f and a is a specifier (or a subject) for f);
- words are introduced as axioms of the form $\vdash w : A$;
- there are (at least) two elimination rules, depending on the flavour of the linear implication:

$$\frac{\vdash a : A \qquad \vdash f : A \multimap_S B}{\vdash [^S a\ f] : B}[\multimap_S\ e] \quad \frac{\vdash f : A \multimap_C B \qquad \vdash a : A}{\vdash [f\ a^C] : B}[\multimap_C\ e]$$

9.2.3 *An example*

Lexicon:

$$\vdash \text{Chris, Chris'} \ : \ NP, e \dashv$$
$$\vdash \text{everyone, everyone'} \ : \ NP, {e_t}^t \dashv$$
$$\vdash \text{likes, like'} \ : \ NP \multimap_c (NP \multimap_s S), e \to (e \to t) \dashv$$

Derivation: See Fig. 9.1.

9.3 Labelled Linear Grammars

LLG (Anoun, 2007; Anoun and Lecomte, 2006) is based on *restricted access to hypothetical reasoning* and the use of exponentials for flexible access to resources. Objects which are required to move in a sentence are not represented by tensor products, like in the type-logical version of MGs seen in the previous chapter, but by lists made of a constant object with a convenient type (generally of high order) followed by a mandatory list of assumptions, each of the same type and corresponding to the various licensee features. In the calculus, only the assumptions involved in the lexical entries can be made. Of course, these assumptions are destined to be discharged, but by the rule of introduction of the linear implication, and not by the elimination of the tensor like it was the case in the type-logical version previously seen. This rule of introduction of the linear implication is itself encapsulated inside a derived rule which is the same as the semantic rule $[\to\ i] + [\to\ e]$ presented in Section 8.4. This results in a system where we may consider *only one* derivation, two interpretations of which are respectively the phonetic and the semantic interpretations. This version is therefore more in accordance with the usual paradigm of the Curry–Howard homomorphism developed in categorial grammars.

$$
\dfrac{
  \vdash \text{Chris}, \text{Chris'} \;:\; NP, e \dashv
  \qquad
  \dfrac{
    \vdash \text{likes}, \text{like}' \;:\; NP \multimap_c (NP \multimap_s S), e \rightarrow (e \rightarrow t) \dashv
    \qquad
    \dfrac{\vdash \text{everyone}, \text{everyone'} \;:\; NP, {e_t}^t \dashv}{\vdash \text{everyone}, x \;:\; NP, e \dashv \text{everyone}_x \;:\; {e_t}^t}\,C
  }{
    \vdash (\text{likes everyone}^c), (\text{like}'\ x) \;:\; NP \multimap_s S, e \rightarrow t \dashv \text{everyone}_x \;:\; {e_t}^t
  }
}{
  \dfrac{\vdash ({}^s\text{Chris (likes everyone}^c)), ((\text{like}'\ x)\ \text{Chris'}) \;:\; S, t \dashv \text{everyone}_x \;:\; {e_t}^t}{\vdash ({}^s\text{Chris (likes everyone}^c)), (\text{everyone}_x\ ((\text{like}'\ x)\ \text{Chris'})) \;:\; S, t \dashv}\,R
}
$$

Fig. 9.1 A derivation in convergent grammars

The rules, as given by Anoun, are

Axioms:

$$\frac{(\vdash A \dashv l_{hyps})^i}{\vdash A}\, Head^i \qquad \frac{(\vdash A \dashv l_{hyps})^i \quad (C \vdash C)^j \in l^i_{hyps}}{C \vdash C}\, Ax^{ij}$$

Logical rules:

$$\frac{\Gamma \vdash A \multimap B \quad \Delta \vdash A}{\Gamma, \Delta \vdash B}\,[\multimap\ e]_1 \qquad \frac{(\Gamma \vdash A \multimap B \dashv l_{hyps})^i \quad \Delta \vdash A}{(\Gamma, \Delta \vdash B \dashv l_{hyps})^i}\,[\multimap\ e]_2$$

$$\frac{\Gamma \vdash A \quad \Delta, A \vdash B}{\Gamma, \Delta \vdash B}\,[cut] \qquad \frac{\Gamma \vdash (A \multimap B) \multimap C \quad \Delta, A \vdash B}{\Gamma, \Delta \vdash C}\,[\multimap\ i + e]^i_c$$

Structural rules:

$$\frac{\Gamma, A \vdash B}{\Gamma, !A \vdash B}\,[!W] \qquad \frac{\Gamma, !A, !A \vdash B}{\Gamma, !A \vdash B}\,[!C]$$

- Axioms may be interpreted as follows: given lexical entries, which are either *free* (where $l_{hyps} = \emptyset$) or *bound* (where $l_{hyps} \neq \emptyset$), it is always possible to extract the "head", which is a non-logical axiom consisting of a constant object associated with this entry. But if the entry is bound, it is also possible to extract from it one of the hypotheses belonging to its list. All the axioms are indexed according to some *enumeration* in the Chomskyan sense: *tokens* are extracted from a lexicon and are given each a different integer i: this integer indexes the axiom $Head$. Moreover, each lexical token given in an enumeration contains several assumptions to be made (from 0 to an arbitrary k), these instances are indexed by a second integer j so that Ax^{ij} is doubly indexed: by the rank of the entry in the enumeration, and by the rank of the assumption taken in consideration. The system is such that each axiom instance may be used once and only once. Moreover, for a given i, the Ax^{ij} must be used in decreasing order relative to the index j: that means that all the assumptions on the list l_{hyps} are scanned from right to left.
- Structural rules are used in order to bind two different instances of an axiom belonging to the same entry. If the list of hypotheses contains $[x : C]$ and $[x' : C]$, there will be a way to contract them in such a manner that x

and x' become bound. It is mandatory that *all the hypotheses attached to the same entry be discharged in one and only one step*, therefore requiring one or several steps of $[!W]$.

- The elimination rule for $\multimap$ is given in two variants; the second one ($[\multimap\ e]_2$) concerns the case where the axiom *Head* has not been used previously and the list of hypotheses is kept after the application of the rule. The index of the token is also kept during the application of the rule. An example of application of this rule is the case of a determiner like *every* which is bound to a hypothesis of type $\mathtt{d}$, but which has to consume an $\mathtt{n}$ before the hypothesis enters into the derivation. In the logical rule $[\multimap\ i+e]^i_c$, the index i also refers to a precise token in the enumeration. The hypothesis $x : A$ and the head $w : (A \multimap B) \multimap C$ must come from the same lexical token. It is the same for $[\multimap\ i+e]^i_s$.

Let us take as an example the sentence *John congratulated every student*, with the following enumeration:

Word	**Head**	**List of hyp.**	**Index**
John	$(\mathtt{d_k} \multimap \mathtt{s}) \multimap \mathtt{s}$	$x : \mathtt{d} \vdash x : \mathtt{d}$	(e_1)
congratulated	$\mathtt{d_{acc}} \multimap (\mathtt{d_{nom}} \multimap \mathtt{s})$		(e_2)
every	$\mathtt{n} \multimap ((\mathtt{d_k} \multimap \mathtt{s}) \multimap \mathtt{s})$	$y : \mathtt{d} \vdash y : \mathtt{d}$	(e_3)
student	$\mathtt{n}$		(e_4)

The beginning of the derivation is

$$\dfrac{\dfrac{\dfrac{\dfrac{(e_4)}{\vdash \mathtt{n}} \quad (\vdash \mathtt{n} \multimap ((\mathtt{d_k} \multimap \mathtt{s}) \multimap \mathtt{s}) \dashv y : \mathtt{d})^3}{(\vdash (\mathtt{d_k} \multimap \mathtt{s}) \multimap \mathtt{s} \dashv y : \mathtt{d})^3}[\multimap\ e]_2}{y : \mathtt{d} \vdash y : \mathtt{d}}Ax^{31} \qquad \dfrac{(e_2)}{\vdash \mathtt{d_{acc}} \multimap (\mathtt{d_{nom}} \multimap \mathtt{s})}}{y^3 : \mathtt{d_k} \vdash \mathtt{d_{nom}} \multimap \mathtt{s}}$$

The continuation is

$$\dfrac{\dfrac{(\vdash (\mathtt{d_k} \multimap \mathtt{s}) \multimap \mathtt{s} \dashv x : \mathtt{d})^1}{x^1 : \mathtt{d} \vdash x : \mathtt{d}}Ax^{11} \qquad y^3 : \mathtt{d_k} \vdash \mathtt{d_{nom}} \multimap \mathtt{s}}{x^1 : \mathtt{d_k}, y^3 : \mathtt{d_k} \vdash \mathtt{s}}$$

At the end, we may use $[\multimap\ i+e]^3_c$ and $[\multimap\ i+e]^1_c$:

$$\dfrac{\dfrac{}{(\mathtt{d_k} \multimap \mathtt{s}) \multimap \mathtt{s}^1}\,Head^1 \qquad \dfrac{\dfrac{}{(\mathtt{d_k} \multimap \mathtt{s}) \multimap \mathtt{s}^3}\,Head^3 \qquad x^1 : \mathtt{d_k}, y^3 : \mathtt{d_k} \vdash \mathtt{s}}{x^1 : \mathtt{d_k} \vdash \mathtt{s}}\,[\multimap\ i+e]^3_c}{\vdash \mathtt{s}}\,[\multimap\ i+e]^1_c$$

The tree representation makes this logical derivation more readable:

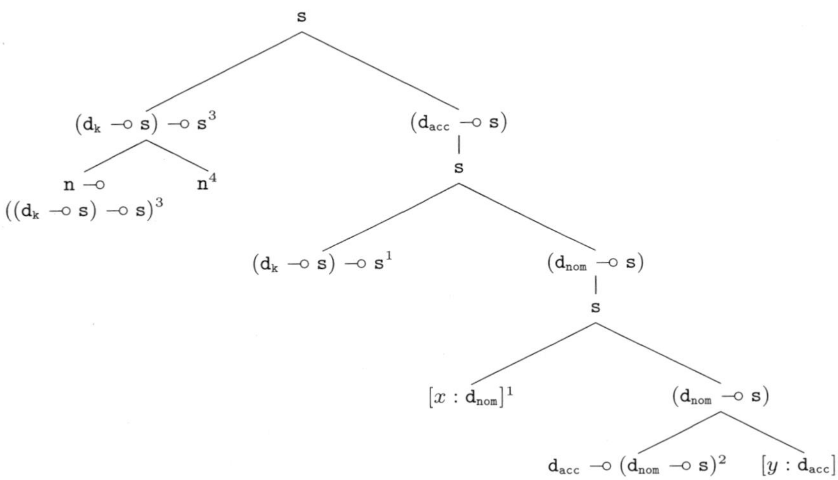

Note that the parameter $\mathtt{k}$ which occurs as an index to some $\mathtt{d}$ is a kind of variable which takes its values only in $\{\mathtt{nom}, \mathtt{acc}, \mathtt{obl}\}$. We denote by superscripts the index related to each lexical entry, as it is preserved by the given rules above.

Let us observe some special properties of this approach:

- Discharging rules, here represented by unary branches, may be applied in any order. This exactly translates the various orders of quantifier raising in generative grammar and enables us to get all the readings due to scope ambiguities.
- The ability to insert lexical items (or phrases built from them) consisting of several parts opens the field to taking *discontinuity* into account (like in the case of verbs with separable particles, like *ring ... up*).

Let us concentrate on the way such trees are interpreted, phonologically and semantically. This requires us to clarify some notions by introducing some formal definitions. The term *abstract syntax* is used to refer to the building of proofs based on syntactic and linear types like $\mathtt{d_{acc}} \multimap (\mathtt{d_{nom}} \multimap \mathtt{s})$; it corresponds to what is called *narrow syntax* in the generative terms. In contrast, *concrete syntax* means the composition of terms on the two interpretative levels: the phonological and the semantic ones. They correspond to the levels of PF (phonetic form) and LF (logical form) in generative grammar. There are therefore two concrete syntaxes, one for the strings and the other for the terms expressing meaning. Terms of the first component are called ϕ-terms, terms of the second one are our familiar λ-terms. Both kinds are *typed.* The types for the ϕ-terms are built from a unique primitive type: *str* (for *string*) with a unique functor $\multimap$. The types for the λ-terms are built, like in Montague grammar, from two primitive types $\mathbf{e}$ and $\mathbf{t}$, using only the intuitionistic arrow $\rightarrow$. Now, the abstract syntactic types are defined starting from a set of primitive types

$$A = \{\mathtt{d_{nom}}, \mathtt{d_{acc}}, \mathtt{d_{obl}}, \mathtt{d_{k}}, \mathtt{s}, \mathtt{c}, \mathtt{c_{wh}}, \mathtt{i}\}$$

and using $\multimap$ and $\mathbf{!}$ as constructors.

Definition 9.1. Let $T(A)$ be the abstract types on the atomic types A. We have

$$T(A) := A \mid T(A) \multimap T(A) \mid !T(A)$$

T_ϕ and T_λ are then defined:

Definition 9.2.

$$T_\phi := str \mid T_\phi \multimap T_\phi$$
$$T_\lambda := \mathbf{e} \mid \mathbf{t} \mid T_\lambda \rightarrow T_\lambda$$

The global architecture is in fact the one of *abstract categorial grammars* (de Groote, 2001a). Two homomorphisms τ_ϕ and τ_λ are defined, the former from $T(A)$ to T_ϕ and the latter from $T(A)$ to T_λ:

Definition 9.3.

(1) $\forall \mathtt{a} \in A, \tau_\phi(\mathtt{a}) = str$
(2) $\forall A, B, \tau_\phi(A \multimap B) = \tau_\phi(A) \multimap \tau_\phi(B)$
(3) $\forall A, \tau_\phi(!A) = \tau_\phi(A)$

Definition 9.4.

(1) $\tau_\lambda(\mathtt{s}) = \tau_\lambda(\mathtt{c}) = \tau_\lambda(\mathtt{i}) = \mathbf{t}$
(2) $\tau_\lambda(\mathtt{d_k}) = \mathbf{e}$ (also for $\mathtt{k} = \mathtt{nom}, \mathtt{acc}, \mathtt{obl}$)

(3) $\tau_\lambda(\mathtt{n}) = \mathbf{e} \rightarrow \mathbf{t}$
(4) $\forall A, B, \tau_\lambda(A \multimap B) = \tau_\lambda(A) \rightarrow \tau_\lambda(B)$
(5) $\forall A, \tau_\lambda(!A) = \tau_\lambda(A)$

Phonological interpretation: ϕ-terms were primarily defined by Dick Oehrle (1994, 1995). Let Σ be a set of phonetic constants ($\approx$ words), and $(\Sigma^*, ^\frown, \epsilon)$ the associated monoid. The set of ϕ-terms is defined as follows:

Definition 9.5.

- Each element $t \in \Sigma^*$ is a ϕ-term of type s;
- each ϕ-variable x^α is a ϕ-term of type α;
- if x^α is a ϕ-variable of type α and A is a ϕ-term of type β where x is free in A and occurs once and only once, then $\lambda x.A$ is a ϕ-term of type $\alpha \multimap \beta$;
- if u is a ϕ-term of type $A \multimap B$ and v a ϕ-term of type A, and if u and v do not share any free variable, then $u(v)$ is a ϕ-term of type B.

The set of λ-terms is defined as usual. Let Δ be a set of semantic constants, provided with their types (σ_Δ is the function which associates each constant with its λ-type).

The relation of β-reduction is the same in both calculi. Interpretations in both dimensions are then calculated using the Curry–Howard homomorphism, which allows us to enrich our rules in the following way:

Axioms:

$$\frac{(\vdash u : A \dashv l_{hyps})^i}{\vdash u : A}\ Head^i \qquad \frac{(\vdash u : A \dashv l_{hyps})^i \quad (x : C \vdash x' : C)^j \in l^i_{hyps}}{x : C \vdash x' : C}\ Ax^{ij}$$

Logical rules:

$$\frac{\Gamma \vdash u : A \multimap B \quad \Delta \vdash v : A}{\Gamma, \Delta \vdash u(v) : B}\ [\multimap\ e]_1 \qquad \frac{(\Gamma \vdash u : A \multimap B \dashv l_{hyps})^i \quad \Delta \vdash v : A}{(\Gamma, \Delta \vdash u(v) : B \dashv l_{hyps})^i}\ [\multimap\ e]_2$$

$$\frac{\Gamma \vdash u^i : \phi \quad \Delta, x^i : \phi \vdash v : B}{\Gamma, \Delta \vdash v[x := u] : B}\ [cut]$$

$$\frac{\Gamma \vdash w^i : (A \multimap B) \multimap C \quad \Delta, x^i : A \vdash v : B}{\Gamma, \Delta \vdash w(\lambda x.v) : C}\ [\multimap\ i + e]^i_c$$

Structural rules:

$$\frac{\Gamma, x : A \vdash w : B}{\Gamma, x : !A \vdash w : B}[!W] \qquad \frac{\Gamma, x : !A, y : !A \vdash w : B}{\Gamma, z : !A \vdash w[x := z, y := z] : B}[!C]$$

It must be noticed that the terms associated with the abstract types may be read as ϕ- as well as λ-terms! If we decide to feed types with ϕ-terms coming from lexical entries, then we get the ϕ-interpretation of the sentence, and if we decide to feed them with λ-terms, then we get the semantic interpretation. Differences between both kinds of terms therefore come only from the lexicon. As a continuation of our example, let us complete our lexicon:

(1) Phonological component:

Word	**Head**	**ϕ-term**
John	$(\mathtt{d_{nom}} \multimap \mathtt{s}) \multimap \mathtt{s}$	$\lambda X.\mathtt{john}^\frown X(\epsilon)$
John	$(\mathtt{d_{acc}} \multimap \mathtt{s}) \multimap \mathtt{s}$	$\lambda X.X(\mathtt{john})$
congratulated	$\mathtt{d_{acc}} \multimap (\mathtt{d_{nom}} \multimap \mathtt{s})$	$\lambda X.\lambda Y.Y^\frown \mathtt{cong}^\frown X$
every	$\mathtt{n} \multimap ((\mathtt{d_{nom}} \multimap \mathtt{s}) \multimap \mathtt{s})$	$\lambda U.\lambda V.(\mathtt{ev}^\frown U)^\frown V(\epsilon)$
every	$\mathtt{n} \multimap ((\mathtt{d_{acc}} \multimap \mathtt{s}) \multimap \mathtt{s})$	$\lambda U.\lambda V.V(\mathtt{ev}^\frown U)$
student	$\mathtt{n}$	$\mathtt{student}$

Word	**List of hyp.**
John	$Z : \mathtt{d_{nom}} \vdash Z : \mathtt{d_{nom}}$
John	$Z : \mathtt{d_{acc}} \vdash Z : \mathtt{d_{acc}}$
congratulated	
every	$W : \mathtt{d_{nom}} \vdash W : \mathtt{d_{nom}}$
every	$W : \mathtt{d_{acc}} \vdash W : \mathtt{d_{acc}}$
student	

(2) Semantic component:

Word	**Head**	**λ-term**
John	$(\mathtt{d_k} \multimap \mathtt{s}) \multimap \mathtt{s}$	$\lambda P.P(john)$
congratulated	$\mathtt{d_{acc}} \multimap (\mathtt{d_{nom}} \multimap \mathtt{s})$	$\lambda x.\lambda y.cong(x, y)$
every	$\mathtt{n} \multimap ((\mathtt{d_k} \multimap \mathtt{s}) \multimap \mathtt{s})$	$\lambda Q.\lambda P.\forall x Q(x) \Rightarrow P(x)$
student	$\mathtt{n}$	$\lambda z.student(z)$

Word	**List of hyp.**
John	$x : \mathtt{d} \vdash x : \mathtt{d}$
every	$y : \mathtt{d} \vdash y : \mathtt{d}$

After decorating the rules with their Curry–Howard corresponding term, we get from the derivation

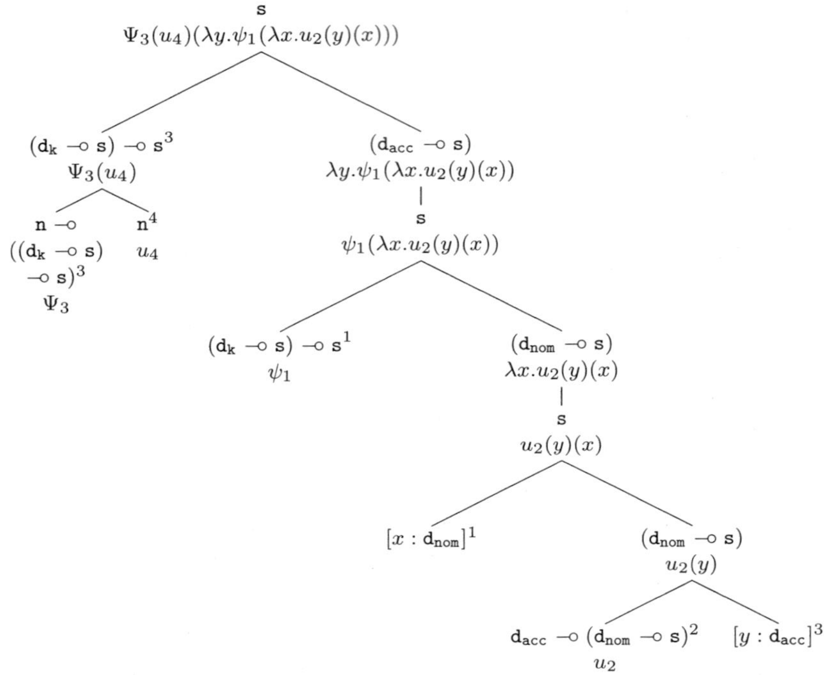

The two interpretations are obtained by substituting terms of the appropriate dimension to the variables in the final expression, $\Psi_3(u_4)(\lambda y.\psi_1 (\lambda x.u_2(y)(x)))$.

In the phonological component, we have the equations

$$\begin{aligned} \Psi_3 &= \lambda U.\lambda V.V(\mathtt{ev}^\frown U) \\ u_4 &= \mathtt{student} \\ u_2 &= \lambda X.\lambda Y.Y^\frown \mathtt{cong}^\frown X \\ \psi_1 &= \lambda X.\mathtt{john}^\frown X(\epsilon) \end{aligned}$$

and therefore the following β-reduction:

$$\begin{array}{lr} \lambda U.\lambda V.V(\mathtt{ev}^\frown U)(\mathtt{student})(\lambda y.\lambda X.\mathtt{john}^\frown X(\epsilon)(\lambda x.\lambda X.\lambda Y.Y^\frown \mathtt{cong}^\frown X(y)(x))) & \rightarrow \\ \lambda U.\lambda V.V(\mathtt{ev}^\frown U)(\mathtt{student})(\lambda y.\lambda X.\mathtt{john}^\frown X(\epsilon)(\lambda x.x^\frown \mathtt{cong}^\frown y)) & \rightarrow \\ \lambda U.\lambda V.V(\mathtt{ev}^\frown U)(\mathtt{student})(\lambda y.\mathtt{john}^\frown (\lambda x.x^\frown \mathtt{cong}^\frown y)(\epsilon)) & \rightarrow \\ \lambda U.\lambda V.V(\mathtt{ev}^\frown U)(\mathtt{student})(\lambda y.\mathtt{john}^\frown \epsilon^\frown \mathtt{cong}^\frown y) & \rightarrow \\ \lambda V.V(\mathtt{ev}^\frown \mathtt{student})(\lambda y.\mathtt{john}^\frown \epsilon^\frown \mathtt{cong}^\frown y) & \rightarrow \\ (\lambda y.\mathtt{john}^\frown \epsilon^\frown \mathtt{cong}^\frown y)(\mathtt{ev}^\frown \mathtt{student}) & \rightarrow \\ \mathtt{john}^\frown \epsilon^\frown \mathtt{cong}^\frown \mathtt{ev}^\frown \mathtt{student} & \end{array}$$

In the semantic component, we have

$$\begin{aligned}\Psi_3 &= \lambda Q.\lambda P.\forall x(Q(x) \Rightarrow P(x))\\ u_4 &= \lambda z.student(z)\\ u_2 &= \lambda x.\lambda y.cong(y,x)\\ \psi_1 &= \lambda U.U(john)\end{aligned}$$

and therefore the following β-reduction:

$$\begin{aligned}&(\lambda Q.\lambda P.\forall x(Q(x) \Rightarrow P(x))(\lambda z.student(z))(\lambda y.\lambda U.U(john)(\lambda x.\lambda x.\lambda y.cong(y,x)(y)(x))) \rightarrow\\ &(\lambda Q.\lambda P.\forall x(Q(x) \Rightarrow P(x))(\lambda z.student(z))(\lambda y.\lambda U.U(john)(\lambda x.cong(x,y))) \rightarrow\\ &(\lambda Q.\lambda P.\forall x(Q(x) \Rightarrow P(x))(\lambda z.student(z))(\lambda y.(\lambda x.cong(x,y)(john))) \rightarrow\\ &(\lambda Q.\lambda P.\forall x(Q(x) \Rightarrow P(x))(\lambda z.student(z))(\lambda y.cong(john,y)) \rightarrow\\ &(\lambda P.\forall x(\lambda z.student(z)(x) \Rightarrow P(x))(\lambda y.cong(john,y)) \rightarrow\\ &(\lambda P.\forall x(student(x) \Rightarrow P(x))(\lambda y.cong(john,y)) \rightarrow\\ &\forall x(student(x) \Rightarrow \lambda y.cong(john,y)(x)) \rightarrow\\ &\forall x(student(x) \Rightarrow cong(john,x))\end{aligned}$$

We must mention that, with regard to the question of scope ambiguities, because of the ϕ- and λ-terms assigned to the words, the various syntactical derivations we can have (depending on the order of the abstractions over the nominative and the accusative DPs) lead of course to different semantic readings but they yield the same phonetic form. It is important to notice here that we got this result without mentioning any quantifier raising operation, nor any special feature, and not even any distinction between weak and strong features!

9.4 Binding in LLG

9.4.1 *Pronouns as pronounced variables*

LLG is able to manage binding issues in a straightforward way. It is there that the constructor "!", that we introduced in our rules and have not yet used, will prove useful, since it allows a sharing of resources, which is exactly what is needed in the treatment of the anaphora relation. In this relation, actually, the same entity is shared between an antecedent and the anaphoric element: a trace or pronoun. We have already seen that traces are translated into hypotheses, that is, variables: they have an empty content with regard to the phonological component, but they have a content in the semantic one, even if we need an assignment to determine what it is. Pronouns, on their side, will be treated similarly as traces, that is, as ordinary variables, except that they have a non-empty content on the phonological component (and they have what is known as "ϕ-features"), being expressed by words like *he*, *she*, *it*, *they* or *him*, *her*, *them*, etc. In some sense, they are *pronounced* variables. But on the semantic side, they will be treated exactly as variables.

There is an idea due to Kayne (2002), according to which we can dispense with some binding principle like Condition C if we take *doubling constituents* into account. Analyzing

Example 9.1. John thinks he's smart

he proposes the following intuitive derivation:

Example 9.2. thinks [John, he] is smart $\rightarrow$ John$_i$ thinks [t_i, he] is smart

In this derivation, as Kayne says, "the double *John* moves from within the doubling constituent $[John, he]$ into a higher theta-position", and he similarly treats the phenomenon of control by assuming, for instance,

Example 9.3. John tried to solve the problem

is derived from

Example 9.4. tried to [John, PRO] solve the problem

Thus, stating that "extraction of a phrase from within a doubling constituent like [*John, he*] is limited to extraction of the Spec", Kayne shows that a sentence like *He thinks that John is smart* cannot be derived from Example 9.2, and that therefore Condition C is *not* a primitive of universal grammar.

Such an analysis is very well suited to LLG, where it suffices to introduce parallel specific entries like

$$\vdash \begin{pmatrix} \lambda P.\mathbf{John}^\frown P(\epsilon) \\ \lambda P.P(\mathbf{John}) \end{pmatrix} : (!\mathtt{d}_{\mathtt{nom}} \multimap \mathtt{c}) \multimap \mathtt{c} \tag{9.1}$$

$$[(x_{\Phi 1}, x_{\lambda 1}) : \mathtt{d}_{\mathtt{nom}} \vdash (\mathtt{x}_{\Phi 1}, \mathtt{x}_{\lambda 1}) : \mathtt{d}_{\mathtt{nom}} \tag{9.2}$$
$$[(x_{\Phi 2}, x_{\lambda 2}) : \mathtt{d}_{\mathtt{nom}} \vdash (\mathtt{he}, \mathtt{x}_{\lambda 2}) : \mathtt{d}_{\mathtt{nom}}$$

where the first line gives the abstract syntactic type (notice the exponential), and where the last two lines are the two successive hypotheses which must be introduced into the derivation before the high-order terms of both

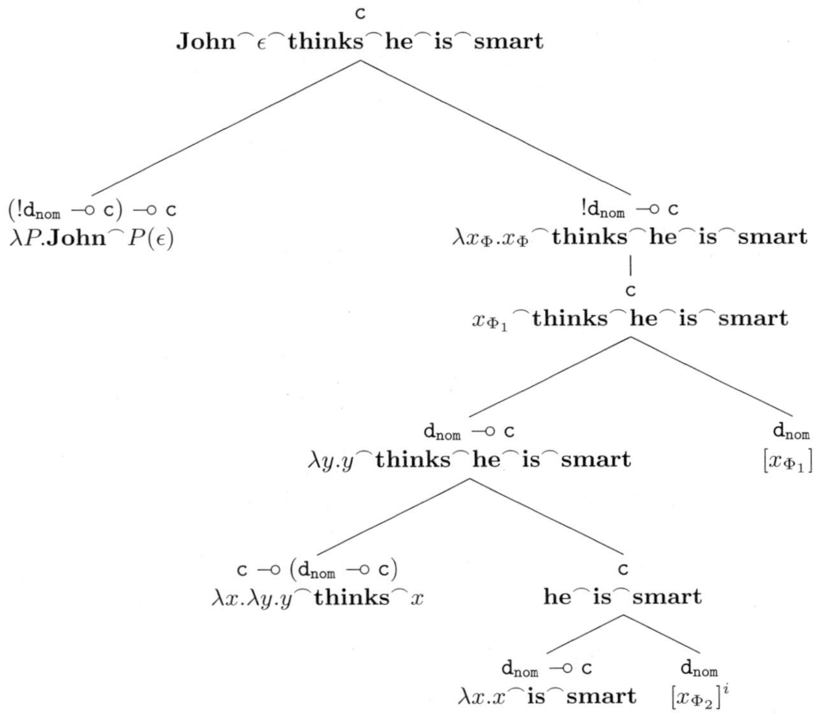

Fig. 9.2 Derivation of Example 9.1 (phonology)

dimensions (phonological and semantic) can apply. The derivation is given in Figs. 9.2 (phonology) and 9.3 (semantics). It must be noticed that the two hypotheses $x_{\lambda 1}$ and $x_{\lambda 2}$ are discharged at the same time when applying the unary rule, thus giving x_λ, but because of this action, the syntactic abstract type resulting from this non-linear abstraction is $!\mathtt{d}_{\mathtt{nom}} \multimap \mathtt{c}$, and not $\mathtt{d}_{\mathtt{nom}} \multimap \mathtt{c}$. It is the reason why the doubling entry corresponding to *John* must be $(!\mathtt{d}_{\mathtt{nom}} \multimap \mathtt{c}) \multimap \mathtt{c}$ and not $(\mathtt{d}_{\mathtt{nom}} \multimap \mathtt{c}) \multimap \mathtt{c}$. In the phonological component, the mechanism is the same except that the abstraction is over x_Φ which occurs only once in the phonological form since an occurrence has been replaced by the phonological constant *he*.

Cases where a nominative antecedent is linked to an accusative pronoun, like in Example 9.5 below, require some extension.

Example 9.5. John believes that Mary loves him

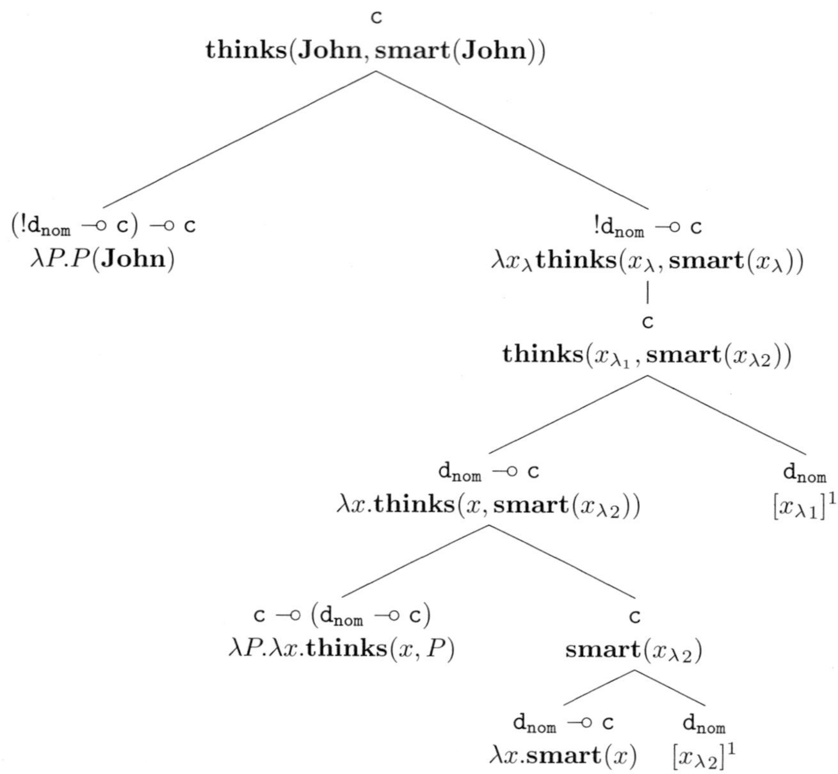

Fig. 9.3 Derivation of Example 9.1 (semantics)

Types are assumed partially ordered by an "*is_a*" relation, according to which

$$\mathtt{d_{nom}}\ is\_a\ \mathtt{d} \tag{9.3}$$

$$\mathtt{d_{acc}}\ is\_a\ \mathtt{d} \tag{9.4}$$

From now on we assume that *specific* axioms (non-logical ones) are attached to the nominal entries, like

$$x : \mathtt{d} \vdash x : \mathtt{d_{nom}} \tag{9.5}$$

$$x : \mathtt{d} \vdash x : \mathtt{d_{acc}} \tag{9.6}$$

By doing so, the system is supposed to answer positively to a demand of type $\mathtt{d_{nom}}$, for instance by introducing a hypothesis of type $\mathtt{d}$. This allows us to unify variables which do not necessarily have the same subtype (that is,

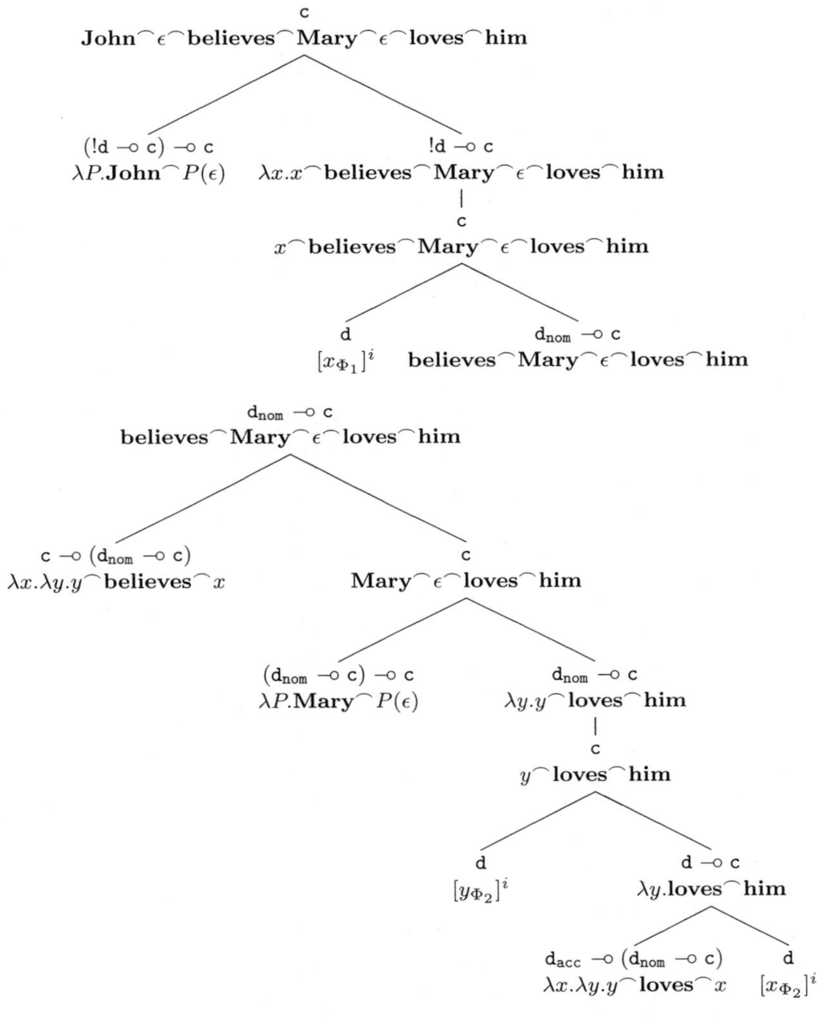

Fig. 9.4 Derivation of Example 9.5 (phonology)

here, the same case). This unification of course happens by the contraction rule applied after the introduction of "!" on the two (or more) ds.[2] This allows us for instance to obtain the derivation in Fig. 9.4.

[2]In this case, the head itself is also changed, becoming of type (!d ⊸ c) ⊸ c instead of, for instance, (!$\mathtt{d}_{\mathtt{nom}}$ ⊸ c) ⊸ c.

9.4.2 *On reflexives*

Reflexives are bound inside their clause, which means that the mechanism of building reflexive structures is at work before the verbal cluster is closed. Anoun (2007) suggests having entries like

$$\begin{pmatrix} \lambda P.\lambda x_\Phi.P(proself, x_\Phi) \\ \lambda P.\lambda x_\lambda.P(x_\lambda, x_\lambda) \end{pmatrix} : (\mathtt{d}_{\mathtt{acc}} \multimap \mathtt{d}_{\mathtt{nom}} \multimap \mathtt{c}) \multimap (\mathtt{d}_{\mathtt{nom}} \multimap \mathtt{c}) \qquad (9.7)$$

This solution requires us to use the regular form associated with the axioms, that is, true logical axioms, and not of the form of Example 9.6.

Example 9.6. Peter likes himself

The derivation of Example 9.6 implies making at first two assumptions, one of type $\mathtt{d}_{\mathtt{acc}}$, another of type $\mathtt{d}_{\mathtt{nom}}$, discharging both in the right order so that we obtain a phrase of type $(\mathtt{d}_{\mathtt{acc}} \multimap \mathtt{d}_{\mathtt{nom}} \multimap \mathtt{c})$, which can be taken as an argument by the reflexive, which suppresses one of the requirements, thus giving a phrase of type $\mathtt{d}_{\mathtt{nom}} \multimap \mathtt{c}$. After that, the derivation may proceed as usual, taking an assumption $\mathtt{d}_{\mathtt{nom}}$ before the rule $[\multimap i + e]$ applies, thus introducing the subject *Peter* as the first word in the phonological form, and as a constant which substitutes x_λ in the logical form.

9.5 On Phases

There is nothing in the previous section preventing undesirable bindings like

Example 9.7. *John$_1$ likes him$_1$

that is, a sentence which is semantically interpreted like *John*$_1$ *likes himself*$_1$ even if obtained by different means. This can be solved by referring to *phases*, a device recently introduced by Chomsky (2001) to regulate the computational flow. The minimalist framework now assumes that there is only one computational cycle, which builds structures until a phase has been created. When a phase has been created, transfer takes place: the phase is sent to the conceptual-intentional and sensori-motor systems, where it is interpreted. There are two levels (or sub-phases) in one phase: the discursive one, which contains the illocutary force, CP, and the predicative one, which contains the argument structure, vP. When a phase is complete, its domain of c-command is sent to the interpretive components and is no longer accessible. Only the periphery, that is, the specifier, is still accessible, as part of a new tableau which gives the input of the following phase. If we wish to prevent cases like Example 9.7, we stipulate that not only the Ax^{ij} (i: the rank of the entry in the lexical enumeration, j: the rank of the hypothesis in the entry) are used once and only once and by decreasing order with regard to j, but also *each in one and only one periphery of a CP-phase* and in such a way that all the Ax^{ij} for the same i occupy successive phases. This is the way LLG satisfies the so-called *phase impenetrability condition* (PIC).

9.6 Comparing CVG and LLG

The LLG framework proves to be very close to convergent grammars, except on the following points:

- Convergent grammars are not oriented towards the "minimalist" viewpoint, for instance there is no question of formal features triggering checking and movements. It is therefore not possible to predict the future steps of a derivation taking simply into account the features which are waiting for checking at some stage.
- In syntax, linear implication is enriched by various "flavours" which indicate the kind of relation there is between two constituents (complement, subject or adjunct), but later versions suggest we can get rid of these flavours (in any case, in other frames these composition modes are also indicated even if by other means).
- Questions of scope taking are taken into consideration via the pure mechanism of Cooper storage we presented in Chapter 3, but later versions have shown the equivalence of this mechanism with a rule similar to the $[\multimap\ i + e]_c$ rule of LLG.
- Whereas syntactic and semantic types in LLG remain isomorphic (for instance to a syntactic type $(\mathbf{d_k} \multimap \mathbf{s}) \multimap \mathbf{s}$ corresponds a semantic type $(\mathbf{e} \rightarrow \mathbf{t}) \rightarrow \mathbf{t}$) in order to use the Curry–Howard isomorphism, they can behave differently in convergent grammars, keeping for instance the syntactic type NP for noun phrases while, semantically, they have the higher-order type. It is claimed that syntax must stay as simple as possible. In this way, Pollard may really distinguish between syntactic and intrinsically semantic ambiguities: scope ambiguities, for instance, are intrinsically semantic and it is a mistake, according to him, to treat them by means of different syntactic analysis, as was done by Montague as well as by LLG.

• Pollard insists on the non-functionality of the relation between syntax and semantics. The approaches based on the Curry–Howard isomorphism are condemned to see this relation as a *function*. (This question will be dealt with differently later on when continuations have been introduced, showing that the choice of a strategy of evaluation for a given semantic analysis may lead to various readings.)

9.7 Concluding Remarks

This chapter introduced two relatively close frameworks, convergent grammars and labelled linear grammars. The former is fundamentally based on the conception of two derivations held in parallel: this allows Pollard to insist on the relational character (as opposed to functional) of the relation between syntax and semantics. The latter is more traditional in the categorial framework since it heavily relies on the Curry–Howard correspondence. Nevertheless, we must mention that this correspondence is here incomplete, and limited to the correspondence between logical rules and their translation in terms of the λ-calculus. The deductive grammar in itself is not a logical system properly speaking since it lacks many logical properties. For instance, it is not complete; some rules are missing or their application is so constrained that there is of course no hope of getting results like normalization. If normalization exists inside the logic in which the system is embedded (multiplicative linear logic with exponential), there is no evidence that normalization preserves the properties of the linguistic derivations. This is a consequence of the discrepancy between logic and grammar which we pointed out in the introduction. One is symmetric, the other not. We may speculate that if it is so, it is because language has to cope with empirical properties of the system of speech. For instance, considerations of economy are intervening in language. It is an open (and fascinating) question to motivate the limitations we encounter in our grammatical models with regard to logic, by means of such considerations.

Now, compared to other models, like multimodal categorial grammar, we suggest that constraints on proofs (which are always processes, as we shall emphasize in the next chapters) can be more easily motivated, by referring to properties of the mind, than special operators like structural modalities, for which there is of course no empirical evidence in the faculty of language.

Chapter 10

Continuations and Contexts

10.1 The Use of Continuations in Semantics

10.1.1 *Continuations and quantification*

The notion of *continuation*[1] plays an important role in computer science, and it could play as big a role in linguistics. What do we mean by "continuation"? It may be explained briefly in the following way. Executing a program consists in successively applying functions, such that each function applies to the result given by the previous one. In the simplest case, we have a mere composition of two functions, let us say f and ϕ. ϕ is applied to an integer n and then the result is given as an argument to f, so that the result is $f(\phi(n))$. We shall say in this case that when ϕ has been applied, its regular continuation consists in applying f. From a certain point inside a program, a continuation is the future of that point. Another way of expressing this amounts to saying that f provides some context into which $\phi(n)$ has to be calculated. Now let us imagine that one wants to change this future (this case occurs for instance when, because of an error, a program has to escape towards another function, signalling an error at this point for instance). We have two ways to proceed: either we enter the program and change the whole future at the point under examination, or we see the process in a reverse way and consider the context itself as an argument, while the object to which f had to apply becomes the function.

We may write

$$\text{context}: \; f[\;] \qquad \text{term}: \; \phi(n)$$

[1]This notion was introduced the first time in theoretical computer science by Strachey and Wadsworth (1974) and later studied by Plotkin (1975), Griffin (1990), etc.

where *term* means the term to compute in some context. In the ordinary case of programming, the context, which may here be also represented by a λ-term $\lambda x.f[x]$, is applied to the computed value $\phi(n)$.

Now, we may look for a new function, associated with ϕ, that we will denote by $\overline{\phi}$, also defined for integers, such that

$$\overline{\phi}(n)(\lambda x.f[x]) = \lambda x.f[x](\phi(n))$$

The context has become an object, and the previous term has now been transformed into a new context. The advantage of this new way of programming is that if we want to change the continuation, we have only to apply the new function, associated with the old term, to another context. Thus *terms* (or *programs*, or *functions*) and *contexts* are treated on a par.

Of course the first question is whether it is always possible to associate such $\overline{\phi}$ to an arbitrary ϕ, and if it is, how to compute it? Programming using the notion of continuation is known as *continuation passing style* (CPS). There are techniques which allow automatic transformation of a program written in the normal style into the continuation passing style. They rest on so-called CPS transforms.

Concerning the use of continuations in natural language processing, we shall first refer to works by Barker (2002).

One of the main motivations still concerns the question of quantifier scoping. We saw that several solutions were tried but all lack uniformity (in the treatment of subjects and objects, for instance). To have a uniform treatment would suppose that the operation by which the scope of an operator is taken always rests on the same mechanism and has always the same effect whatever its location. This will be the case if all nominal phrases denote operators which manipulate their continuations, and ambiguities will result then from various ways to combine these continuations.

10.1.2 *Continuizing a grammar*

Let us start from an elementary grammar like the one given in Fig. 10.1. The semantic objects (between brackets [[.]]) have all fixed semantic types:

[[S]]	t
[[NP]]	e
[[VP]]	$e \to t$
[[TV]]	$e \to (e \to t)$

[[N]]	$e \to t$
[[Det]]	$(e \to t) \to ((e \to t) \to (e \to t))$

Let us consider an object of a given type α in the context of a sentence (for instance the noun *John* in the context of the sentence *John saw Mary*). Its context is implicitly typed $\alpha \to t$, and if we wish to make it apply to its context, we have then to raise its type to $(\alpha \to t) \to t$. As already seen, in the case of *John*, we shall have as an associated semantic recipe $\lambda X.X(j)$. In this new semantic representation, the variable bound by λ, X, is a variable of type $e \to t$; in other words, it is a variable for the continuation of a term of type e (here, a variable for a continuation of *John*). We shall therefore admit from now on that besides ordinary variables (of any type α), there are *continuation variables* of type $\alpha \to t$. For an ordinary variable x, we shall denote by $\overline{x}$ its counterpart as a continuation variable. Thus, for instance, the continuized semantics of *John* may be written as $\lambda\overline{x}.\overline{x}(j)$.

When continuizing a grammar, we wish each expression to be a function of its continuation. For instance, concerning the unary rules of the given grammar above, we shall have

$NP \to John$	$[[\text{NP}]]^c = \lambda\overline{x}.\overline{x}(\mathbf{j})$
$NP \to Mary$	$[[\text{NP}]]^c = \lambda\overline{x}.\overline{x}(\mathbf{m})$
$VP \to left$	$[[\text{VP}]]^c = \lambda\overline{P}.\overline{P}(\lambda x.\mathbf{left}(x))$
$TV \to saw$	$[[\text{TV}]]^c = \lambda\overline{R}.\overline{R}(\lambda x.\lambda y.\mathbf{saw}(y,x))$
$N \to boy$	$[[\text{N}]]^c = \lambda\overline{P}.\overline{P}(\lambda x.\mathbf{boy}(x))$

$S \to NP\ VP$	$[[\text{S}]] = [[\text{VP}]]([[\text{NP}]])$
$NP \to NP$	$[[\text{NP}]] = [[\text{NP}]]$
$NP \to Det\ N$	$[[\text{NP}]] = [[\text{Det}]]([[\text{N}]])$
$VP \to TV\ NP$	$[[\text{VP}]] = [[\text{TV}]]([[\text{NP}]])$
$NP \to John$	$[[\text{NP}]] = \mathbf{j}$
$NP \to Mary$	$[[\text{NP}]] = \mathbf{m}$
$VP \to left$	$[[\text{VP}]] = \lambda x.\mathbf{left}(x)$
$TV \to saw$	$[[\text{TV}]] = \lambda x.\lambda y.\mathbf{saw}(y,x)$
$N \to boy$	$[[\text{N}]] = \lambda x.\mathbf{boy}(x)$
$Det \to every$	$[[\text{Det}]] = \lambda P.\lambda Q.\forall x.(P(x) \Rightarrow Q(x))$

Fig. 10.1 An elementary grammar

As we may see, not only the NPs have now a lifted type: the raising operation now becomes the rule, or, put differently, all the phrases (whatever their location) are uniformly raised. It remains of course to say how we shall proceed with binary rules, that is, how the application of some λ-term to another one is transformed by CPS transform.[2]

Let M and N be two λ-terms, the former of type $\alpha \to \beta$ and the latter of type α. They are respectively transformed into M^c and N^c of respective types $((\alpha \to \beta) \to t) \to t$ and $(\alpha \to t) \to t$. By performing $(M\ N)$, we consider N put in the context $M[\,]$ and $(M\ N)$ put in the empty context $[\,]$. Therefore, finally, $(M\ N)$ should be transformed into

$$N^c(M^c[\,])$$

which should be itself an expression of a continuized type (here $(\beta \to t) \to t$), where any continuation of type t may be put inside the hole $[\,]$. If the sentence stops here, we shall simply consider that this continuation, of type $t \to t$, is the identity function.

Nevertheless, the types of expressions do not allow a direct functional application, and in the same way a proof gives the λ-term associated with the continuization of a proper name by means of the Curry–Howard isomorphism, it is also a proof which allows us to extract the type of combination associated with the continuization of functional application.

How in effect to prove $(\beta \to t) \to t$ from $((\alpha \to \beta) \to t) \to t$ and $(\alpha \to t) \to t$? Here is a proof in natural deduction:

$$\dfrac{\dfrac{\dfrac{\dfrac{\dfrac{\dfrac{[u:\beta\to t]^3 \quad \dfrac{[m:\alpha\to\beta]^2[x:\alpha]^1}{(m\ x):\beta}}{(u\ (m\ x)):t}}{\lambda x.(u\ (m\ x)):\alpha\to t}\,(1) \quad N^c:(\alpha\to t)\to t}{(N^c\ \lambda x.(u\ (m\ x))):t}}{\lambda m.(N^c\ \lambda x.(u\ (m\ x))):(\alpha\to\beta)\to t}\,(2) \quad M^c:((\alpha\to\beta)\to t)\to t}{(M^c\ \lambda m.(N^c\ \lambda x.(u\ (m\ x)))):t}}{\lambda u.(M^c\ \lambda m.(N^c\ \lambda x.(u\ (m\ x)))):(\beta\to t)\to t}$$

[2]We encounter here a problem similar to the one Jacobson had to cope with (cf. Section 6.3.2), when pronouns are not treated as e-type entities but as $e \to e$-type ones.

Hence, the conclusion we draw:

$$(M\ N)^c = \lambda u.(M^c\ \lambda m.(N^c\ \lambda x.(u\ (m\ x))))$$

Let us apply this to the sentence *John left.* It follows that

$$\begin{array}{ll}
\lambda u.(left^c\ \lambda m.(john^c\ \lambda x.(u\ (m\ x)))) & \longrightarrow \\
\lambda u.([\lambda \overline{P}.\overline{P}(\lambda z.left(z))]\ \lambda m.(\lambda \overline{x}.\overline{x}(j)\ \lambda x.(u\ (m\ x)))) & \longrightarrow \\
\lambda u.([\lambda \overline{P}.\overline{P}(\lambda z.left(z))]\ \lambda m.(\lambda x.(u\ (m\ x)))(j)) & \longrightarrow \\
\lambda u.([\lambda \overline{P}.\overline{P}(\lambda z.left(z))]\ \lambda m.(u\ (m\ j))) & \longrightarrow \\
\lambda u.(\lambda m.(u\ (m\ j))\ (\lambda z.left(z))) & \longrightarrow \\
\lambda u.(u\ (\lambda z.left(z)\ j)) & \longrightarrow \\
\lambda u.(u\ left(j)) &
\end{array}$$

And we check that in effect, by applying this expression to the identity function, we just obtain *left*(j).

But we must notice that the previous proof in natural deduction was not unique, since this is another one:

$$\dfrac{\dfrac{\dfrac{\dfrac{\dfrac{[u:\beta\to t]^3 \qquad \dfrac{[m:\alpha\to\beta]^2[x:\alpha]^1}{(m\ x):\beta}}{(u\ (m\ x)):t}}{\lambda m.(u\ (m\ x)):(\alpha\to\beta)\to t}\,(2) \qquad M^c:((\alpha\to\beta)\to t)\to t}{(M^c\ \lambda m.(u\ (m\ x))):t}}{\lambda x.(M^c\ \lambda m.(u\ (m\ x))):\alpha\to t}\,(1) \qquad N^c:(\alpha\to t)\to t}{\dfrac{(N^c\ \lambda x.(M^c\ \lambda m.(u\ (m\ x)))):t}{\lambda u.(N^c\ \lambda x.(M^c\ \lambda m.(u\ (m\ x)))):(\beta\to t)\to t}}$$

which gives a second translation of the application $(M\ N)$ into the continuized version:

$$(M\ N)^c_n = \lambda u.(N^c\ \lambda x.(M^c\ \lambda m.(u\ (m\ x))))$$

These two possibilities express two distinct evaluation orders. In the former, what occurs inside M (in our example the VP semantics) has a wide scope with regard to what occurs inside N (the subject of the VP), while in the latter, relations are inverted: the subject has the wide scope and the VP the narrow one. In a simple case as *John left*, this is immaterial, but as we shall see, as soon as we introduce quantified NPs, that will matter. Given that each translation is as valid as the other, both can be used, resulting in distinct logical representations of sentences. Let us end this section by giving the continuized binary rules of our toy grammar.

$$
\begin{array}{ll}
S \to NP\ VP & [[S]]^c = \lambda u.([[VP]]^c\ \lambda m.([[NP]]^c\ \lambda x.(u\ (m\ x)))) \\
 & [[S]]^c = \lambda u.([[NP]]^c\ \lambda x.([[VP]]^c\ \lambda m.(u\ (m\ x)))) \\
NP \to Det\ N & [[NP]]^c = \lambda u.([[Det]]^c\ \lambda m.([[N]]^c\ \lambda x.(u\ (m\ x)))) \\
 & [[NP]]^c = \lambda u.([[N]]^c\ \lambda x.([[Det]]^c\ \lambda m.(u\ (m\ x)))) \\
VP \to TV\ NP & [[VP]]^c = \lambda u.([[TV]]^c\ \lambda m.([[NP]]^c\ \lambda x.(u\ (m\ x)))) \\
 & [[VP]]^c = \lambda u.([[NP]]^c\ \lambda x.([[TV]]^c\ \lambda m.(u\ (m\ x))))
\end{array}
$$

10.1.3 *Call-by-value and call-by-name*

In fact what we have seen has much to do with some traditional approaches in theoretical computer science, which deserve to be surveyed in order to shed some light on the question of *evaluation*. After exploring these issues we will be able to come back to continuation semantics for natural language. In computer science, generally two evaluation strategies are distinguished for symbolic forms. Let us consider for instance the evaluation of the following λ-expression:

$$(\lambda x.(\lambda y.(x\ y)\ u)\ z)$$

β-reduction may be performed in two ways. The *normal* reduction starts from the periphery:

$$(\lambda x.(\lambda y.(x\ y)\ u)\ z) \longrightarrow (\lambda y.(z\ y)\ u) \longrightarrow (z\ u)$$

But the reduction *by value* starts from the inside:

$$(\lambda x.(\lambda y.(x\ y)\ u)\ z) \longrightarrow (\lambda x.(x\ u)\ z) \longrightarrow (z\ u)$$

Both lead to the same result, as is expected from the Church–Rosser theorem (Girard *et al.*, 1989). Nevertheless, there may be advantages of using one strategy preferably to the other, and in the case of a non-confluent system, that would give distinct evaluations.

We may notice that the reduction by value deserves its name since it amounts to first evaluating the λ-terms before performing their composition. Let us take this other example:

$$(\lambda M.(M\ u)\ (\lambda k.\lambda y.(k\ y)\ g))$$

If we proceed by value, we make the following reduction:

$$\begin{array}{ll}(\lambda M.(M\ u)\ (\lambda k.\lambda y.(k\ y)\ g)) & \longrightarrow\\ (\lambda M.(M\ u)\ \lambda y.(g\ y)) & \longrightarrow\\ (\lambda y.(g\ y)\ u) & \longrightarrow\\ (g\ u) & \end{array}$$

This amounts to first calculating the value of the argument, here $(\lambda k.\lambda y.(k\ y)\ g)$, before passing it to the function (as a value of M). In other words, in this strategy, it is only possible to reduce an $(M\ N)$ term after N is evaluated. N may be a variable, a constant, a λ-abstraction, but in no case an application $(M'\ N')$ not yet evaluated.

The other strategy amounts to not evaluating the argument first, and therefore evaluating it as late as possible (something which is sometimes named *freezing* the evaluation of the argument). We have

$$\begin{array}{ll}(\lambda M.(M\ u)\ (\lambda k.\lambda y.(k\ y)\ g)) & \longrightarrow\\ ((\lambda k.\lambda y.(k\ y)\ g)\ u) & \longrightarrow\\ (\lambda y.(g\ y)\ u) & \longrightarrow\\ (g\ u) & \end{array}$$

In this case, the argument is passed to M without evaluation: this is referred to as passing the parameter *by name*.

In continuation semantics, an expression is evaluated after it receives a continuation. For instance, $(\lambda P.P(j)\ \lambda x.\mathit{left}(x))$ may be evaluated since a continuation has been given to $\lambda P.P(j)$, in the form of the VP *left*. We get $\mathit{left}(j)$ of type t. If not, it amounts to an instruction: $\lambda P.P(j)$ is an expression waiting for a continuation of type $e \to t$ in order to be evaluated.

In call-by-value, a function M of type $\alpha \to \beta$ will be interpreted in continuation semantics as an expression which takes for its argument a *value* of type α, and then returns a result (of type t) if it is given a continuation of a β-type term, therefore something of type $\alpha \to ((\beta \to t) \to t)$. The continuized type of this function is

$$M_v^c : (\alpha \to ((\beta \to t) \to t) \to t) \to t$$

and like previously, an argument N of type α will be CPS-transformed into

$$N_v^c : (\alpha \to t) \to t$$

$$
\dfrac{\dfrac{\dfrac{\dfrac{\dfrac{\dfrac{\dfrac{[n:\alpha]^1 [m:\alpha \to \beta^*]^2}{(m\ n):\beta^*} \qquad [k:\beta \to t]^3}{((m\ n)\ k):t}}{\lambda n.((m\ n)\ k):\alpha \to t}\,(1) \qquad N^c_v:\alpha^*}{N(\lambda n.((m\ n)\ k)):t}}{\lambda m.N(\lambda n.((m\ n)\ k)):(\alpha \to \beta^*)^*}\,(2) \qquad M:(\alpha \to \beta^*)^*}{M(\lambda m.N(\lambda n.((m\ n)\ k))):t}}{\lambda k.M(\lambda m.N(\lambda n.((m\ n)\ k))):\beta^*}
$$

Fig. 10.2 CPS transform by call-by-value where σ^* is a shorthand for $(\sigma \to t) \to t$

By a reasoning similar to that in the previous section, a proof in natural deduction is able to give the "call-by-value" CPS transform of the application $(M\ N)$, that is,

$$(M\ N)^c_v = \lambda k.(M\ \lambda m.(N\ \lambda n.((m\ n)\ k)))$$

A proof is provided in Fig. 10.2. In call-by-name, we shall get something slightly different. This time, a function of type $\alpha \to \beta$ does not take as its argument a value but a "calculus" (not yet performed), but the rest stays the same. We get

$$(M\ N)^c_n = \lambda k.(M\ \lambda m.(N\ m)\ k)$$

To sum up, in call-by-value, an expression of type $\alpha \to \beta$ is interpreted as mapping an evaluated form (of type α) to a form to evaluate in some context (which itself waits for a β-type value). Its continuized form is a form to evaluate inside the kind of context which waits for an expression of type $\alpha \to \beta$, hence the continuized type of a function of type $\alpha \to \beta$ in the call-by-value regime is $(\alpha \to ((\beta \to t) \to t) \to t) \to t$.

In call-by-name, an expression of type $\alpha \to \beta$ is interpreted as mapping a form to evaluate in some context which waits for a value of type α to a form to evaluate in a context which waits for a value of type β. Its continuized version is itself a form to evaluate in the kind of context which waits for an expression of type $\alpha \to \beta$, hence the continuized type of a function of type $\alpha \to \beta$ in the call-by-name regime is $(((\alpha \to t) \to t) \to ((\beta \to t) \to t) \to t) \to t$.

$$\dfrac{\dfrac{M : (\alpha^* \to \beta^*)^* \qquad \dfrac{\dfrac{\dfrac{N : \alpha^* \quad [m : \alpha^* \to \beta^*]^1}{(N\ m) : (\beta \to t) \to t} \quad [k : \beta \to t]^2}{((N\ m)\ k) : t}}{\lambda m.((N\ m)\ k) : (\alpha^* \to \beta^*) \to t}\,(1)}{M(\lambda m.((N\ m)\ k)) : t}}{\lambda k.M(\lambda m.((N\ m)\ k)) : \beta^*}$$

Fig. 10.3 CPS transform by call-by-name where σ^* is a shorthand for $(\sigma \to t) \to t$

If we distinguish three sorts of objects:

- the *values* of type α, forming a set V_α,
- the *continuations* for a term of type α, forming a set K_α,
- and the *calculi* which provide α-type terms with contexts in order to evaluate them, that is, the forms to evaluate in contexts waiting for expressions of type α, forming a set C_α,

then, in call-by-value, the expressions of type $\alpha \to \beta$ are interpreted as functions from V_α to C_β, and in call-by-name, as functions from C_α to C_β. Let us remark that the elements of C_α are functions from K_α to R, where R is the type of answer, that we have until now taken to be t.

In call-by-name, like in call-by-value, we have a proof in natural deduction giving the CPS transform of the application, as shown in Fig. 10.3. We may notice now something amazing. Replacing the "answer" type t with $\perp$ (the absurdity), $(\alpha \to t) \to t$ (or α^* in our abbreviation) becomes $(\alpha \to \perp) \to \perp$, that is, $\neg\neg\alpha$, the *double negation* of α. Thus, using continuations leads us to consider negations and double negations in logic. Let us recall then that, whereas in classical logic, we have $\neg\neg\alpha \equiv \alpha$, this is no longer the case in intuitionistic logic (where we have only $\alpha \vdash \neg\neg\alpha$ and not the reciprocal relation). On the other hand, intuitionistic logic is known to be *constructive* contrarily to classical logic. So how to conciliate a calculus using an implicit involutive negation with the requirement of constructivity? In fact, there is an old result due to Gödel, according to which if we encode classical formulae by means of their double negation, then it is always possible to prove in intuitionistic logic the encoding of any classical formula. In fact, this is precisely what we have done here: we encoded α by $\neg\neg\alpha$ and $\alpha \to \beta$ by $\neg\neg(\alpha \to \beta)$, so that we got a proof (even two proofs!) of $\neg\neg\beta$ in intuitionistic logic from these two encoded premises. The constructivity of intuitionistic logic shows up in that these two proofs provide us with two λ-terms expressing the composition of the λ-terms associated with the premises. We can now ask whether there could exist other possible translations of classical formulae which would provide the same effects.

10.2 Symmetric Calculi

10.2.1 *Towards the $\lambda\mu$-calculus*

Introducing the notion of continuation allows us to conceive changes of evaluation contexts, that is, "jumps" from an instruction to another one. This assumes distinct instructions are labelled (like when using a *goto* instruction), or "named". From a logical point of view, this corresponds to formulae endowed with a *name*. We have then two types of variables: λ-variables (for standard use with regard to the λ-calculus) and new μ-variables which serve to name formulae. In a sequent, there are therefore two series of formulae: those which are labelled by λ-variables (put on the left of the sequent) and those which are labelled by μ-variables (put on the right-hand side) (see Parigot, 1992).

To give a name to a formula is a condition to access it directly in the future, apart from any given sequential order. When a term t (associated with a formula) is named by a label α, this is denoted by $[\alpha]t$. When a named formula becomes active, this is rendered by the μ-abstraction over the name: $\mu\alpha.e$. That amounts to choosing a formula among all the available ones and storing its name in memory before activating it. When a $\lambda\mu$-term $\mu\alpha.M$ is evaluated in a context N, this context is transmitted to all the subterms of M named by α, hence the following reduction rule:

$$(\mu\alpha.M)N \longrightarrow \mu\alpha.M[N/^\star\alpha]$$

where the substitution ("$/^\star$") is such that

(1) if M is a λ-variable $x : x[N/^\star\alpha] = x$;
(2) if M is a λ-term $\lambda x.M$, $(\lambda x.M)[N/^\star\alpha] = \lambda x.M[N/^\star\alpha]$;
(3) if M and M$'$ are λ-terms, then $(MM')[N/^\star\alpha] = (M[N/^\star\alpha]M'[N/^\star\alpha])$;
(4) if M is a μ-term $\mu\beta.M$, $(\mu\beta.M)[N/^\star\alpha] = \mu\beta.M[N/^\star\alpha]$;
(5) if M is $[\alpha]M$, $([\alpha]M)[N/^\star\alpha] = [\alpha](M[N/^\star\alpha]N)$;
(6) if M is $[\beta]M$, $([\beta]M)[N/^\star\alpha] = [\beta](M[N/^\star\alpha])$ if $\beta \neq \alpha$.

Clause (5) must be noticed. Let $[\alpha]M$ be a term; substituting N inside this term amounts to keeping the label α, but for another term, which is obtained by applying M (where N has been substituted) to N.

For instance the $\lambda\mu$-term $\lambda y.\mu\alpha.[\beta](y \ \ \lambda x.\mu\delta.[\alpha]x)$ gives

$$
\begin{array}{ll}
((((\lambda y.\mu\alpha.[\beta](y \ \ \lambda x.\mu\delta.[\alpha]x) \ \ u) \ \ a) \ \ b) \rightarrow & \\
((\mu\alpha.[\beta](u \ \ \lambda x.\mu\delta.[\alpha]x) \ \ a) \ \ b) & [\beta - \mathit{reduction}] \\
(\mu\alpha.[\beta](u \ \ \lambda x.\mu\delta.[\alpha](x \ \ a)) \ \ b) & [\mu - \mathit{reduction}] \\
\mu\alpha.[\beta](u \ \ \lambda x.\mu\delta.[\alpha]((x \ \ a) \ \ b)) & \mathit{idem}
\end{array}
$$

Notice that at step 2, $(\mu\alpha.[\beta](u \ \ \lambda x.\mu\delta.[\alpha]x) \ \ a)$ is reduced by applying the above rules, which give

$$
\begin{array}{ll}
(\mu\alpha.[\beta](u \ \ \lambda x.\mu\delta.[\alpha]x) \ \ a) & \rightarrow \\
\mu\alpha.([\beta](u \ \ \lambda x.\mu\delta.[\alpha]x) \ \ a)[a/^{\star}\alpha]) & \rightarrow \\
\mu\alpha.[\beta]((u \ \ \lambda x.\mu\delta.[\alpha]x) \ \ a)[a/^{\star}\alpha]) & \rightarrow \\
\mu\alpha.[\beta]((u[a/^{\star}\alpha] \ \ \lambda x.\mu\delta.[\alpha]x[a/^{\star}\alpha])) \ \ a) & \rightarrow \\
\mu\alpha.[\beta]((u \ \ \lambda x.\mu\delta.[\alpha]x[a/^{\star}\alpha])) \ \ a) & \rightarrow \\
\mu\alpha.[\beta](u \ \lambda x.\mu\delta.[\alpha](x \ \ a)) &
\end{array}
$$

Every term of the form $[\alpha]x$ is replaced by $[\alpha](x \ a)$. At the end, for any number of arguments, we have

$$((((\lambda y.\mu\alpha.[\beta](y\lambda x.\mu\delta.[\alpha]x) \ u) \ v_1) \ v_2) \ v_n)$$

which reduces to $\mu\alpha.[\beta](u \ \lambda x.\mu\delta.[\alpha]((x \ v_1) \ \ v_n))$.

The fact that the label α is preserved allows us to apply such a $\lambda\mu$-term to any sequence of terms: we are not obliged to plan the exact number of terms to which the operator applies, like it is the case in the standard λ-calculus.

It may be useful to look at the behaviour of an operator, known as *Felleisen's*:

$$E[C(M)] \rightarrow M(\lambda z.E[z])$$

It consists in extracting a term M from its environment $E[\ ]$ to transform it into a function taking this environment as its argument.

Let α and β be arbitrary types. Let us assume E is of type β and the hole in it is waiting for a term of type α, then $\lambda z.E[z]$ is of type $\alpha \rightarrow \beta$. Since the two sides of the reduction are of type β, M must be of type $(\alpha \rightarrow \beta) \rightarrow \beta$, but $C(M)$ must be of type α since it fills in E's hole, therefore C maps an object of type $(\alpha \rightarrow \beta) \rightarrow \beta$ onto an object of type α: it is a *type-lowering* operator.

Griffin (1990) noticed that if N is a ground term of type β, then it is always possible to make any kind of vacuous abstraction on it, for instance over a variable of type $\alpha \to \beta$, for any α, and then it would be enough to apply C to $\lambda d.N$ (where d would be of type $\alpha \to \beta$) in order to get a term of type α. Thus, any type α would be provable starting from any type β! To avoid this pitfall, Felleisen's operator should only apply when β is empty, that is, when it is $\perp$. Hence the idea that the type of "observables" (the resolved calculi) is $\perp$. In this case, d is of type $\neg\alpha$ and $\lambda d.N$ of type $\neg\neg\alpha$, and we get the fact that α is provable if $\neg\neg\alpha$ is, that is, an application of the rule of double negation. The operator C happens therefore to be associated with the elimination of the double negation; its type is $\neg\neg\alpha \to \alpha$.

It becomes then appealing to realize a calculus allowing a kind of Curry–Howard isomorphism. It would be necessary for that to give a set of classical rules, with the corresponding operations of the $\lambda\mu$-calculus. It is what was done by Parigot (1992).

The rules different from those of the λ-calculus are (where a proof term labels a sequent and where all the formulae are labelled except the active one)

$$\frac{t : (\Gamma \vdash A, \Delta)}{[\alpha]t : (\Gamma \vdash \alpha : A, \Delta)} [naming] \qquad \frac{e : (\Gamma \vdash \alpha : A, \Delta)}{\mu\alpha.e : (\Gamma \vdash A, \Delta)} [activation]$$

The system of rules includes structural ones but they are left implicit. It is assumed that labelled formulae are subject to weakening and contraction. Thus, Δ of the left rule may already contain $\alpha : A$, in which case, the active A is contracted with this A when applying naming. In the same way, the $\alpha : A$ in the premise of the right rule may not be present: A will have been obtained by weakening. These rules make it possible to demonstrate the validity of the following derived rule:

$$\frac{t : (\Gamma \vdash A, \beta : B, \Delta)}{\mu\beta[\alpha]t : (\Gamma \vdash B, \alpha : A, \Delta)}$$

It allows us to switch from an active formula to another one. Let us still notice that the structural rules are implicit. For instance, $\beta : B$ may not be present on the right-hand side of the premise, which gives a case of weakening, or $\alpha : A$ may be already present in Δ, in which case we get a contraction.

It is also possible to unearth the interpretation of negation in this calculus by using the fact that the formula $\perp$ itself may be named and activated. We thus have the following deductions:

$$\dfrac{\dfrac{t : (\Gamma, x : A \vdash \Delta)}{\mu\alpha.t : (\Gamma, x : A \vdash \bot, \Delta)}}{\lambda x.\mu\alpha.t : (\Gamma \vdash \neg A, \Delta)} \qquad \dfrac{\dfrac{t : (\Gamma \vdash A \Rightarrow \bot, \Delta) \quad u : (\Gamma' \vdash A, \Delta')}{(t\ u) : (\Gamma, \Gamma' \vdash \bot, \Delta, \Delta')}}{[\beta](t\ u) : (\Gamma, \Gamma' \vdash \beta : \bot, \Delta, \Delta')}$$

If we admit by convention that the labelled $\bot$ do not appear (are implicit) (a convention already admitted in the left deduction where $\mu\alpha$ comes from an implicit $\alpha : \bot$), we get the two rules

$$\dfrac{t : (\Gamma, x : A \vdash \Delta)}{\lambda x.\mu\alpha.t : (\Gamma \vdash \neg A, \Delta)} \qquad \dfrac{t : (\Gamma \vdash \neg A, \Delta) \quad u : (\Gamma' \vdash A, \Delta')}{[\beta](t\ u) : (\Gamma, \Gamma' \vdash \Delta, \Delta')}$$

It is then possible to name the application of a *co-term* (that is, a term of type $\neg A$) to a term, inside a computation, in order to go back to it afterwards (like it is obviously the case in a reduction to the absurdity, where the detection of a contradiction forces us to go backward, until the wrong hypothesis is met). It is also possible to construct a co-term, by activating $\bot$ and abstracting on a variable. Let us notice also that the classical negation allows the transfer of all the formulae to the same side of $\vdash$. Thus a sequent like

$$x_1 : A_1, x_2 : A_2, ..., x_n : A_n \vdash \alpha_1 : B_1, \alpha_2 : B_2, ..., \alpha_m : B_m$$

may be transformed into

$$x_1 : A_1, ..., x_n : A_n, \alpha_1 : \neg B_1, ..., \alpha_j : \neg B_j, ..., \alpha_m : \neg B_m \vdash \bot$$

The μ-variables occur then as labelling *negative formulae.* To activate a formula amounts to passing it to the right by a negation, and to name a formula amounts to passing it to the left:

$$\dfrac{\dfrac{e : (\Gamma, \alpha : \neg A \vdash \bot)}{\Gamma \vdash \neg\neg A}[\neg\ D]}{\mu\alpha.e : (\Gamma \vdash A)}[\neg\neg\ elim] \qquad \dfrac{t : (\Gamma \vdash A)}{[\alpha]t : (\Gamma, \alpha : \neg A \vdash \bot)}[naming]$$

From the natural deduction viewpoint (hence in terms of introduction and elimination rules), the two rules become

$$\dfrac{\Gamma, \alpha : \neg A \vdash t : \bot}{\Gamma \vdash \mu\alpha.t : A}[\bot\ elim] \qquad \dfrac{\Gamma \vdash \alpha : \neg A \qquad \Gamma' \vdash t : A}{\Gamma, \Gamma' \vdash [\alpha]t : \bot}[\bot\ intro]$$

It is worth coming back to μ-reduction, in order to see where it comes from. We know that reduction is, on the calculus side, the equivalent of normalization on the proof-theoretic side. Let us therefore examine a case of normalization, when for instance we suppress a detour caused by the application of the naming rule (making an $[\alpha]$ appear) followed by a step of activation (which makes a binder $\mu\alpha$ appear). The proof is the following:

$$\dfrac{\dfrac{\begin{array}{c}\dfrac{\alpha : \neg(A \Rightarrow B) \vdash \alpha : \neg(A \Rightarrow B) \qquad \vdash t : A \Rightarrow B}{\alpha : \neg(A \Rightarrow B) \vdash [\alpha]t : \bot} \\ \vdots \\ \alpha : \neg(A \Rightarrow B) \vdash u : \bot\end{array}}{\vdash \mu\alpha.u : A \Rightarrow B} \qquad \vdash v : A}{\vdash (\mu\alpha.u\ v) : B}.$$

It may be replaced by the following proof, which is more straightforward:

$$\dfrac{\begin{array}{c}\dfrac{\beta : \neg B \vdash \beta : \neg B \qquad \dfrac{\vdash t : A \Rightarrow B \qquad \vdash v : A}{(t\ v) : B}}{\beta : \neg B \vdash [\beta](t\ v) : \bot} \\ \vdots \\ \beta : \neg B \vdash u[[\alpha]t := [\beta](t\ v)] : \bot\end{array}}{\vdash \mu\beta.u[[\alpha]t := [\beta](t\ v)] : B}.$$

Substitution comes from that the very same proof is made in the first proof than in the second one, starting from a term of type $\bot$, but in the first case, the term is $[\alpha]t$, while in the second, it is $[\beta](t\ v)$.

We may also have

$$\dfrac{\vdash v : (A \Rightarrow B) \Rightarrow B \qquad \dfrac{\begin{array}{c}\dfrac{\alpha : \neg(A \Rightarrow B) \vdash \alpha : \neg(A \Rightarrow B) \qquad \vdash t : A \Rightarrow B}{\alpha : \neg(A \Rightarrow B) \vdash [\alpha]t : \bot} \\ \vdots \\ \alpha : \neg(A \Rightarrow B) \vdash u : \bot\end{array}}{\vdash \mu\alpha.u : A \Rightarrow B}}{\vdash (v\ \mu\alpha.u) : B}.$$

which reduces to

$$\dfrac{\begin{array}{c}\dfrac{\beta : \neg B \vdash \beta : \neg B \qquad \dfrac{\vdash v : (A \Rightarrow B) \Rightarrow B \qquad \vdash t : A \Rightarrow B}{(v\ t) : B}}{\beta : \neg B \vdash [\beta](v\ t) : \bot} \\ \vdots \\ \beta : \neg B \vdash u[[\alpha]t := [\beta](v\ t)] : \bot\end{array}}{\vdash \mu\beta.u[[\alpha]t := [\beta](v\ t)] : B}.$$

which then gives the rule (μ'). The standard $\lambda\mu$-calculus is the one which sticks to the rule (μ), only. De Groote (2001b) nevertheless showed a direct application of this calculus, endowed with the two reduction rules, to the multiple scope ambiguity problem.

10.2.2 *Applications of the $\lambda\mu$-calculus to the problem of scope construal*

De Groote showed that expressions of a restricted language may be typed by means of the classical types of the $\lambda\mu$-calculus, and that, thanks to this typing, it was possible to freely get scope ambiguities. For this aim, let us define the following typing:

$$\begin{array}{lll} x & ::= & e \\ \alpha & ::= & e \rightarrow t \\ \textbf{individual} & ::= & e \rightarrow t \\ \Rightarrow & ::= & t \rightarrow (t \rightarrow t) \\ \forall & ::= & e \rightarrow (t \rightarrow t) \end{array}$$

where t is here used in place of $\bot$ and where, consequently, we should be allowed to write

$$\begin{array}{lll} \alpha & ::= & \neg e \\ \textbf{individual} & ::= & \neg e \\ \Rightarrow & ::= & \neg\neg\bot \\ \forall & ::= & e \rightarrow \neg\bot \end{array}$$

This typing may be extended to complex semantic expressions like

$$\mu\alpha.\forall x.((individual\ x) \Rightarrow [\alpha]x)$$

This is made by means of the following deduction:

$$\dfrac{\dfrac{\dfrac{\dfrac{x : e \vdash x : e \qquad \vdash individual : e \rightarrow t}{x : e \vdash (individual\ x) : t} \qquad \dfrac{x : e \vdash x : e \qquad \alpha : e \rightarrow t \vdash \alpha : e \rightarrow t}{x : e, \alpha : e \rightarrow t \vdash [\alpha]x : t}}{x : e, \alpha : \neg e \vdash ((individual\ x) \Rightarrow [\alpha]x) : t}}{\alpha : \neg e \vdash \forall x.((individual\ x) \Rightarrow [\alpha]x) : t}}{\vdash \mu\alpha.\forall x.((individual\ x) \Rightarrow [\alpha]x) : e}$$

This deduction is obtained by using, in passing, rules we did not yet give, for the introduction and the elimination of quantifiers. We shall notice that we use indifferently $e \rightarrow t$, $\neg e$ or $e \rightarrow \bot$. That means that a name used to name a term may also be viewed as a variable predicate. $[\alpha]x$ may be read as $(\alpha\ x)$: the name is a property. However, the peculiarity of this system is that it gives as a result of this deduction a term of type e, and not of type $(e \to t) \to t$ like in Montague grammar. This may be interpreted by saying that the expression denotes some kind of *generic individual*, which suggests that at the level of the model, a place is given to such individuals, which have the type e, like others. In the same way, for an existential expression such as

$$\mu\alpha.\exists x.((individual\ x) \wedge [\alpha]x)$$

the calculus leads us to introduce the assumption that individuals may be interpreted as *anonymous constants*, according to a tradition of thought which goes back to the ϵ-calculus of David Hilbert (1922).

Let us then give the syntactic typing, with for each item, in addition to its syntactic type, its semantic representation in the form of a $\lambda\mu$-term (and no longer only of a λ-term):

everybody	:	$\mu\alpha.\forall x.((individual\ x) \Rightarrow [\alpha]x)$	:	NP
someone	:	$\mu\alpha.\exists x.((individual\ x) \wedge [\alpha]x)$	:	NP
loves	:	$\lambda x.\lambda y.((love\ y)\ x)$	:	$(NP\backslash s)/NP$

The syntactic derivation assembles the $\lambda\mu$-terms in the following way:

$$((\lambda x.\lambda y.((love\ y)\ x)\ \ \mu\alpha.\exists x.((individual\ x) \wedge [\alpha]x)) \\ \mu\alpha.\forall x.((individual\ x) \Rightarrow [\alpha]x)) \quad (10.1)$$

thus producing a new $\lambda\mu$-term which may be reduced by the rules (μ) and (μ').

It is worth noting that we shall be also allowed to add a rule "of simplification" (σ) which aims at getting rid of the operator $\mu\alpha$ when there is neither a context containing an $[\alpha]$ to which the μ-term could be applied, nor a context of this kind into which it could apply. The rule is then

$$(\sigma) \qquad \mu\alpha.u \longrightarrow u[[\alpha]t := t]$$

The reduction may be obviously seen as the normalization of a proof which contains only semantic types. In this sense, this view could be in accordance with the purpose of a grammar as presented by Chomsky in the Minimalist Program: the first phase of the analysis (corresponding to the pre-"spell out") only uses syntactic types (also known as formal and categorial features), and the second phase, which in minimalism corresponds to the covert moves, would be equivalently expressed here by a normalization of the proof only based on semantic types.

The two reductions are shown in Fig. 10.4. It is important here to note that this divergence comes from the evaluation of a form (often called a *critical pair*) $(\mu\alpha.u \quad \mu\beta.v)$. We have (simplifying in some way the formulae) the bifurcation shown in Fig. 10.5.

10.2.3 $\lambda\mu\tilde{\mu}$-calculus

If the $\lambda\mu$-calculus has a privileged representation in natural deduction, the $\lambda\mu\tilde{\mu}$-calculus (Herbelin, 2005; Curien and Herbelin, 2000), that we shall see now, has one in the sequent calculus.

In the $\lambda\mu$-calculus, we had ordinary sequents with only one distinguished formula on the right-hand side (the active formula). The $\lambda\mu\tilde{\mu}$-calculus symmetrizes things in order to have distinguished formulae on the right and on the left. We are then led to distinguish several sorts of sequents:

- sequents with a distinguished formula on the right, which are *sequents* properly speaking,
- sequents with a distinguished formula on the left, or *co-sequents.*

In the computational interpretation, an active formula on the right-hand side corresponds to a term to be built, while an active formula on the left corresponds to a co-term to be built (or a *context*). μ stays the operator which activates a right formula, but it is accompanied by another one, $\tilde{\mu}$, which serves to activate a left formula.

If a term represents a program (or a variable, or a function), like it is always the case in the λ-calculus, a co-term represents a context in which the program is executed.

Thus, from now on, we will have a mechanism for building contexts as complex as the one used for building programs! This mechanism includes *co-variables* as well as operations of *co-application* and *co-abstraction.*

We shall therefore consider the following rules, in a sequent presentation:

- two variants of the identity axiom, one for a left active formula, and one for a right active formula:

$$\Gamma; A \vdash A, \Delta \qquad \Gamma, A \vdash A; \Delta$$

- rules for implication:

$$\frac{\Gamma \vdash A; \Delta \quad \Gamma; B \vdash \Delta}{\Gamma; A \to B \vdash \Delta} [\to \ L] \qquad \frac{\Gamma, A \vdash B; \Delta}{\Gamma \vdash A \to B; \Delta} [\to \ R]$$

$$
\begin{array}{ll}
\quad ((\lambda x.\lambda y.((love\ y)\ x)\ \ \mu\alpha.\exists x.((individual\ x) \wedge [\alpha]x))\ \ \mu\alpha.\forall x.((individual\ x) \Rightarrow [\alpha]x)) & \\
\longrightarrow ((\lambda y.((love\ y)\ \mu\alpha.\exists x.((individual\ x) \wedge [\alpha]x)))\ \ (\mu\alpha.\forall x.((individual\ x) \Rightarrow [\alpha]x))) & (\beta) \\
\longrightarrow ((love\ (\mu\alpha.\forall x.((individual\ x) \Rightarrow [\alpha]x)))\ \ \mu\alpha.\exists x.((individual\ x) \wedge [\alpha]x)) & (\beta) \\
\longrightarrow (\mu\beta.\forall x.((individual\ x) \Rightarrow [\beta](love\ x))\ \ \mu\alpha.\exists x.((individual\ x) \wedge [\alpha]x)) & (\mu') \\
\longrightarrow \mu\beta.\forall x.((individual\ x) \Rightarrow ([\beta](love\ x)\ \ \mu\alpha.\exists x.((individual\ x) \wedge [\alpha]x))) & (\mu) \\
\longrightarrow \forall x.((individual\ x) \Rightarrow ((love\ x)\ \ \mu\alpha.\exists x.((individual\ x) \wedge [\alpha]x))) & (\sigma) \\
\longrightarrow \forall x.((individual\ x) \Rightarrow (\mu\alpha.\exists y.((individual\ y) \wedge [\alpha]((love\ x)\ y)))) & (\mu') \\
\longrightarrow \forall x.((individual\ x) \Rightarrow (\exists y.((individual\ y) \wedge ((love\ x)\ y)))) & (\sigma)
\end{array}
$$

$$
\begin{array}{ll}
\quad ((\lambda x.\lambda y.((love\ y)\ x)\ \ \mu\alpha.\exists x.((individual\ x) \wedge [\alpha]x))\ \ \mu\alpha.\forall x.((individual\ x) \Rightarrow [\alpha]x)) & \\
\longrightarrow ((\lambda y.((love\ y)\ \mu\alpha.\exists x.((individual\ x) \wedge [\alpha]x)))\ \ (\mu\alpha.\forall x.((individual\ x) \Rightarrow [\alpha]x))) & (\beta) \\
\longrightarrow ((love\ (\mu\alpha.\forall x.((individual\ x) \Rightarrow [\alpha]x)))\ \ \mu\alpha.\exists x.((individual\ x) \wedge [\alpha]x)) & (\beta) \\
\longrightarrow (\mu\beta.\forall x.((individual\ x) \Rightarrow [\beta](love\ x))\ \ \mu\alpha.\exists x.((individual\ x) \wedge [\alpha]x)) & (\mu') \\
\longrightarrow \mu\alpha.\exists y.((individual\ y) \wedge [\alpha](\mu\beta.\forall x.((individual\ x) \Rightarrow [\beta](love\ x))\ \ y)) & (\mu') \\
\longrightarrow \exists y.((individual\ y) \wedge (\mu\beta.\forall x.((individual\ x) \Rightarrow [\beta](love\ x))\ \ y)) & (\sigma) \\
\longrightarrow \exists y.((individual\ y) \wedge (\mu\beta.\forall x.((individual\ x) \Rightarrow [\beta]((love\ x)\ \ y)))) & (\mu) \\
\longrightarrow \exists y.((individual\ y) \wedge (\forall x.((individual\ x) \Rightarrow ((love\ x)\ \ y)))) & (\sigma)
\end{array}
$$

Fig. 10.4 Two evaluations for the sentence *Everyone loves someone*

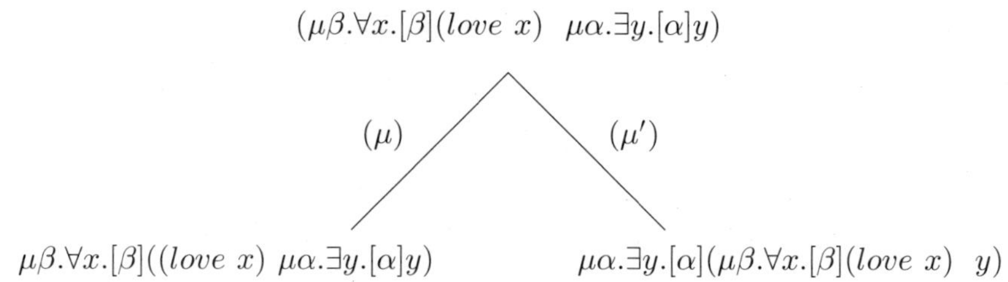

Fig. 10.5 A critical pair

- a cut rule:

$$\frac{\Gamma \vdash A; \Delta \quad \Gamma'; A \vdash \Delta'}{\Gamma, \Gamma' \vdash \Delta, \Delta'} [cut]$$

Thus, given a term of type A and a context of type B, it is possible to build a context of type $A \to B$, which is a particular context that when filled with the A-type term, gives a context of type B. The right rule gives, like it is the case in the λ-calculus, a term of type $A \to B$ by means of an assumption of type A, and which allows us to get a term of type B.

The cut rule makes sequents appear which have active formulae neither on the left nor on the right! Hence a third type of sequents, in order that we can finally distinguish between

- sequents, or right sequents;
- co-sequents, or left sequents;
- centred sequents.

Moreover, as soon as we get centred sequents in a proof, we no longer have rules to handle them! This compels us to introduce particular rules, called *focus* rules:

$$\frac{\Gamma, A \vdash \Delta}{\Gamma; A \vdash \Delta} [\tilde{\mu}] \qquad \frac{\Gamma \vdash A, \Delta}{\Gamma \vdash A; \Delta} [\mu]$$

The contraction rule remains to be expressed. Now, let us imagine a left sequent with some distinguished $A : \Gamma, A; A \vdash \Delta$. The fact that A is distinguished (or occurs in what Girard calls the *stoup*) will allow us to contract it. Let us make in effect the cut with a right axiom $\Gamma, A \vdash A; \Delta$ then there will come $\Gamma, A \vdash \Delta$, that is, the contraction will be done! The same is true, symmetrically, for the sequents $\Gamma \vdash A; A, \Delta$ and the left axiom $\Gamma; A \vdash A, \Delta$. The conclusion we draw from this is that the contraction rule only applies to formulae, an instance of which occurs in the stoup. The stoup therefore involves linearity constraints. As we shall see later on, the constructors build linear terms (values) and linear co-terms (environments).

The computational interpretation of the system, when limited to μ, $\tilde{\mu}$ and *cut*, is then provided by means of the following labelling:

$$\Gamma; \alpha : A \vdash \alpha : A, \Delta \qquad \Gamma, x : A \vdash x : A; \Delta$$

$$\frac{c : (\Gamma, x : A \vdash \Delta)}{\Gamma; \tilde{\mu}x.c : A \vdash \Delta}[\tilde{\mu}] \qquad \frac{c : (\Gamma \vdash \alpha : A, \Delta)}{\Gamma \vdash \mu\alpha.c : A; \Delta}[\mu]$$

$$\frac{\Gamma \vdash v : A; \Delta \quad \Gamma'; e : A \vdash \Delta'}{(v \bullet e) : (\Gamma, \Gamma' \vdash \Delta, \Delta')}[cut]$$

In the first axiom (on the left), a co-term is built by means of the co-variable α while in the second (on the right), a term is built by means of the variable x. The rule $[\tilde{\mu}]$ is interpreted as making A active on the left, storing the variable x which labelled this resource, and the rule $[\mu]$ as making A active on the right, storing the context α which labelled it. The cut rule is interpreted as evaluating the result of the application of the program v in the context e. We may incidentally notice that the notation $(v \bullet e)$ is used to represent the fact that v is put inside e, which is also sometimes written $e[v]$, but a symmetrical notation must be preferred.

Rules the conclusion of which is a right sequent build *values*, rules the conclusion of which is a left sequent build *contexts*. Centred sequents are interpreted as *commands* or *executions*.

- A value is either a variable or a value constructor.
- An evaluation context is either a co-variable or a context constructor.

In the $\mu\tilde{\mu}$-calculus (limited to axioms, cut and $[\mu]$ and $[\tilde{\mu}]$), we have the following array:

$$\begin{array}{lll} \textit{commands} & : c & ::= (v \bullet e) \\ \textit{values} & : V & ::= x \\ \textit{terms} & : \tau & ::= V \mid \mu\alpha.c \\ \textit{contexts} & : E & ::= \alpha \\ \textit{co-terms} & : e & ::= \tilde{\mu}x.c \mid E \end{array}$$

The reduction rules are

$$(\mu) \quad (\mu\alpha.c \bullet e) \longrightarrow c[\alpha ::= e]$$

$$(\tilde{\mu}) \quad (v \bullet \tilde{\mu}x.c) \longrightarrow c[x ::= v]$$

This precisely means that an execution of the term $\mu\alpha.c$ "against" the context e gives as a result the substitution of e in c everywhere the co-variable α occurs in c; it evaluates the subterms named by α in the context e. Similarly, an execution of the term v in the context $\tilde{\mu}x.c$ substitutes v for x in this context everywhere x occurs. It is simply an instruction *let-in*.

$$(v \bullet \tilde{\mu}x.c) = (let\ x = v\ in\ c)$$

As we can see, $\mu\alpha.c$ is a term waiting for a continuation, like $\tilde{\mu}x.c$ is a context waiting for a term. Curien and Herbelin show that this part of the system with only the operator μ is equivalent to the $\lambda\mu$-calculus in the call-by-name regime.

10.2.4 *Call-by-name vs call-by-value*

When faced with a critical pair, like in Fig. 10.5, a problem arises. It is what happens during the evaluation of the following expression:

$$(\mu\alpha.(y_1 \bullet \beta_1) \bullet \tilde{\mu}x.(y_2 \bullet \beta_2))$$

If we give priority to (μ), we get

$$(\mu\alpha.(y_1 \bullet \beta_1) \bullet \tilde{\mu}x.(y_2 \bullet \beta_2)) \longrightarrow (y_1 \bullet \beta_1)[\alpha ::= \tilde{\mu}x.(y_2 \bullet \beta_2)] = (y_1 \bullet \beta_1)$$

and if we give priority to $(\tilde{\mu})$, we get

$$(\mu\alpha.(y_1 \bullet \beta_1) \bullet \tilde{\mu}x.(y_2 \bullet \beta_2)) \longrightarrow (y_2 \bullet \beta_2)$$

The system is therefore non-deterministic. In order to solve this problem, we may try to "dissymmetrize" the calculus by distinguishing two strategies, which are precisely what we have named call-by-name and call-by-value.

- In the first case, $\tilde{\mu}$ is systematically preferred; this corresponds to call-by-name.
- In the second case, μ is systematically preferred; this corresponds to call-by-value.

We may see the role of $\tilde{\mu}$ in the implementation of the call-by-value strategy: to impose this strategy amounts to *freezing* the application of a function until its argument is evaluated. To enforce the calculus to adopt this strategy in evaluating $(M\ N)$ may be done by using the construction $let\ x = N\ in\ (M\ x)$. Let us notice that this construction exactly corresponds to $(N \bullet \tilde{\mu}x.(M\ x))$.

On the contrary, the call-by-name strategy may be seen as one which freezes the evaluation of the argument until it has been passed as a parameter to the function. This amounts to having another kind of *let-in*, let us denote it by *let′-in*, which, instead of operating on terms, operates on contexts with a syntax of the kind $let'\ \alpha = e\ in\ c$, but this is exactly what is given by μ. It is therefore possible to write, symmetrically with regard to $\tilde{\mu}$, $let'\ \alpha = e\ in\ c \equiv (\mu\alpha.c \bullet e)$. By using this construction, we enforce the environment to be evaluated before the argument is given to the function. The conflict case may be represented in *let-in* terms:

$$let'\ \alpha = (let\ x = [\ ]\ in\ (y_2 \bullet \beta_2))\ in\ (y_1 \bullet \beta_1)$$

versus

$$let\ x = (let'\ \alpha = [\ ]\ in\ (y_1 \bullet \beta_1))\ in\ (y_2 \bullet \beta_2)$$

If we give priority to the reduction of μ-terms (that is, of the *let′-in*), we get the second form, where $let'\ \alpha = [\ ]\ in\ (y_1 \bullet \beta_1)$ is first evaluated (x is not allowed to have as its value a μ-term), and we get call-by-value.

If we give priority to the reduction of the $\tilde{\mu}$-terms, that is, of the *let-in*, we get the first form, where $let\ x = [\ ]\ in\ (y_2 \bullet \beta_2)$ is first evaluated (α is not allowed to have as its co-value a $\tilde{\mu}$-term), and we get call-by-name.

This leads in fact to two different calculi, a CBN and a CBV calculus. In order to distinguish them, we introduce more notation:

$$\begin{array}{lll} commands & : c & ::= (\tau \bullet e) \\ constructors\ of\ values & : V_c & \\ terms & : \tau & ::= \mu\alpha.c \mid x \mid V_c \\ constructors\ of\ linear\ contexts & : E_c & \\ linear\ contexts & : E & ::= \alpha \mid E_c \\ contexts & : e & ::= \tilde{\mu}x.c \mid E \end{array}$$

and for CBN, the reduction rules are

$$(\mu_n) \quad (\mu\alpha.c \bullet E) \longrightarrow_n c[\alpha ::= E]$$

$$(\tilde{\mu}) \quad (\tau \bullet \tilde{\mu}x.c) \longrightarrow_n c[x ::= \tau]$$

This calculus can be proven confluent.

If we now apply the CBV strategy, things are alike, but symmetrically. We get

$$\begin{array}{lll} commands & : c & ::= (\tau \bullet e) \\ constructors\ of\ values & : V_c & \\ values & : V & ::= x \mid V_c \end{array}$$

$$\begin{array}{llll} \textit{terms} & : \tau & ::= \mu\alpha.c \mid V \\ \textit{constructors of linear contexts} & : E_c & \\ \textit{contexts} & : e & ::= \tilde{\mu}x.c \mid \alpha \mid E_c \end{array}$$

and for CBV, the reduction rules are

$$(\mu) \quad (\mu\alpha.c \bullet e) \longrightarrow_v c[\alpha ::= e]$$

$$(\tilde{\mu}_v) \quad (V \bullet \tilde{\mu}x.c) \longrightarrow_v c[x ::= V]$$

In other terms, in CBV, [*let* $x = \tau$ *in* c] reduces to c where x may only be replaced (by substitution) by values and not by terms waiting for their continuation.

10.2.5 *Back to implication*

10.2.5.1 *Rules and reductions*

The main interest of this system lies in its ability to include other logical constructors. We get then interesting tools for building terms and co-terms and analyzing their interactions. We already saw above the case of implication, which we recall here, with proper decoration by terms and co-terms:

$$\frac{\Gamma, x : A \vdash v : B; \Delta}{\Gamma \vdash \lambda x.v : A \to B; \Delta}[\to\ R] \qquad \frac{\Gamma \vdash v : A; \Delta \quad \Gamma; e : B \vdash \Delta}{\Gamma; v@e : A \to B \vdash \Delta}[\to\ L]$$

As said above, the rule $[\to\ R]$ is nothing else than the usual rule for forming a λ-term, by abstraction. The rule $[\to\ L]$ has nevertheless a new interpretation, which comes from the fact that the distinguished formulae in the premises are not on the same side. A term v and a context e are used for building a new context (and not a term): $v@e$.

The first rule is a value constructor, the second one a context constructor, therefore $\lambda x.v$ is a value (this term counts as a value when transmitted by a *let-in* in the call-by-value strategy), while $v@e$ is a co-value (it counts as a context when transmitted by a *let′-in* in the call-by-name strategy).

How to interpret this constructor "@"? We may say that $v@e$ is the context obtained by applying v, putting the result in the context e, something we may denote by $e[([\ ]\ v)]$.

The reduction rule associated with these two new constructions is

$$(\lambda x.v \bullet v'@e) \longrightarrow (v' \bullet \tilde{\mu}x.(v \bullet e))$$

or in other terms

$$e[(\lambda x.v\ v')] \longrightarrow let\ x = v'\ in\ e[v]$$

Another way of saying it is

> if v is of type A and if the hole of the context $e[\ ]$ is of type B, then the hole of the context $e[[\ ]v]$ is of type $A \to B$ (Curien and Herbelin, 2000).

With the implication, our "dictionary" of objects (commands, values, environments, etc.) is enriched. We are now able to assign expressions to value constructors and constructors of linear contexts:

$$\begin{array}{lll} commands & : c & ::= (\tau \bullet e) \\ value\ constructors & : V_c & ::= \lambda x.v \\ values & : V & ::= x \mid V_c \\ terms & : \tau & ::= \mu\alpha.c \mid x \mid V_c \\ constructors\ of\ linear\ contexts & : E_c & ::= v@e \\ linear\ contexts & : E & ::= \alpha \mid E_c \\ contexts & : e & ::= \tilde{\mu}x.c \mid E \end{array}$$

10.2.6 *Substraction*

The two previous rules have their symmetrical counterparts, which define a new connective, that we denote by "$-$" and name *substraction.*[3]

$$\frac{\Gamma; A \vdash B, \Delta}{\Gamma; A - B \vdash \Delta}[-\ L] \qquad \frac{\Gamma \vdash A; \Delta \quad \Gamma; B \vdash \Delta}{\Gamma \vdash A - B; \Delta}[-\ R]$$

We may still decorate its rules with appropriate terms and co-terms:

$$\frac{\Gamma; e : A \vdash \alpha : B, \Delta}{\Gamma; \tilde{\lambda}\alpha.e : A - B \vdash \Delta}[-\ L] \qquad \frac{\Gamma \vdash v : A; \Delta \quad \Gamma; e : B \vdash \Delta}{\Gamma \vdash v - e : A - B; \Delta}[-\ R]$$

The reduction rule is dual with regard to the one which concerns implication:

$$(v - e' \bullet \tilde{\lambda}\alpha.e) \longrightarrow (\mu\alpha.(v \bullet e) \bullet e')$$

[3]Substraction corresponds to $\oslash$ and $\otimes$, in the non-permutational variant of the calculus which was presented in Section 6.6.3.

This still allows us to enrich our dictionary:

$$
\begin{array}{lll}
\textit{commands} & : c & ::= (\tau \bullet e) \\
\textit{value constructors} & : V_c & ::= \lambda x.v | v - e \\
\textit{values} & : V & ::= x \mid V_c \\
\textit{terms} & : \tau & ::= \mu\alpha.c \mid x \mid V_c \\
\textit{constructors of linear contexts} & : E_c & ::= v@e|\tilde{\lambda}\alpha.e \\
\textit{linear contexts} & : E & ::= \alpha \mid E_c \\
\textit{contexts} & : e & ::= \tilde{\mu}x.c \mid E
\end{array}
$$

10.2.7 *Back to links between logic and duality*

10.2.7.1 *Duality*

To get CPS translations of terms and co-terms introduced by the $\lambda\mu\tilde{\mu}$-calculus by means of the rules of classical logic amounts to translating classical logic into intuitionistic logic. We will finally get λ-terms. There exist several known translations. They in fact correspond to various evaluation strategies.

It is for instance possible to show that if we use the following translation between types:

$$p^* = p$$

$$(A \to B)^* = A^* \to \neg\neg B^*$$

we get such an embedding.

Theorem 10.1. *If π is a proof of $\Gamma \vdash \alpha$ in classical implicational logic, then there exists a proof π^* of $\Gamma^* \vdash \neg\neg\alpha^*$ in intuitionistic implicational logic.*

To use this translation amounts, on the computational aspect, to translating a term of type A that we build on the right-hand side of a sequent into a term of type $\neg\neg A$, in other words into a function over a context (which is of type $A \to \bot$), where we retrieve all the intuitions behind the continuation calculus. This leads us to interpret every sequent with an active right formula as what we shall call a *computation*, that is, a higher-order term (taking a continuation as its argument) and therefore of the type of a double negation.

Generally speaking, we shall assume that we have always some kind of answer to be expected from our programs, which we decide to represent by $\bot$ (sometimes R, sometimes **t**, etc.). This kind of object represents the level of output of a program (the *top level*). A *continuation* of co-type A is a function of objects of type A (programs) towards this particular answer type. As said by Wadler (2003), it consumes values but returns nothing, so that, like we hypothesized for the $\lambda\mu$-calculus, a continuation of co-type A is in fact a *co-term* of type $A \to \bot$, or $\neg A$, or $V_A \to \bot$, where V_A explicitly denotes a *value* of type A.

In a call-by-value calculus, the value of $(v \bullet e)$ is obtained by applying v to e so that we get

$$(v \bullet e)^v = (v^v \; e^v)$$

where the superscript v means a particular translation (associated with call-by-value) into the λ-calculus. That is to say that the programs (terms) are viewed as functions of the continuations towards the distinguished type $\bot$, therefore as objects of type $(A \to \bot) \to \bot$, which is perfectly in agreement with the *-translation above. By introducing the variables k_α as continuation variables, we get, for variables and co-variables,

$$\begin{aligned} x^v &= \lambda k_\alpha (k_\alpha \; x) \\ \alpha^v &= k_\alpha \end{aligned}$$

Actually, this is what we get by means of the following proposition:

Proposition 10.1.

$$\begin{aligned} \Gamma \vdash v : A; \Delta &\Longrightarrow \Gamma^*, \neg\Delta^* \vdash v^* : \neg\neg A^* \\ \Gamma; e : A \vdash \Delta &\Longrightarrow \Gamma^*, \neg\Delta^* \vdash e^* : \neg A^* \\ c : (\Gamma \vdash \Delta) &\Longrightarrow \Gamma^*, \neg\Delta^* \vdash c^* : \bot \end{aligned}$$

In the axiom case, we get in effect

$$x : (x : A \vdash A) \; \textit{translated into} \; x^* : (x : A \vdash \neg\neg A)$$

according to the following proof:

$$\cfrac{\cfrac{A \vdash A \quad \bot \vdash \bot}{A, A \to \bot \vdash \bot}[\to L]}{A \vdash (A \to \bot) \to \bot}[\to R]$$

which may be decorated according to

$$\cfrac{\cfrac{x : A \vdash x : A \quad \bot \vdash (k\; x) : \bot}{x : A, k : A \to \bot \vdash (k\; x) : \bot}[\to L]}{x : A \vdash \lambda k.(k\; x) : (A \to \bot) \to \bot}[\to R]$$

showing thus that $x^* = \lambda k.(k\ x)$.

In a call-by-name calculus, symmetrically, contexts are obtained by means of values. For instance, in $(v \bullet e)$, v is put inside the context $e[\ ]$. We get

$$(v \bullet e)^n = e^n[v^n]$$

In this case, contexts are functions of values towards the distinguished type $\perp$, and their type is $A \to \perp$. We get for variables and co-variables

$$\begin{aligned} x^n &= x \\ \alpha^n &= \lambda x.(x\ k_\alpha) \end{aligned}$$

Let us look particularly at the functional objects of type $A \to B$. In call-by-value, the rule $[\to\ R]$ shows that we build a term (of type $((A \to B) \to \perp) \to \perp$ according to our translation) by means of a value (x of type A) giving a term v of type B (hence of type $(B \to \perp) \to \perp$ according to our translation). We shall say therefore that in call-by-value, *a value of type $A \to B$ is a function of a value of type A to a computation of type B.*

We write

$$V_{A \to B} = A \to ((B \to \perp) \to \perp)$$

The CPS translation of a term $\lambda x.v$ of type $A \to B$ is also determined by the above proposition. The rule $[\to\ R]$ becomes, in effect,

$$\frac{\Gamma^*, \neg\Delta^*, x : A \vdash v^* : \neg\neg B^*}{\Gamma^*, \neg\Delta^* \vdash (\lambda x.v)^* : ((A^* \to \neg\neg B^*) \to \perp) \to \perp}$$

and in order to determine $(\lambda x.v)^*$, we simply have to make the proof of this rule in intuitionistic logic and label it with λ-terms. It follows that

$$\frac{\dfrac{\dfrac{\Gamma^*, \neg\Delta^*, x : A \vdash v^* : \neg\neg B^*}{\Gamma^*, \neg\Delta^* \vdash \lambda x.v^* : A \to \neg\neg B^*} \quad k(\lambda x.v^*) : \perp \vdash k(\lambda x.v^*) : \perp}{\Gamma^*, \neg\Delta^*, k : (A \to \neg\neg B^*) \to \perp \vdash k(\lambda x.v^*) : \perp}}{\Gamma^*, \neg\Delta^* \vdash \lambda k.k(\lambda x.v^*) : ((A \to \neg\neg B^*) \to \perp) \to \perp}$$

and therefore

$$(\lambda x.v)^* = \lambda k.k(\lambda x.v^*)$$

Let us consider now the left rule for $\to$.

$$\frac{\Gamma \vdash v : A; \Delta \quad \Gamma; e : B \vdash \Delta}{\Gamma; v@e : A \to B \vdash \Delta}[\to\ L]$$

translates into

$$\frac{\Gamma^*, \neg\Delta^* \vdash v^* : \neg\neg A^* \quad \Gamma^*, \neg\Delta^* \vdash e^* : \neg B}{\Gamma^*, \neg\Delta^* \vdash (v@e)^* : \neg(A \to B)^*}$$

or

$$\frac{\Gamma^*, \neg\Delta^* \vdash v^* : \neg\neg A^* \quad \Gamma^*, \neg\Delta^* \vdash e^* : \neg B}{\Gamma^*, \neg\Delta^* \vdash (v@e)^* : \neg(A^* \to \neg\neg B^*)}$$

We make then the following proof:

$$\dfrac{\dfrac{\Gamma^*, \neg\Delta^* \vdash \neg\neg A^* \quad \dfrac{\Gamma^*, \neg\Delta^* \vdash \neg B^* \quad \Gamma^*, \neg\Delta^*, \bot \vdash \bot}{\Gamma^*, \neg\Delta^*, (B^* \to \bot) \to \bot \vdash \bot}}{\Gamma^*, \neg\Delta^*, \neg\neg A^* \to ((B^* \to \bot) \to \bot) \vdash \bot}}{\Gamma^*, \neg\Delta^* \vdash (\neg\neg A^* \to ((B^* \to \bot) \to \bot)) \to \bot}$$

which we may decorate as follows:

$$\dfrac{\dfrac{\Gamma^*, \neg\Delta^* \vdash v^* : \neg\neg A^* \quad \dfrac{\Gamma^*, \neg\Delta^* \vdash e^* : \neg B^* \quad \Gamma^*, \neg\Delta^*, \bot \vdash (y\ e^*) : \bot}{\Gamma^*, \neg\Delta^*, y : (B^* \to \bot) \to \bot \vdash (y\ e^*) : \bot}}{\Gamma^*, \neg\Delta^*, x : \neg\neg A^* \to ((B^* \to \bot) \to \bot) \vdash ((x\ v^*)\ e^*) : \bot}}{\Gamma^*, \neg\Delta^* \vdash \lambda x.((x\ v^*)\ e^*) : (\neg\neg A^* \to ((B^* \to \bot) \to \bot)) \to \bot}$$

We get the fact that the CPS transform in CBV of $v@e$ is $\lambda x.((x\ v^*)\ e^*)$ (to apply v to a hole and then to put the result in the context e).

We know that in CBV, the context requires that the argument be a value before being substituted for x; in other words, we crucially make the distinction between values (V) and non-values (terms v which are not $V's$). Similarly, in CBN, the term requires its environment to be particular. E is a co-variable α or a constructor of linear context E_c. Therefore in this case, the distinction between V and v no longer has much importance; it is the one existing between E and e which counts: it is necessary to distinguish the linear contexts from others.

In CBV, we distinguish V and v, then draw the following array for a value:

$$\begin{aligned} \alpha^v &= k_\alpha \\ V^v &= \lambda k.(k\ V^{vl}) \\ (v \bullet e)^v &= (v^v\ e^v) \end{aligned}$$

with

$$\begin{aligned} x^{vl} &= x \\ (\lambda x.v)^{vl} &= \lambda x.v^v \end{aligned}$$

Otherwise, in CBV, by definition, only $V@e$ has a meaning: it consists in putting a value V inside the environment e. It is in effect not allowed to put something other than a value in an environment before evaluating it. It is then possible to add to the array the following line:

$$(V@e)^v = \lambda x.((x\ V^{vl})\ e^v)$$

With these distinctions, Proposition 10.1 can be reformulated in a more fine-grained way in CBV:

Proposition 10.2.

$$\begin{aligned} \Gamma \vdash V : A; \Delta &\Longrightarrow \Gamma^v, \neg\Delta^v \vdash V^{vl} : A^v \\ \Gamma \vdash v : A; \Delta &\Longrightarrow \Gamma^v, \neg\Delta^v \vdash v^v : \neg\neg A^v \\ \Gamma; e : A \vdash \Delta &\Longrightarrow \Gamma^v, \neg\Delta^* \vdash e^v : \neg A^v \\ c : (\Gamma \vdash \Delta) &\Longrightarrow \Gamma^v, \neg\Delta^v \vdash c^v : \bot \end{aligned}$$

The above translation is nevertheless not unique. For example, by "decurryfying", we get $A \to ((B \to \bot) \to \bot) = (A \times (B \to \bot)) \to \bot$ and therefore

$$V_{A\to B} = R^{V_A \times K_B}$$

where K_B denotes a continuation of type B. Logically speaking, this amounts to defining

$$A \to B = \neg(A \wedge \neg B)$$

The translation of the implication is no longer the same. We have now

$$\begin{aligned} (A \to B)^v &= \neg(A^v \wedge \neg B^v) \\ (A - B)^v &= B^v \wedge \neg A^v \end{aligned}$$

For instance, to calculate the CPS transform of $\lambda x.V$, we are led to prove

$$\frac{\Gamma^v, \neg\Delta^v, x : A \vdash V^{vl} : B^v}{\Gamma^v, \neg\Delta^v \vdash (\lambda x.V)^v : \neg\neg\neg(A^v \wedge \neg B^v)}$$

We get

$$\dfrac{\dfrac{\dfrac{\dfrac{\Gamma^v, x : A^v \vdash V^{vl} : B^v \qquad (V^{vl}\ \alpha) : \bot \vdash (V^{vl}\ \alpha) : \bot}{\Gamma^v, x : A^v, \alpha : B^v \to \bot \vdash (V^{vl}\ \alpha) : \bot}}{\Gamma^v, (x, \alpha) : A^v \wedge (B^v \to \bot) \vdash (V^{vl}\ \alpha) : \bot}}{\Gamma^v \vdash \lambda(x, \alpha).(V^{vl}\ \alpha) : (A^v \wedge (B^v \to \bot)) \to \bot} \qquad \bot \vdash k(\lambda(x,\alpha).(\alpha\ V^{vl})) : \bot}{\dfrac{\Gamma^v, k : ((A^v \wedge (B^v \to \bot)) \to \bot) \to \bot \vdash k(\lambda(x, \alpha).(V^{vl}\ \alpha)) : \bot}{\Gamma^v \vdash \lambda k.k(\lambda(x, \alpha).(V^{vl}\ \alpha)) : (((A^v \wedge (B^v \to \bot)) \to \bot) \to \bot) \to \bot}}$$

which gives the CPS translation

$$(\lambda x.V)^v = \lambda k.k(\lambda(x, \alpha).(V^{vl}\ \alpha))$$

Let us consider now the left rule for implication:

$$\frac{\Gamma \vdash v : A; \Delta \quad \Gamma; e : B \vdash \Delta}{\Gamma; v@e : A \to B \vdash \Delta}\ [\to\ L]$$

We know that $K_{A\to B} = R^{R^{V_A \times R^{V_B}}}$, therefore we can calculate the CPS translation of $v@e$. Since it is a context, its value is the one of a continuation of type $K_{A\to B}$, where e^v is supposed to be a continuation of type B, hence belonging to K_B, or R^{V_B}. Let y be a variable of type A, the pair (y, e^v) of type $V_A \times R^{V_B}$, and if k is of type $R^{V_A \times R^{V_B}}$, $(k\ (y,\ e^v))$ is of type $\bot$, $\lambda y.(k\ (y,\ e^v))$ of type $A \to \bot$ and $(v^v\ \lambda y.(k\ (y\ e^v))$ of type $\bot$, and $\lambda k.(v^v\ \lambda y.(k\ (y,\ e^v))$ of type $R^{R^{V_A \times R^{V_B}}}$, that is, exactly of the type of $v@e$.

Hence the translation

$$(v@e)^v = \lambda k.(v^v\ \lambda y.(k\ (y,\ e^v))$$

For the translations of μ and $\tilde{\mu}$-terms, we get

$$\frac{e : (\Gamma, A \vdash \Delta; \beta : B)}{\mu\beta.(x \bullet e) : (\Gamma; x : A \vdash \Delta, B)} \qquad \frac{\tau : (\Gamma; x : A \vdash \Delta, B)}{\tilde{\mu}x.(\tau \bullet \beta) : (\Gamma, A \vdash \Delta; \beta : B)}$$

In the first proof, the premise is a co-sequent (which builds a context e) and the conclusion is a sequent, which builds a term by using the variable x. In order to do that, the variable x is put inside the given context e, and the formula with name B is activated in order to put the result in the right place. The value of the constructed term is therefore a computation of type B^* which uses a co-variable β, that is, a continuation of type B $(B \to \bot)$, and gives as an output a result $(\bot)$ obtained by putting x in the context e.

Therefore the CPS translation in the call-by-value regime of $\mu\beta.(x \bullet e)$ is $\lambda\beta.e[x]$. Similarly, in the second proof, we get the value of $\tilde{\mu}x.(\tau \bullet \beta)$ as $\lambda x.\beta[\tau]$.

In call-by-name, we have only to exchange the roles between variables and co-variables, between terms and contexts.

The symmetries of the $\lambda\mu\tilde{\mu}$-calculus are actually obvious. Take the four rules $[\rightarrow\ L]$, $[\rightarrow\ R]$, $[-\ L]$ and $[-\ R]$. Change the variables x into co-variables α and a sequent $\Gamma \vdash \Delta$ becomes its symmetric $\Delta \vdash \Gamma$. The rules $[\rightarrow\ R]$ and $[-\ L]$ are also exchanged, and therefore abstraction $(\lambda x.v)$ and co-abstraction $(\tilde{\lambda}\alpha.e)$. In the same way, still by left-right symmetry, values and co-values (contexts) are exchanged, and therefore the rules $[\rightarrow\ L]$ and $[-\ R]$, so that the application $v@e$ leads to the co-application $e - v$, and reciprocally. Let us denote by ^{o} this transform by symmetry (see Moortgat, 2007); what we just said simply expresses as

$$\begin{aligned} X^o &= X \\ (A \rightarrow B)^o &= B^o - A^o \\ (B - A)^o &= A^o \rightarrow B^o \end{aligned}$$

with the following exchanges:

$$\begin{aligned} x &\leftrightarrow \alpha \\ \mu\beta &\leftrightarrow \tilde{\mu}x \\ \lambda x &\leftrightarrow \tilde{\lambda}\alpha \end{aligned}$$

Since going from CBV to CBN amounts to exchanging roles, we can deduce from these symmetries the CBN translations from the CBV ones. This is made by embedding the dual calculus into intuitionistic logic, which amounts to defining a new translation by setting

$$(\phi)^n = (\phi^o)^*$$

We get then

$$\begin{aligned} X^n &= X \\ (A \rightarrow B)^n &= B^n \wedge \neg A^n \\ (B - A)^n &= \neg(A^n \wedge \neg B^n) \end{aligned}$$

where, as we may see, the translation is interpreted as a *negation* $((A \rightarrow B)^n = \neg B \wedge \neg\neg A = A \wedge \neg B = \neg(A \rightarrow B))$.

Finally, in call-by-name, the implication is defined by

$$A \rightarrow B = \neg A \vee B$$

We may also write

$$(A \to B) \to \bot = (B \to \bot) \wedge ((A \to \bot) \to \bot)$$

When translated into computational terms, this reads: *a continuation of type $A \to B$ is obtained by a continuation of type B and a computation of type A*. That is,

$$K_{A \to B} = C_A \times K_B$$

or, by taking the negation

$$C_{A \to B} = (C_A \times K_B) \to \bot$$

which also gives

$$C_{A \to B} = C_A \to C_B$$

That means that, in call-by-name, the strategy does not take values as inputs, but computations. We have the following theorem.

Theorem 10.2. *If π is a proof of $\Gamma \vdash \alpha$ in the classical implicational logic, then there is a proof π^n of $\Gamma^n \vdash \neg\alpha^n$ in the intuitionistic implicational fragment.*

Otherwise, $C_A = R^{K_A}$ and therefore

$$K_{A \to B} = R^{K_A} \times K_B$$

but we have also

$$K_{A \to B} = R^{R^{V_A} \times R^{V_B}}$$

It is then easy to get by symmetry the interpretation of the substraction:

$$V_{A-B} = R^{V_B} \times V_A$$

$$K_{A-B} = R^{K_A \times C_B}$$

which, from a logical point of view, means that

$$A - B = A \wedge \neg B$$

$$\neg(A - B) = \neg(\neg A \wedge \neg\neg B)$$

or, in other terms, in call-by-value, $A - B$ must be interpreted as the logical difference $A \wedge \neg B$, and in call-by-name as $\neg A \wedge \neg\neg B$.

If we now seek the CPS translation in call-by-name of $v@e$, we shall use the translation which transforms $A \to B$ into $\neg A^n \wedge B^n$ (a computation of type A and a continuation of type B). In this case, we have the following proposition:

Proposition 10.3.

$$\Gamma \vdash v : A; \Delta \Longrightarrow \neg\Gamma^n, \Delta^n \vdash v^v : \neg A^n$$

$$\Gamma; e : A \vdash \Delta \Longrightarrow \neg\Gamma^n, \Delta^n \vdash e^n : \neg\neg A^n$$

$$c : (\Gamma \vdash \Delta) \Longrightarrow \neg\Gamma^n, \Delta^n \vdash c^n : \bot$$

so that the left rule for implication translates into

$$\frac{\neg\Gamma^n, \Delta^n \vdash v^n : \neg A^n \quad \neg\Gamma^n, \Delta^n \vdash e^n : \neg\neg B^n}{\neg\Gamma^n, \Delta^n \vdash (v@e)^n : \neg\neg(A \to B)^n}$$

or

$$\frac{\neg\Gamma^n, \Delta^n \vdash v^n : \neg A^n \quad \neg\Gamma^n, \Delta^n \vdash e^n : \neg\neg B^n}{\neg\Gamma^n, \Delta^n \vdash (v@e)^n : \neg\neg(\neg A^n \wedge B^n)}$$

We get then, as the call-by-name CPS translation of $v@e$, $\lambda k.(\lambda y.k\ (v^n, y)\ e^n)$.

Finally, we get the following array of CPS translations:

$\lambda\mu\tilde{\mu}$-terms	Call-by-value	Call-by-name
α	α	$\lambda x.(x\ \alpha)$
x	$\lambda k.(k\ x)$	x
$\lambda x.v$	$\lambda k.k(\lambda(c,\alpha).(v^v\ k_\alpha))$	$\lambda(x,k).(v^n\ k)$
$v@e$	$\lambda k.(v^v\ \lambda x.k(x, e^v))$	$\lambda k.(\lambda y.k\ (v^n, y)\ e^n)$
$\tilde{\lambda}\alpha.e$	$\lambda(x,\alpha).(e^v\ x)$	$\lambda k.k(\lambda(c,\alpha).(\alpha\ e^n))$
$e - v$	$\lambda k.(v^v\ \lambda x.k(x, e^v))$	$\lambda k.(\lambda\alpha.k(v^n,\alpha)\ e^n)$
$\mu\alpha.c$	$\lambda\alpha.c^v$	$\lambda\alpha.c^n$
$\tilde{\mu}x.c$	$\lambda x.c^v$	$\lambda x.c^n$
$v \bullet e$	$(v^v\ e^v)$	$(e^n\ v^n)$

This array makes the symmetries of the $\lambda\mu\tilde{\mu}$-calculus clear.

To sum up, the CPS translations in call-by-name and call-by-value are dual to each other (see particularly for that Wadler, 2003). This duality is based on the one which exists between the two translations of classical logic towards intuitionistic logic that we saw, call-by-value being based on the equivalence

$$A \to B \equiv \neg(A \wedge \neg B)$$

and call-by-name on

$$A \to B \equiv \neg A \vee B$$

The first of these translations leads, in the literature, to the works by Plotkin and Fisher, the second to those of Lafont, Hofmann and Streicher.

It is also worth recalling that, in call-by-value, we have

$$V_{A \to B} = V_A \to C_B$$

and in call-by-name

$$C_{A \to B} = C_A \to C_B$$

In other words, call-by-value transforms values into computations and call-by-name computations into computations. Moreover, we have, in call-by-name,

$$K_{A \to B} = K_B \times C_A$$

that is, in call-by-name, a continuation of type $A \to B$ is provided by a pair of a computation of type A and a continuation of type B, while in call by value, a value of type $A \to B$ is a function of a value of type A towards a computation of type B.

10.2.8 *From the $\lambda\mu$-calculus to the $\lambda\mu\tilde{\mu}$-calculus*

Curien and Herbelin gave a translation of the $\lambda\mu$-calculus into the $\lambda\mu\tilde{\mu}$-calculus:

$$\begin{aligned} x^\dagger &= x \\ (M\ N)^\dagger &= \mu\alpha.(N^\dagger \bullet \tilde{\mu}x.(M^\dagger \bullet x@\alpha)) \\ (\lambda x.M)^\dagger &= \lambda x.M^\dagger \\ ([\alpha]M)^\dagger &= (M^\dagger \bullet \alpha) \\ (\mu\beta.c)^\dagger &= \mu\beta.c^\dagger \end{aligned}$$

A judgement $\Gamma \vdash M : A; \Delta$ of the $\lambda\mu$-calculus translates then into a judgement of the $\lambda\mu\tilde{\mu}$-calculus: $\Gamma \vdash M^\dagger : A; \Delta$.

10.2.9 *A small linguistic example*

Let us take, as an elementary example, the analysis of the sentence *Alice left*, concentrating only on the semantic part of the analysis (leaving aside for the time being the use of the Lambek–Grishin calculus presented in

Subsection 6.6.3). Let us assume the following typing:

$$\begin{array}{lcl} \text{alice} & := & a : np \\ \text{left} & := & l : np \to s \end{array}$$

The goal is to build a term associated with the sequent

$$a : np \ \circ \ l : np \to s \vdash \ ? : s$$

The proof in the intuitionistic linear logic is the following:

$$\frac{np \vdash np \quad s \vdash s}{np \ \circ \ np \to s \vdash s}\,[\to L]$$

Nevertheless, in our calculus, things do not exactly go like this. The rule for $\to$ has premises, one of which has an active term, and the other an active co-term. The active formulae are those without labels. We therefore get

$$\frac{a : (a : np \vdash np) \quad \alpha : (s \vdash \alpha : s)}{a@\alpha : (a : np \ \circ \ np \to s \vdash \alpha : s)}\,[\to L]$$

But this builds a co-term, and not a term! To build a term, we have to go from an active co-term to an active term. The proof then goes on by a new step:

$$\frac{a@\alpha : (a : np \ \circ \ np \to s \vdash \alpha : s)}{\mu\alpha.(l \bullet (a@\alpha)) : (a : np \ \circ \ l : np \to s \vdash s)}\,[switch]$$

Finally, the result term is $\mu\alpha.(l\bullet(a@\alpha))$, which is not easily decodable. That is why we proceed to a translation into λ-terms. We have two strategies at our disposal: call-by-value and call-by-name.

The translation in call-by-value is such that

$$\mu\alpha.(l \bullet (a@\alpha))^* = \lambda k_\alpha.((a@k_\alpha)^* \ l)$$

but, as we have seen,

$$(a@k_\alpha)^* = \lambda k.(a^* \ \lambda x.(k \ \langle x, k_\alpha\rangle))$$

We get therefore

$$\begin{aligned} \mu\alpha.(l \bullet (a@\alpha))^* &= \lambda k_\alpha.(a^* \ \lambda x.(l \ \langle x, k_\alpha\rangle)) \\ &= \lambda k_\alpha.(\lambda k.(k \ a) \ \lambda x.(l \ \langle x, k_\alpha\rangle)) \\ &= \lambda k_\alpha.((\lambda x.(l \ \langle x, k_\alpha\rangle))a) \\ &= \lambda k_\alpha.((l \ \langle a, k_\alpha\rangle)) \end{aligned}$$

We observe then that in call-by-value, contexts are substituted in the first place in order to get computations from values. In the call-by-name regime, things go differently: the terms are substituted in the first place in order to get computations from continuations. A co-term is therefore built in place of a term. We may make another proof, or we may dualize the term associated with the proof in call-by-value, according to the symmetries put in evidence above, and we get

$$\begin{aligned}(\mu\alpha.(l \bullet (a@\alpha))^\circ &= \tilde{\mu}x.((a@\alpha)^\circ \bullet l^\circ)\\ &= \tilde{\mu}x.((\alpha^\circ - a^\circ) \bullet l^\circ)\\ &= \tilde{\mu}x.((x - a^\circ) \bullet l^\circ)\end{aligned}$$

that we may transform according to

$$\begin{aligned}(\tilde{\mu}x.((x - a^\circ) \bullet l^\circ))^* &= \lambda x.((x - a^\circ)^*\ l^\circ)\\ &= \lambda x.(\lambda k.(x^*\ \lambda z.(k\ \langle z, a^\circ\rangle))\ l^\circ)\\ &= \lambda x.(x^*\ \lambda z.(l^\circ\ \langle z, a^\circ\rangle))\\ &= \lambda x.((\lambda k.(k\ x))\ \lambda z.(l^\circ\ \langle z, a^\circ\rangle))\\ &= \lambda x.((\lambda z.(l^\circ\ \langle z, a^\circ\rangle))\ x)\\ &= \lambda x.((l^\circ\ \langle x, a^\circ\rangle))\end{aligned}$$

But the lexical entries must themselves be adapted to these two regimes. For instance, the term associated with *left* must be taken according to its translation in call-by-name or in call-by-value; the same for the constant a (*Alice*). In call-by-value, since *left* has a functional type (from np to s), its translation is the one of an implication $A \to B$ like we have already seen, that is, it takes as its argument a pair formed by a value np and a continuation s to give a result $\perp$. The semantic representation is therefore $\lambda(x,k).(k\ (left\ x))$. Similarly, always in call-by-value, the constant *alice* stays *alice*. Therefore the semantic representation of the sentence *Alice left* becomes

$$\lambda k_\alpha.(\lambda(x,k).(k\ (left\ x))\ \langle alice, k_\alpha\rangle)$$

which, by β reduction, gives

$$\lambda k.(k\ (left\ alice))$$

which, applied to the trivial continuation (the identity function), gives

$$(left\ alice)$$

In call-by-name, the translation of *Alice* is $\lambda k'.(k'\ alice)$, and the one of the function associated with *left* is a function from a pair formed by a continuation of type np and a computation of type s to $\perp$, that gives

$$\lambda(k,q).(q\ \lambda x.(k\ (left\ x)))$$

The semantic representation of the sentence becomes then

$$\lambda x.((\lambda(k,q).(q\ \lambda x.(k\ (left\ x)))\ \langle x, \lambda k'.(k'\ alice)\rangle))$$

which gives, still by β-reduction,

$$\lambda k.(k\ (left\ alice))$$

Finally, in a simple sentence both translations coincide. It is left to the reader to show that this is not the case for a sentence with multiple quantifiers.

10.3 Concluding Remarks and Further Works

This chapter presented some calculi which deserve attention for linguistic purposes. As already announced in the introduction, we may expect that ambiguity in language results from various ways of computing meaning. These ways of computing rely on *strategies*, not in the game-theoretic sense, but in the sense given to this word by computing theory. Linguistic expressions are evaluated exactly like programs. In the course of this evaluation, bifurcation points may occur, concerning critical pairs. Intuitionistic logic is oriented towards a deterministic analysis, but classical logic, because of its inherent non-determinism, is suitable for representing bifurcations of meaning. We have not enough space here to go further in the analysis allowed by these calculi. We may notice that some points are still not addressed. For instance, there is a point according to which the phenomenon of ambiguity is in fact limited to some constructions. There is no ambiguity for instance in a sentence like *Each boy saw a teacher* since it is clear that there is necessarily a teacher for each boy, and not one and the same teacher for all. *Each* is "distributive" as compared with *every*, and therefore, the sentence has only one reading. In other terms, sometimes, the two strategies we defined above are not equiprobable. It may happen that only one may be used, according to the context and to the words which are used.

Another issue is that the preceding views only concern top-level phenomena, that is, phenomena of interpretation involving the whole sentence.

Moreover, it would be worth exploring the relationship between this continuation semantics and the other frameworks we presented earlier in the book, like Cooper storage, Jacobson's variable-free semantics and the notions of grammatical reasoning.

Much has already been done on this by Shan (2004, 2008). He showed in particular how to solve the first issues by means of operators of *delimited control*, that is, operators which explicitly manipulate continuations, like SHIFT and RESET. For instance, the semantic representation of *Somebody saw everybody* is supposed to be, before evaluation,

$$\texttt{reset}((\texttt{shift}\ f.\exists x.fx)^{<}(saw\ ^{>}(\texttt{shift}\ g.\forall y.gy)))$$

where $^{<}$ and $^{>}$ indicate backward and forward application. As stated by Shan (2008), "the occurrences of `shift` in this translation arise from the lexical entries for the quantifiers *somebody* and *everybody*, whereas the occurrence of `reset` is freely available at every clause boundary". The program so obtained may then be evaluated, and the evaluation gives the following sequence of steps:

$$\rightarrow \texttt{reset}(\exists x.(\lambda x.\texttt{reset}(x^{<}(saw\ ^{>}(\texttt{shift}\ g.\forall y.gy))))x) \tag{10.2}$$

$$\rightarrow \texttt{reset}(\exists x.\texttt{reset}(x^{<}(saw\ ^{>}(\texttt{shift}\ g.\forall y.gy)))) \tag{10.3}$$

$$\rightarrow \texttt{reset}(\exists x.\texttt{reset}(\forall y.(\lambda y.\texttt{reset}(x^{<}(saw\ ^{>}y)))y)) \tag{10.4}$$

$$\rightarrow \texttt{reset}(\exists x.\texttt{reset}(\forall y.\texttt{reset}(x^{<}(saw\ ^{>}y)))) \tag{10.5}$$

$$\rightarrow \texttt{reset}(\exists x.\texttt{reset}(\forall y.(x^{<}(saw\ ^{>}y)))) \tag{10.6}$$

$$\rightarrow \texttt{reset}(\exists x.\forall y.(x^{<}(saw\ ^{>}y))) \tag{10.7}$$

$$\rightarrow \exists x.\forall y.(x^{<}(saw\ ^{>}y)) \tag{10.8}$$

`reset` is defined as an operator which does not perform any action by itself:

$$C[\texttt{reset}\ V] \rightarrow C[V] \tag{10.9}$$

`shift` removes the surrounding context *up to the nearest enclosing* `reset` into a variable. For instance, $1 + \texttt{reset}(2.\texttt{shift}\ f.(3.f(f(5))))$ (the example is due to Shan) will result in

$$\rightarrow 1 + \texttt{reset}(3.(\lambda x.\texttt{reset}(2.x))((\lambda x.\texttt{reset}(2.x))(5)))$$

$$\rightarrow 1 + \texttt{reset}(3.(\lambda x.\texttt{reset}(2.x))(\texttt{reset}(2.5)))$$

$$\rightarrow 1 + \texttt{reset}(3.(\lambda x.\texttt{reset}(2.x))(\texttt{reset}10))$$

$$\rightarrow 1 + \texttt{reset}(3.(\lambda x.\texttt{reset}(2.x))\ 10)$$

$$\rightarrow 1 + \texttt{reset}(3.\texttt{reset}(2.10))$$

$$\rightarrow 1 + \texttt{reset}(3.\texttt{reset}\ 20)$$
$$\rightarrow 1 + \texttt{reset}(3.20)$$
$$\rightarrow 1 + \texttt{reset}60$$
$$\rightarrow 1 + 60 \rightarrow 61$$

Therefore we have, as a formal definition of `shift`,

$$C[\texttt{reset}\ D[\texttt{shift}\ f.E]] \rightarrow C[\texttt{reset}\ E\{[f := \lambda x.\texttt{reset}\ D[x]\}] \qquad (10.10)$$

Operators of delimited control have also been introduced by De Groote (2006) to formalize Kamp's discursive representation theory (DRT).

Chapter 11

Proofs as Meanings

11.1 From Intuitionistic Logic to Constructive Type Theory

11.1.1 *Proof processes and proof objects*

One of the main advances in the conception of meanings as (sets of) proofs has been provided by Martin-Löf's (1984) type theory, which has been used by Ranta (1994) for a linguistic purpose. Ranta has made two kinds of presentation of his views concerning meaning. His first conception, which appeared in his seminal book *Type-Theoretical Grammar*, is that logical structures are abstractly built, using formation and equality rules starting from *primitive meanings*, like propositional functions associated with verbs, nouns and other categories, and then these structures are *sugared* into concrete sequences of words respecting the syntax of the language at hand. Since 1998, Ranta has slightly changed his viewpoint and found it more convenient (in particular for computer applications like *parsing*) to develop a level of abstract syntax with two functions, one going to logical structures and the other to concrete syntax (Ranta, 2004). The first one is an *interpretation* function and the second one is called a *linearization* function. This three-layered model is more in accordance with other frameworks currently studied like Muskens's (2003) lambda grammars and De Groote *et al.*'s abstract categorial grammars (De Groote, 2001a). The abstract objects are actually *trees* which may contain various indications about the operators applied to abstract units in order to build sentences.

Ranta (2004) shows that Montague grammar may be totally embedded in type theory.[1] We shall have a look at this embedding in the next section,[2] but our main concern here is to ask how a proof-theoretical conception of meaning is possible.

Ranta (1994, Chapter 3) studies what he calls *formalizations* of English sentences (even if no mechanical rules are given) by means of logical operators. He more specifically studies sentences which may be formalized as propositions of Σ and Π forms. Σ is the operator which translates in English into words like *some*, *a(n)*, *a certain*, and Π the operator which translates into *every*, *each*.... When applied to a set A, which is the domain of some common noun, they provide quantifier phrases like *every man* or *a man*. $\Sigma(A)$ (resp. $\Pi(A)$) is in fact better written $(\Sigma x : A)$ (resp. $(\Pi x : A)$). Such a notation better evokes the usual ones like those used in standard mathematics ($(\exists x \in A)$ or $(\forall x \in A)$).

With this notation, *a man walks* is formalized by

$$(\Sigma x : man)(x\ walks) \tag{11.1}$$

where, necessarily, *man* denotes a set and $(x\ walks)$ denotes a proposition, provided that x belongs to the previous set.

It is worth noticing that saying "*man* denotes a set" does not mean that *man* denotes the set of *men* as it is the case in usual denotational theories. It is better here to interpret such a claim as "we have a set of *proofs* for *man*". These "proofs" may be simply seen as *methods* for finding a man.[3] Briefly speaking, an item of the domain of *man* is any way of referring to a man (by pointing it out, by describing it, by calling it by his name and so on) and not a "real" man.

How can such a proposition be obtained? We have in fact *formation rules*, that is, rules which are employed to explicitly say how to introduce logical operators in order to get a new proposition. The first rules that we will see are the ΣF and ΠF rules. They are similar, and we shall summarize

[1] Nevertheless, the framework does not have unlimited "quantifying in".

[2] The conception of syntax internal to this theory will not be discussed.

[3] Ranta (private communication) prefers to say that canonical proofs are... men themselves, whereas "methods for finding" are non-canonical ones. From our viewpoint, we prefer to stay in a less realistic position. Proofs initiate in our perception, and what we interpret as "men" are perceptive stimulations coming from "real" men of course, but their roots are mental.

them by the following meta-rule, where Q is either Σ or Π:

$$\frac{A : set \qquad \begin{array}{c}[x : A] \\ \vdots \\ B(x) : set\end{array}}{(Qx : A)B(x) : set}\ QF$$

The notation $[x : A]$, like in natural deduction, represents a hypothesis. A ground proposition is obtained when all hypotheses have been discharged. $(x\ walks)$ as said above is a proposition provided that x is given in some domain and corresponds effectively to an entity which may walk. $walks$ is therefore seen as a propositional function which maps any proof of an element of a domain A (that is, any way to point towards it) onto a proof of the fact that this element $walks$. It is therefore not a function from a set I of individuals to the set of truth values (as usual in denotational frames), but a function *from a set of proofs to a set of proofs*. We may give the entry for *walks* as

$$\begin{array}{c}[x : animate] \\ \vdots \\ (x : walks) : prop\end{array}$$

A more convenient notation is the linearization of this vertical scheme:

$$(x\ walks) : prop \quad (x : animate) \tag{11.2}$$

where what occurs on the right of the resulting type *prop* (here $(x : animate)$) is a pre-requisite for the judgement expressed by the assignment of the type *prop* to the expression $(x\ walks)$.

We may now assume an ontology of types and subtypes according to which any entity explicitly given in discourse which has the feature *animate* may discharge the assumption $(x : animate)$ (that is, it may be an argument for the functional proposition associated with *walks*). The formation of *a man walks* is then explained by the following proof:

$$\frac{man : set \qquad \dfrac{\begin{array}{c}[x : animate] \\ \vdots \\ (x\ walks) : prop\end{array} \qquad [x : man]}{(x\ walks) : prop}}{(\Sigma x : man)(x\ walks) : prop}\ \Sigma F$$

where the hypothesis $[x : animate]$ is replaced by the hypothesis $[x : man]$, and then discharged. This is obviously itself a proof! The important feature of this framework is that starting from objects (proofs) which inhabit our domains, we develop other proofs which may be dealt with in the same way as our previous objects. The only thing to do is to pass from the notion of *proofs as processes* to the one of *proofs as objects*. As soon as we have considered our previous proof as an object z we may make the following reasoning: we know that a proof of a proposition $(\Sigma x : A)B(x)$ is necessarily given by a pair (a, b) where a is a "proof" of A (that is, some element of it) and b a proof of $B(a)$. This may be applied to the previous proof. By taking p as the first projection (in the space of proofs of such a formula) and q as the second one, we get $p(z)$ as "the man" we are able to point to as a witness of the fact that there is a man who walks, and $q(z)$ as the proof that he walks.

That results in the proof in Fig. 11.1 (reproduced from Ranta, 1994, p. 66), where the subproof on the left-hand side is retrieved as an object z in the subproof on the right-hand side. To formalize

Example 11.1. If a man walks he whistles

we form the implicatum $(p(z)\ whistles)$ in the context of a variable proof of the implicans, $z : (\Sigma x : man)(x\ walks)$, and combine the two by Π formation into

$$(\Pi z : (\Sigma x : man)(x\ walks))(p(z)\ whistles) : prop \tag{11.3}$$

This approach is very fruitful because, thanks to it, there is no difficulty to take into account sentences which pose many problems to the classical analyses, like so-called *Geach sentences*, also called *donkey sentences*, like

Example 11.2. If a farmer owns a donkey, he beats it

since the context provides the proof z of the first sentence, which may be used for giving the references of the pronouns (as respectively $p(z)$ and $p(q(z))$).

11.1.2 *Judgements*

Martin-Löf's type theory takes judgements as primary forms of expression. At first glance, it distinguishes between several primitive types, like *Set* and *Prop*, but we shall see soon that they can be equated if a proposition is seen

$$
\dfrac{
  \dfrac{
    man : set
    \qquad
    \dfrac{
      \begin{array}{c}(x : man)\\ \vdots \\ (x\ walks) : prop\end{array}
      \quad x : man^{1}
    }{
      (x\ walks) : prop
    }
  }{
    (\Sigma x : man)(x\ walks) : prop
  }\Sigma F, 1
  \qquad
  \dfrac{
    \begin{array}{c}(x : man)\\ \vdots \\ (x\ whistles) : prop\end{array}
    \quad
    \dfrac{
      z : (\Sigma x : man)(x\ walks)^{2}
    }{
      p(z) : man
    }\Sigma E
  }{
    (p(z)\ whistles) : prop
  }
}{
  (\Pi z : (\Sigma x : man)(x\ walks))(p(z)\ whistles) : prop
}\Pi F, 2
$$

Fig. 11.1 From proof as process to proof as object

as the set of its proofs. Starting from the *Set* type, we may have distinct judgements:

$A : Set$	for	"A is of type *Set*"
$a : A$	for	"a is a member of A" (assuming the previous judgement)
$a = b : A$	for	"a and b are equal members of the set A provided A is a set"

and there are of course rules for forming such judgements.

When using the type *Prop*, we also have judgements such as

$A : Prop$	for	"A is of type *Prop*"
$A\ true$	for	"A is a true proposition"

and there are also rules for forming such judgements: they are simply the rules of intuitionistic logic. With the *propositions as types* principle (see Chapter 5), propositions and sets are identified and propositions are interpreted along the following lines (due to Heyting):

proposition	is proved by
$\perp$	no proof
$A \wedge B$	a pair (a, b), with a a proof of A and b a proof of B
$A \vee B$	either a proof of A or a proof of B
$A \Rightarrow B$	a function which associates a proof of B with any proof of A
$\neg A$	a function from the set of proofs of A to the empty set
$(\forall x : A)B(x)$	a function which maps any $a \in A$ onto a proof of $B(a)$
$(\exists x : A)B(x)$	a pair (a, b) where $a \in A$ and b is a proof of $B(a)$

What is striking is of course that an existential judgement appears as a generalization of a conjunction, and a universal one as a generalization of an implication. Assuming this identification, it is easy to give rules for Σ (formation and introduction):

$$\frac{A : set \quad \begin{array}{c}(x : A)\\ B(x) : set\end{array}}{(\Sigma x : A)B(x) : set} \qquad \frac{a : A \quad b : B(a)}{(a, b) : (\Sigma x : A)B(x)}$$

We give also rules for elimination:

$$\frac{c : (\Sigma x : A)B(x)}{p(c) : A} \qquad \frac{c : (\Sigma x : A)B(x)}{q(c) : B(p(c))}$$

and for equality:

$$\frac{a : A \quad b : B(a)}{p((a,b)) = a : A} \qquad \frac{a : A \quad b : B(a)}{q((a,b)) = b : B(a)}$$

We similarly get rules for Π:
Formation and introduction:

$$\frac{A : set \;\; \begin{matrix}(x : A)\\ B(x) : set\end{matrix}}{(\Pi x : A)B(x) : set} \qquad \frac{\begin{matrix}(x : A)\\ b(x) : B(x)\end{matrix}}{\lambda(x \to b(x)) : (\Pi x : A)B(x)}$$

We give also a rule for elimination:

$$\frac{c : (\Pi x : A)B(x) \quad a : A}{app(c,a) : B(a)}$$

and for equality:

$$\frac{\begin{matrix}(x : A)\\ b(x) : B(x)\end{matrix} \;\; a : A}{app(\lambda(x \to b(x)), a) = b(a) : B(a)}$$

These rules are natural. The notation $\lambda(x \to b(x))$ is just a variant of the usual notation $\lambda x.b(x)$; it puts the emphasis on the fact that λ is simply a symbol which indicates that we handle a function which maps x on $b(x)$.

We don't go further into details, simply remaining for a while on the discussion Ranta introduces concerning "non-mathematical propositions". It is true that type theory (like it was for the case for Frege's *Begriffsschrift*) was built for dealing only with mathematical propositions. Its application to ordinary language leads to some puzzles. There is no problem in saying that the proposition

There is a prime number between 212 and 222

is represented by the set of its proofs, namely the list of prime numbers between these two numbers. But it is of course not the same for an ordinary existential sentence like

There is a railway from Moscow to Vladivostok

except if we accept to answer with the set of railways between these two towns, but that consists in a (perhaps premature) jump outside the mental processes. What kind of proof has a sentence like

Noam very well knows an African language

It depends of course on what we mean by "knowing" here, and even on what we mean by "African language". In cases like this, we are not confronted with proofs properly speaking but with ways of *arguing*, which makes a fundamental difference. If a proof is a finite object, an argumentation is a potentially infinite one, to the extent that a discussion may continue indefinitely on what we mean by some judgements in particular cases. We have only to remember hard discussions in scientific fields to notice that.[4]

Ranta proposes several solutions to this kind of problem:

- one is to use types instead of sets, in which case, he says, "exhaustive introduction rules are not needed", but it does not solve the problem that in ordinary language, we handle possibly infinite objects (objects which are defined by a potentially infinite list of properties);
- a second solution consists in developing techniques of approximation;
- and a third one consists in limiting our pretensions to delimited models of language use, like it is the case in many applications concerning AI.

In fact the solution we shall tentatively suggest in the next chapter is concerned with the second one, but by totally assuming non-finite objects as potential argumentations.

Another problem pointed out by Ranta is the *modification* problem. Let us assume we wish to represent an expression like

$$man\ who\ walks$$

The usual representation is

$$(\Sigma x : man)(x\ walks)$$

which means the set of pairs of a man and a proof that he walks.

This makes use of the function represented by $(x\ walks)$. It is a *propositional* function, that is, something we usually represent by

$$\begin{gathered}(x : man) \\ (x : walks) : prop\end{gathered}$$

Suppose we have now the sentence

$$Every\ man\ walks$$

[4]Such points will be addressed in the next chapter.

we get

$$(\Pi x : man)(x\ walks)$$

but if we have

Every old man walks

we get

$$(\Pi z : ((\Sigma x : man)(x\ is\ old))(z\ walks))$$

where the function (x *walks*) is defined in the first case on the set of *men* and in the second case on the set of *men who are old*, while we would like to have a more polymorphic function. Such a problem is solved by the use of *projections*. We have for *Every old man walks*

$$(\Pi z : (\Sigma x : man)(x\ is\ old))(p(z)\ walks)$$

It is this solution which can be reused in the case of donkey sentences.

11.1.3 *Dependent types*

Another feature which gives much interest to the constructive type theory is its admission of *dependent* types. Such a type is defined every time an object has a type which varies on another one, of another type. A classical example is provided by the type *el* (for *element*) which varies on the set it is a member of. For instance, instead of $a : A(A : set)$, we may write $a : el(A)(A : set)$. Every member of a set is an object of the type *element of this set*. Another easy example is provided by the type *day of the month*, which depends on the month: if $x = june$, the *day of the month* varies from 1 to 30, and if $x = july$, it varies from 1 to 31. For a day j we may then write

$$j : day(x)\ (x : month)$$

Dependent types have a crucial use in linguistics since linguistic categories often depend on some parameters, for example values of features, like agreement features: gender, number, person, etc. If there is a *pronoun* type, for instance, it will be suitable to write

$$\begin{aligned} I &: pron(sing, 1st\ person) \\ you &: pron(sing, 2nd\ person) \\ he &: pron(sing, masc, 3rd\ person) \\ she &: pron(sing, fem, 3rd\ person) \end{aligned}$$

and generally, for a pronoun a,

$$a : pron(n, g, p)(n : number)(g : gender)(p : person)$$

Semantic selection features (like *human, animate, comestible...*) may also be encoded by using dependent types.

In fact, dependent types provide a powerful device by means of which the definitions of type theory can be rephrased in the *higher-level type theory*, where all the definitions which appeared as rules may now be seen as judgements themselves, thus lowering the level of the language. For instance, considering the operator Σ seen above, we may formulate its formation rule as a function *Sigma* with a functional type:

$$Sigma : (A : set) \rightarrow (El(A) \rightarrow set) \rightarrow set$$

which exactly expresses that the expression $\Sigma(A)(B)$ which we can form is a set, with A a set and B a family of sets over A.

The Σ-introduction rule tells us that an element of a Σ set is a pair of an element of the domain and a proof of the propositional function as applied to that element.

$$\begin{aligned} pair : (A : set) \rightarrow (B : El(A) \rightarrow set) \rightarrow (a : El(A)) \rightarrow El(B(a)) \\ \rightarrow El(Sigma(A)(B)) \end{aligned}$$

The elimination rules introduce projection functions:

$$\begin{aligned} &p : (A : set) \rightarrow (B : El(A) \rightarrow set) \rightarrow El(Sigma(A)(B)) \rightarrow El(A) \\ &q : (A : set) \rightarrow (B : El(A) \rightarrow set) \rightarrow (c : El(Sigma(A)(B)) \\ &\quad \rightarrow El(B(p(A)(B)(c)))) \end{aligned}$$

and equality rules specify that

$$\begin{aligned} p(\_)(\_)(pair(\_)(\_)(a)(\_)) = a \\ q(\_)(\_)(pair(\_)(\_)(\_)(b)) = b \end{aligned}$$

By means of these tools, we see how to generate a sentence like *A man walks*. It suffices to take *man* as a set and *walks* as a functional proposition, then to use the Σ-formation rule which yields $\Sigma(man)(walks)$, more readable as $(\Sigma x : man)(x\ walks)$. The Σ-introduction rule gives us a pair (a, b) where $a : man$ and $b : a\ walks$, and the elimination rules are such that for any element c of $(\Sigma x : man)(x\ walks)$, we have

$$\begin{aligned} p(c) &: man \\ q(c) &: p(c)\ walks \end{aligned}$$

Then a sugaring function transfoms the abstract form $(\Sigma x : man)(x\ walks)$ into A MAN WALKS. Ranta (1994) provides a set of sugaring rules, like

$$\begin{array}{ll} S((\Sigma x : A)B(x)) \rhd S(A)\ and\ S(B(x)) & (C) \\ \ S(B(indef(N(A)))) & (Q) \end{array}$$

(where "S" is sugaring as a sentence, and "N" as a noun) corresponding respectively to *There is a man and he walks* and *A man walks.*

By means of projections, we can get discourse continuations like

A man walks. He whistles

since for any proof of *A man walks* (that is, any pair c obtained by the Σ-introduction rule) there is an element $p(c)$ obtained by the first projection, which is available as a referent for *he* in the second sentence. Of course the sentence *Every man walks* does not give such a possibility since it has not a pair as its proof but a function $\lambda(x : man \rightarrow walks(x))$ which does not allow access to a particular man (but only an application of that function to a particular man, allowing for instance *Every man walks. Bill is a man, therefore Bill walks*).

Of course producing such sentences, with pronouns, needs more explanations. Pronouns are introduced by means of rules which express their dependency with regard to sets occurring in their context. For instance, if there is a man in the context (obtained by a projection, say) we may introduce a pronoun which refers to this man. The rule which is used is then the following one:

$$\frac{a : man}{\begin{array}{l} he(a) : man \\ he(a) = a : man \end{array}}$$

This rule expresses the fact that the pronoun *he* is dependent on a constant (already introduced) which is of type *man*; it is itself of type *man* and its denotation is exactly equal to that constant (the pronoun is an identity mapping). Another way of providing this information, in the lower-level format, consists in saying that *Pron* is a function

$$(A : set) \rightarrow El(A) \rightarrow A$$

which, to every set A, associates a function from A to A, this function being the identity mapping, as it is expressed by the equation

$$Pron(\_)(a) = a$$

Things therefore go as follows. Suppose we want to formalize the sentence

Example 11.3. If a man walks he talks

We first formalize the sentence *A man walks*, as $(\Sigma x : man)(x\ walks)$, that is, we suppose we have a proof of that sentence (something we got by the Σ-introduction rule), that we call z:

$$z : (\Sigma x : man)(x\ walks)$$

This provides a context where *he talks* may be formalized, since we have

$$p(z) : man$$

and therefore

$$he(p(z)) : man$$

By means of the propositional function $(x\ talks)$, we get by Π-formation

$$(\Pi z : (\Sigma x : man)(x\ walks))(he(p(z))\ talks)$$

which is equal to

$$(\Pi z : (\Sigma x : man)(x\ walks))(p(z)\ talks)$$

by identity mapping, but is exactly sugared as

IF A MAN WALKS HE TALKS

11.1.4 *Contexts*

Notice that the proposition $(\Pi z : (\Sigma x : man)(x\ walks))(p(z)\ talks)$ is obtained by forming

- first, the proposition $(\Sigma x : man)(x\ walks)$, which itself amounts to forming man (given) and then the proposition $x\ walks$ under the assumption that $x : man$, that is, in the context consisting of a judgement $(x : man)$;
- second, the proposition $y\ talks$ under the assumption that $y : (\Sigma x : man)(x\ walks)$, that is, in the context consisting of this judgement.

Putting together these formation rules we get that a proposition like *he talks* depends on a context which contains

$$(z : (\Sigma x : man)(x\ walks))$$

We may write

$$he(p(z))\ talks : prop \quad (z : (\Sigma x : man)(x\ walks))$$

thus showing that it really uses the context in the technical sense of type theory. In such a technical sense, a *context* is, in effect, any sequence of judgements which is required to derive an expression, that is, in the logical sense, any sequence of non-discharged hypotheses. Of course some hypotheses may depend on others; it is the reason why we speak of a sequence, and not a set. In the previous example, the hypothesis $(z : (\Sigma x : man)(x\ walks))$ depended on $(x : man)$ for the formation of $(x\ walks)$, but by Σ-introduction, this hypothesis has been discharged, thus leaving only one active hypothesis.

Generally, a context is expressed as a list Γ:

$$\Gamma \;=\; x_1 : A_1, \ldots, x_n : A_n(x_1, \ldots, x_{n-1})$$

and a judgement may be obtained in such a context, which depends on $x_1, \ldots, x_n$. For instance, when forming the proposition which sugars into *If a farmer owns a donkey, he beats it*, we have to form the proposition x *owns* y where x : *farmer* and y : *donkey*:

$$x\ owns\ y : prop \quad (x : farmer)\ (y : donkey)$$

before obtaining $(\Sigma x\ :\ farmer)(\Sigma y : donkey)(x\ owns\ y)$, which provides a proof z which can be used to form the proposition which is sugared into *he beats it*, that is,

$$he(p(z))beats\ it(p(q(z)))$$

An important distinction made by Ranta in his book concerns the opposition between *actual* and *potential*. We quote here Ranta's words:

> We can distinguish between what is *actually given* in the context Γ, namely the variables $x_1\ :\ A_1$ to $x_n\ :\ A_n(x_1, \ldots, x_{n-1})$ themselves, and what is *potentially given*, namely what can be derived by the rules of type theory from what is actually given.

For instance, in the context providing a variable z of type $(\Sigma x : A)B(x)$, z is actual but $p(z)$ is potential. This opposition opens the field to a *dynamical* view on interpretation. In particular, this is helpful to express *presupposition* in a dynamic way.

We have just seen that judgements can be made in context. For example, to express that *John stops smoking*, we may write

$$John\ stops\ smoking : prop \qquad (\text{John smoked } true)$$

which comes from

$$John\ stops\ smoking : prop \quad (x : \text{John smoked})$$

where x is a "proof" of *John smoked*, but because the proof itself is not used, we suppress it and simply write "John smoked *true*". In constructive type theory, dynamicity is expressed by the notion of progressivity. Actually, the main difference between simple type theory (and categorial grammars) and constructive type theory lies in this notion. In simple type theory, the ordinary conjunction *and* has the type $t \to t \to t$, also written $(t)(t)t$ (or (prop)(prop)prop), whereas in constructive type theory, we may add dependent types, and consider Σ as a generalization of conjunction, that is, an operator of type $(A : set)((A)set)set$, or in a more familiar notation: $(A : set) \to (El(A) \to set) \to set$ (or $(A : t) \to (A \to t) \to t$), that is, a function which, to every set A, and every propositional function B over A, associates the set $\Sigma(A)(B)$, a shorthand for $(\Sigma x : A)B(x)$. There is *progressivity* in the sense that, first, a particular set A is chosen and then, another object is selected, which depends on this A. It is as if, in $t \to t \to t$, the second t could not be taken arbitrarily, but selected according to the first one. The superiority of constructive type theory over simple type theory is clear since it allows complex types like $(\Pi x : A)B(x)$ (according to the *proposition as type* principle) that we could not of course have in simple type theory, thus allowing us to endow lexical units with types which express better their behaviour.

It is interesting to look at how Ranta deals with presuppositional verbs like *to stop*. He says verbs are *progressively typed*. In effect, a verb like *to stop* has the following type:

$$\text{stop} : (A : set) \to (B : El(A) \to set) \to (x : El(A)) \to B(x) \to set$$

that is, from a set A and a propositional function B over A, from an element x of A, and from the application of B to that x, we get a set (the set of proofs that "x stopped $B(x)$").

In other words, the proposition that "x stops $B(x)$" is only valid in a context which contains

$$\begin{array}{l} A : set \\ B : El(A) \to set \\ x : El(A) \\ B(x) \end{array}$$

This is exactly what we named a context, and it is progressive in the sense that each element of the context depends on the previous ones. It is then possible to write also

$$\text{x } \textit{stops} \text{ B-ing} : \textit{prop} \quad (A : \textit{set})(B : El(A) \rightarrow \textit{set})(x : El(A))\ B(x)\ \textit{true}$$

which presupposes that $B(x)$ is *true*. We can contrast this with *to start*, which does not presuppose such a truth:

$$\text{x } \textit{starts} \text{ B-ing} : \textit{prop} \quad (A : \textit{set})(B : El(A) \rightarrow \textit{set})(x : El(A))$$

Of course, this is not sufficient and in order to express better the meaning of these verbs we should have to add some temporal parameter. This is done by introducing a time scale *time* : set. A possibility is then

$$\begin{aligned} &\text{x } \textit{stops} \text{ B-ing} : \textit{prop} \quad (A : \textit{set})(B : El(A) \rightarrow \textit{time} \rightarrow \textit{set})(x : El(A)) \\ &\qquad ((\Pi t < \textit{now})B(x,t)) \\ &\text{x } \textit{starts} \text{ B-ing} : \textit{prop} \quad (A : \textit{set})(B : El(A) \rightarrow \textit{time} \rightarrow \textit{set})(x : El(A)) \\ &\qquad ((\Pi t < \textit{now})\neg B(x,t)) \end{aligned}$$

to stop corresponds then to the type

$$\begin{aligned} &\text{stop} : (A : \textit{set}) \rightarrow (B : El(A) \rightarrow \textit{time} \rightarrow \textit{set}) \rightarrow (x : El(A)) \\ &\quad \rightarrow ((\Pi t < \textit{now})B(x,t)) \rightarrow \textit{set} \end{aligned}$$

We can also give this information in the form of a natural deduction rule:

$$\frac{A : \textit{set} \qquad \begin{array}{c}(x : A)\ (t : \textit{time}) \\ B(x,t) : \textit{set}\end{array} \qquad (\Pi t < \textit{now})B(x,t) : \textit{true}}{\textit{stop}(x, B(x, \textit{now})) : \textit{set}}$$

Consequences are then introduced, as parts of the meaning, as mere *a posteriori* rules like

$$\frac{\textit{stop}(x, B(x, \textit{now})) : \textit{set}}{(\Pi t > \textit{now})\neg B(x,t) : \textit{set}}$$

This augments the dynamical aspect of the *proof-theoretic* interpretation. Presuppositions are really prior to what is explicitly said, and consequences are posterior to it.

Another aspect of presupposition can also be taken into account; it is the so-called *projection problem.* How is it that a complex sentence which embeds a presuppositional one, in some circumstances, loses the presupposition of the embedded clause? It is something which happens for conditionals like

If Bill smoked, he stopped smoking

This involves a particular treatment of conditionals, and therefore of words like *if ... then*.

To say it briefly, *if A then B* is a function from a clause A and a propositional function $\lambda z.B(z)$ (where z is a proof variable) to Set, such that the result comes from the application of the second argument to the first one. By doing so, the hypothesis z in B is discharged by means of A, and the proof of *If A then B* is made without the hypothesis, that is, no presupposition.

More precisely, we know that if $s(A)$ and $s(B)$ are sugared forms of propositions A and $\lambda z.B(z)$ (where z is a proof variable), *if* $s(A)$ *then* $s(B)$ comes from the sugaring of $(\Pi z : A)B(z)$, the latter form being represented by a function which to any proof z of A associates a proof of $B(z)$.

In the case of presupposition, the proof of B thus obtained is assumed to come from the normalization process which replaces the hypothesis expressing the presupposition by the proof of it, given by the antecedent of the conditional.

11.2 Formalizing Montague Grammar in Constructive Type Theory

Ranta (2004) shows that Montague grammar may be totally formalized in the grammatical framework (GF), an extension of logical frameworks with built-in support for syntax. The enterprise is motivated by the observation that "what is found in modern linguistics theory under the title of Montague grammar is seldom a uniform theory, but rather a bundle of ideas, inspired by Montague and reused in another context". The architecture of the grammar distinguishes three levels:

- abstract syntax — the level of trees,
- concrete syntax — the level of strings,
- semantic interpretation — the level of formulae.

These levels are mapped to each other by means of functions of

- linearization (from abstract syntax to the concrete one),
- translation (from abstract syntax to the interpretation level).

We shall here mainly concentrate on translation. At the level of abstract syntax, trees are built by means of *construction rules* which owe much to grammatical rules in the context-free format. For instance, the rule $S \rightarrow NP\ VP$ is expressed by a function, written as follows in GF:

$$\textbf{fun}\ Pred : NP \rightarrow VP \rightarrow S$$

As usual in the layout conventions for GF, bolded symbols denote *judgements*. Thus, the previous line expresses a **fun** judgement, that is, a declaration according to which we have a function, named *Pred* which, to each expression of category *NP*, associates a function which to each expression of category *VP* associates an expression of category *S* (sentence). This presupposes another kind of judgement, namely **cat** judgements, according to which some categories are introduced into the grammar. For instance, there are judgements like

$$\textbf{cat}\ NP$$

The lexicon itself is a set of **fun** judgements, that is, an assignment of a category to each lexical item, like

$$\textbf{fun}\ man : CN$$
$$\textbf{fun}\ love : TV$$
$$\textbf{fun}\ walk : VP$$
$$\textbf{fun}\ John : PN$$

Other construction rules are

$$\textbf{fun}\ Compl : TV \rightarrow NP \rightarrow VP$$
$$\textbf{fun}\ Every : CN \rightarrow NP$$
$$\textbf{fun}\ Indef : CN \rightarrow NP$$
$$\textbf{fun}\ Raise : PN \rightarrow NP$$

According to such rules, tree structures are obtained, like

$$Pred\ (Every\ man)(Compl\ love\ (Raise\ John))$$

11.2.1 *Predicate calculus*

Logical formulae are defined according to the definition, in GF, of predicate calculus. It amounts to declaring two categories, one for *propositions* and another one for *entities*, which are the *logical types*. Connectives and quantifiers are then defined by **fun** judgements:

$$\textbf{fun}\ and, or, if : Prop \rightarrow Prop \rightarrow Prop$$
$$\textbf{fun}\ all, exists : (Ent \rightarrow Prop) \rightarrow Prop$$

This syntax (also a "concrete" one, with regard to the trees of abstract syntax) allows structures which do not actually look like usual predicate logic formulae, but which can be easily decrypted.[5] For instance, the following construct is a formula of this level:

$$all\ (\lambda x.if(man\ x)(exists\ (\lambda y.and(man\ y)(love\ x\ y)))) \tag{11.4}$$

which may be linearized into

$$(\forall x)(man(x) \Rightarrow (\exists y)(man(y) \wedge love(x, y))) \tag{11.5}$$

11.2.2 *Translation into predicate calculus*

For each category, there is an interpretation function that takes terms of that category into a logical type. It is the exact correspondent of the Montagovian morphism which maps every syntactic category onto a semantic type built from e and t.

$$\begin{aligned}
&\textbf{fun } iS : S \rightarrow Prop\\
&\textbf{fun } iNP : NP \rightarrow (Ent \rightarrow Prop) \rightarrow Prop\\
&\textbf{fun } iVP : VP \rightarrow Ent \rightarrow Prop\\
&\textbf{fun } iTV : TV \rightarrow Ent \rightarrow Ent \rightarrow Prop\\
&\textbf{fun } iCN : CN \rightarrow Ent \rightarrow Prop\\
&\textbf{fun } iPN : PN \rightarrow Ent
\end{aligned}$$

Each interpretation function has a defining equation, expressed by a **def** judgement:

$$\begin{aligned}
&\textbf{def } iS(Pred\ Q\ F) = iNP\ Q\ (\lambda x.iVP\ F\ x)\\
&\textbf{def } iVP(Compl\ F\ Q)x = iNP\ Q\ (\lambda y.iTV\ F\ x\ y)\\
&\textbf{def } iNP(Every\ A)F = all(\lambda x.if(iCN\ A\ x)(F\ x))\\
&\textbf{def } iNP(Indef\ A)F = exists(\lambda x.and(iCN\ A\ x)(F\ x))\\
&\textbf{def } iNP(Raise\ a)\ F\ =\ F(iPN\ a)
\end{aligned}$$

The sentence *Every man loves John* thus comes from the tree of abstract syntax given above, to which interpretation functions can be

[5] It is also possible, like Ranta does in his article, to define precise linearization rules which apply to these structures in order to give the formulae we are accustomed to.

applied, leading to the formula of the predicate level. The steps are the following ones:

$$
\begin{array}{ll}
iS\ (Pred\ (Every\ man)(Compl\ love\ (Raise\ John))) & \rightarrow \\
iNP\ (Every\ man)\ (\lambda x.iVP\ (Compl\ love\ (Raise\ John))\ x) & \rightarrow \\
all(\lambda y.if(iCN\ man\ y)((\lambda x.iVP\ (Compl\ love\ (Raise\ John))\ x)\ y)) & \rightarrow \\
all(\lambda y.if(man\ y)((\lambda x.iVP\ (Compl\ love\ (Raise\ John))\ x)\ y)) & \rightarrow \\
all(\lambda y.if(man\ y)((\lambda x.iNP\ (Raise\ John)\ (\lambda z.iTV\ love\ x\ z)\ y))) & \rightarrow \\
all(\lambda y.if(man\ y)((\lambda x.iNP\ (Raise\ John)\ (\lambda z.love\ x\ z)\ y))) & \rightarrow \\
all(\lambda y.if(man\ y)((\lambda x.(\lambda z.love\ x\ z)(iPN\ John)\ y))) & \rightarrow \\
all(\lambda y.if(man\ y)((\lambda x.(\lambda z.love\ x\ z)\ John\ y))) & \rightarrow \\
all(\lambda y.if(man\ y)((\lambda x.(love\ x\ John)\ y))) & \rightarrow \\
all(\lambda y.if(man\ y)(love\ y\ John)) &
\end{array}
$$

11.2.3 *Introducing dependent types*

Selectional restrictions are expressed, as said above, by *dependent types*. This means that categories are now relativized to sets. This allows us to say that, in the application of, say, the *Pred* rule, the semantic feature (expressed as a set) of the NP must correspond to the semantic feature of the subject of VP. The interpretation of categories is now enriched:

$$
\begin{array}{l}
\textbf{fun}\ iS : S \rightarrow Set \\
\textbf{fun}\ iNP : (A : Set) \rightarrow NP\ A \rightarrow (El(A) \rightarrow Set) \rightarrow Set \\
\textbf{fun}\ iVP : (A : Set) \rightarrow VP\ A \rightarrow El(A) \rightarrow Set \\
\textbf{fun}\ iTV : (A, B : Set) \rightarrow TV\ A\ B \rightarrow El(A) \rightarrow El(B) \rightarrow Set \\
\textbf{fun}\ iCN : CN \rightarrow Set \\
\textbf{fun}\ iPN : (A : Set) \rightarrow PN\ A \rightarrow El(A)
\end{array}
$$

This implicitly contains the fact that propositions are identified with sets (the sets of their proofs). For instance, when the interpretation function iS is applied to a given sentence, it associates with it a set: the set of the proofs of the proposition that it expresses. Common nouns (*CN*) are also interpreted as sets, so that for instance $a : iCN(man)$ is interpreted as a is a member of the set of men. Nevertheless, we are not obliged to consider such sets as sets of "real" objects, but in analogy with the elements of a set of proofs associated with a sentence, we may interpret a in this case as a mere process of "finding a man".

The relativization is then extended to the construction rules:

$$\begin{aligned}
&\textbf{fun } \mathit{Pred} : (A : \mathit{Set}) \to \mathit{NP}\ A \to \mathit{VP}\ A \to S\\
&\textbf{fun } \mathit{Compl} : (A, B : \mathit{Set}) \to \mathit{TV}\, AB \to \mathit{NP}\ B \to \mathit{VP}\ A\\
&\textbf{fun } \mathit{Every} : (A : \mathit{CN}) \to \mathit{NP}\ (\mathit{iCN}\ A)\\
&\textbf{fun } \mathit{Indef} : (A : \mathit{CN}) \to \mathit{NP}(\mathit{iCN}\ A)\\
&\textbf{fun } \mathit{Raise} : (A : \mathit{Set}) \to \mathit{PN}\ A \to \mathit{NP}\ A
\end{aligned}$$

Comment: A quantificational NP, for instance an *NP* in *every*, is dependent on the set associated with the common noun to which the determiner *every* is applied.

These functions are equationally defined by **def** judgements:

$$\begin{aligned}
&\textbf{def } iS(\mathit{Pred}\ A\ Q\ F) = \mathit{iNP}\ A\ Q\ (\lambda x.iVP\ A\ F\ x)\\
&\textbf{def } iVP\ B\ (\mathit{Compl}\ A\ B\ F\ Q)x = \mathit{iNP}\ B\ Q\ (\lambda y.iTV\ \ A\ B\ F\ x\ y)\\
&\textbf{def } \mathit{iNP}(\mathit{Every}\ A)F = \Pi(\mathit{iCN}\ A)\ F\\
&\textbf{def } \mathit{iNP}(\mathit{Indef}\ A)F = \Sigma(\mathit{iCN}\ A)\ F\\
&\textbf{def } \mathit{iNP}(\mathit{Raise}\ A\ a)F = F(\mathit{iPN}\ A\ a)
\end{aligned}$$

Notice that because dependent types involve operators (Π and Σ) which allow us to build sets which generalize implication and conjunction, the addition of sets as parameters on which categories are now dependent compels us to change the interpretation of quantificational NPs, using operators Π and Σ.

It is time now to define the key functions which allow the framework to deal with dynamical interpretation, namely functions associated with *If*, *and* and with anaphoric expressions.

11.3 Dynamical Interpretation and Anaphoric Expressions

11.3.1 If . . . then *sentences*

If and *And* are defined as quantifiers, interpreted as Π and Σ:

$$\begin{aligned}
&\textbf{fun } \mathit{If} : (A : S) \to (\mathit{El}(iS\ A) \to S) \to S\\
&\textbf{fun } \mathit{And} : (A : S) \to (\mathit{El}(iS\ A) \to S) \to S\\
&\textbf{def } iS(\mathit{If}\ A\ B) = \Pi\ (iS\ A)(\lambda x.iS(B\ x))\\
&\textbf{def } iS(\mathit{And}\ A\ B) = \Sigma\ (iS\ A)(\lambda x.iS(B\ x))
\end{aligned}$$

If we look at these typings with attention, we see that *If* is interpreted as a function of two arguments. The first argument is a sentence A and the second argument is not a sentence but a function which, to each proof of A, associates a second sentence, B. The result is a sentence which may be read as *If* A, B. The same for *And* where the second conjunct is also dependent on a proof of the first one. This is the way GF renders the notion of dynamical conjunction, necessary for the implementation of discourse processes. This intuition is reflected in the formal equational definitions of the function iS in the case of a conditional and of a conjunction:

$$\begin{aligned}&\textbf{def } iS(\mathit{If}\ A\ B) = \Pi(iS\ A)(\lambda x.iS\ (B\ x))\\&\textbf{def } iS(\mathit{And}\ A\ B) = \Sigma(iS\ A)(\lambda x.iS\ (B\ x))\end{aligned}$$

The first definition tells us that the interpretation of an *If*-sentence is the set of proofs of a formula which could be sugared into $(\Pi z : iS(A))(iS(B(z))$, and we know that proofs of such a formula are *functions* which to each proof z of $iS(A)$ associate a proof of $iS(B(z))$. The second definition says something similar: the interpretation of an *and*-sentence is the set of proofs of a formula which could be sugared into $(\Sigma z : iS(A))(iS(B(z))$, that is, *pairs* made of a proof z of $iS(A)$ and a proof of $iS(B(z))$.

Pronouns are defined by

$$\begin{aligned}&\textbf{fun } \mathit{Pron} : (A : \mathit{CN}) \to \mathit{El}(\mathit{iCN}\ A) \to \mathit{PN}(\mathit{iCN}\ A)\\&\textbf{def } \mathit{iPN\_}(\mathit{Pron\_a}) = a\end{aligned}$$

These typings and definitions mean what we saw earlier, but here expressed in a more rigorous way: that is, to each domain of a common noun, a pronoun associates a function which, to each element of this domain, associates something which is considered a proper name depending on that domain, and when this proper name is interpreted, it exactly returns the element a of the domain to which the pronoun was applied. Let us take then the abstract syntax for the sentence *If a man owns a donkey, he beats it.* This tree has the form of an *If*-sentence, with a sentence as the first argument:

$$\begin{aligned}&\mathit{Pred}\ (iCN\ man)(Indef\ man)\\&\qquad (\mathit{Compl}(iCN\ man)(iCN\ donkey)own(Indef\ donkey))\end{aligned}$$

and then, as a second argument a sentential function which associates a sentence to each proof of the first one. Concerning this second sentence, it may be formalized by

$$\begin{array}{lll} \lambda z.\ Pred & & (iCN\ man) \\ & & (Raise(iCN\ man)(Pron\ man(p(iCN\ man)(\_)z))) \\ & Compl & (iCN\ man)(iCN\ donkey)beat \\ & & (Raise\ (iCN\ donkey) \\ & & (Pron\ donkey(p(iCN\ donkey)(\_)(q(iCN\ man)(\_)z)))) \end{array}$$

where all the type information is not mentioned for the sake of clarity.

The interpretation function applied to this abstract tree leads to

$$\begin{array}{l} \Pi\ (iS(Pred\ (iCN\ man)(Indef\ man) \\ \qquad (Compl(iCN\ man)(iCN\ donkey)own(Indef\ donkey)))) \\ \qquad (\lambda z.iS(B\ z)) \end{array}$$

where $iS(B(z))$ is the body of the previous λ-expression, so that finally we get a λz over the interpretation of this body, which leads to (omitting the type information)

$$\begin{array}{llr} iS(Pred & (Raise(Pron\ man(p\ z))) & \\ & Compl\ beat(Raise(Pron\ donkey(p(q\ z))))) & \rightarrow \\ iNP & (Raise(Pron\ man(p\ z)))(\lambda x.iVP & \\ & Compl\ beat(Raise(Pron\ donkey(p(q\ z))))x) & \rightarrow \\ iNP & (Raise(Pron\ man(p\ z))) & \\ & (\lambda x.iNP(Raise(Pron\ donkey(p(q\ z))))\lambda y\ beat\ x\ y) & \rightarrow \\ (\lambda x.iNP & (Raise(Pron\ donkey(p(q\ z))))\lambda y\ beat\ x\ y) & \\ & iPN(Pron\ man(p\ z))) & \rightarrow \\ (\lambda x.\lambda y & beat\ x\ y\ iPN(Pron\ donkey(p(q\ z)))) & \\ & iPN(Pron\ man(p\ z))) & \rightarrow \\ (\lambda x.\lambda y & beat\ x\ y(p(q\ z))(p\ z) & \rightarrow \\ (\lambda x.beat\ x & (p(q\ z))(p\ z) & \rightarrow \\ (beat(p\ z) & (p(q\ z))) & \end{array}$$

Finally the interpretation of the conditional becomes

$$\begin{array}{l} \Pi\ (iS(Pred\ (iCN\ man)(Indef\ man) \\ \qquad (Compl\ (iCN\ man)(iCN\ donkey)own(Indef\ donkey)))) \\ \qquad (\lambda z.(beat(p\ z)(p\ (q\ z)))) \end{array}$$

which gives, from a previous result,

$$\Pi((\Sigma\ man)(\Sigma\ donkey)(\lambda x.\lambda y.own\ x\ y))(\lambda z.(beat(p\ z)(p(q\ z))))$$

which sugars into

$$(\Pi z : ((\Sigma x \ : \ man)(\Sigma y : donkey)(own\ x\ y))(beat(p\ z)(p(q\ z)))$$

11.3.2 *A presuppositional account*

Relations between anaphora and presupposition have often been noticed (see for instance van der Sandt, 1992); it is therefore not a surprise if we can give an account of the presupposition phenomenon in the same frame as for anaphora resolution. We have first to enrich the grammar by means of new construction rules, like

fun $ComplInf\ :\ (A\ :\ Set) \to (B\ :\ El(A) \to Set) \to Vcomp\ A\ B \to Sinf\ A \to El(iSinf\ A) \to VPinf\ A$
fun $PredInf\ :\ (A\ :\ Set) \to Vinf\ A \to El(A) \to Sinf\ A$
fun $smoking\ :\ Vinf$
fun $stop\ :\ Vcomp$

with the following categories:

cat $Sinf\ Set$
cat $Vinf\ Set$
cat $Vcomp\ Set\ (Set \to Set)$

These rules give the abstract tree

$$\begin{array}{l} Pred\ \mathtt{hum}\ (Raise\ \mathtt{hum}\ Peter) \\ \quad (ComplInf\ \mathtt{hum}\ (iVinf\ \mathtt{hum}\ smoke)\ stop \\ \quad\quad (PredInf\ \mathtt{hum}\ smoking)\ z) \end{array}$$

We define interpretation functions as

fun $iSinf : (A : Set) \to Set$
fun $iVPinf : (A : Set) \to Vinf\ A \to El(A) \to Set$
fun $iVinf A : Vinf \to El(A) \to Set$
fun $iVcomp : (A : Set) \to (B : El(A) \to Set) \to Vcomp\ A\ B \to (a\ :\ El(A)) \to (B(a)\ :\ Set) \to El(B(a)) \to Set$

with the following equational definitions:

def $iSinf(PredInf\ A\ F\ a) = iVinf\ A\ F\ a$
def $iVPinf\_(ComplInf\ A\ B\ F\ Q)\ a\ z =$
$\quad iVcomp\ A\ B\ F\ a\ (iSinf\ (Q\ a))\ z$

As an illustration, let us suppose there is a proof z of the sentence Peter smokes, then the interpretation of *Peter stops smoking* leads to the following steps:

$iS(Pred\ \mathtt{hum}\ (Raise\ \mathtt{hum}\ Peter)$
$(ComplInf\ \mathtt{hum}\ (iVinf\ \mathtt{hum}\ smoke)\ stop\ (PredInf\ \mathtt{hum}\ smoking)\ z) \rightarrow$

$$iNP(Raise\ \mathtt{hum}\ Peter)\ (\lambda x.iVP!nf\ (ComplInf\ \mathtt{hum}\ (iVinf\ \mathtt{hum}\ smoke)\ stop\ (PredInf\ \mathtt{hum}\ smoking)\ x)\ z) \rightarrow$$

$(\lambda x.iVPinf\ (ComplInf\ \mathtt{hum}\ (iVinf\ \mathtt{hum}\ smoke)$
$stop\ (PredInf\ \mathtt{hum}\ smoking)\ x)\ z)\ iPN\ \mathtt{hum}\ Peter$
$\rightarrow iVPinf\ (ComplInf\ \mathtt{hum}\ (iVinf\ \mathtt{hum}\ smoke)$
$(stop\ (PredInf\ \mathtt{hum}\ smoking)\ Peter)\ z)$
$\rightarrow iVcomp\ \mathtt{hum}\ (iVinf\ \mathtt{hum}\ smoke)$
$(stop\ (iSinf\ (PredInf\ \mathtt{hum}\ smoke)\ Peter)\ Peter)\ z$
$\rightarrow ((stop\ iVinf\ \mathtt{hum}n(iVinf\ \mathtt{hum}\ smoke)\ (smoking\ Peter)\ Peter)\ z)$
$\rightarrow ((stop\ (smoking\ Peter)\ Peter)\ z)$

A sentence which is expressed as a function of a free proof-variable is said to have a presupposition. The only way to suppress the presupposition is therefore to bind such a variable with a quantifier which ranges over the set of proofs of the presupposed sentence, like

$$\Pi(smokes(Peter))(\lambda z.((stop\ (smoking\ Peter)\ Peter)\ z))$$

which sugars into

$$(\Pi z : (smokes(Peter)))((stop\ (smoking\ Peter)\ Peter)\ z)$$

It is worth noticing that according to the projection properties of presupposition, a conditional the antecedent of which is not the presupposed proposition of the consequent keeps this presupposition, and therefore the free variable linked to the presupposed proposition.

A conditional is built in effect from a sentence A and a function B from $El(iS\ A)$ to S. In case we have a sentence with a presupposition like $((stop(smoking\ Peter)Peter)z)$, this function is in fact $\lambda z'.(((stop(smoking\ Peter)Peter)z)\ z')$ if A is different from the presupposed sentence. It is only in the case where A coincides with the presupposed sentence that the free variable z is used for abstraction, thus yielding $\lambda z.((stop(smoking\ Peter)Peter)z)$. It is something which also happens with anaphora. In the sentence *If a man owns a donkey, he beats it*, it could happen in some context that the anaphoric pronouns refer to other individuals. That entails that there is not a single declaration for *ComplInf*.

11.4 From Sentences to Dialogue

Constructive type theory allows a formalization of texts in a very precise way. A discursive sequence like

Example 11.4. One window is broken. John will repair it

can be formalized into (Ranta, 1994, p. 127):

$$x_1 : (\Sigma x \ : \ window)(x \ is \ broken) \tag{11.6}$$

$$x_2 : (John \ will \ repair(q(x_1)) \ Pron(window, p(x_1))) \tag{11.7}$$

This of course assumes an entry for the verb *repair* which makes it depend on a presuppositional variable expressing that its object is broken:

$$(x \ repair(z) \ y) : prop \quad (x : man, y : window, z : (y \ is \ broken)) \tag{11.8}$$

thus explaining the place of an extra argument.

Every sentence is interpreted in the context of the previous ones. This leads to representing texts as sequences of hypotheses:

$$x_1 : A_1 \quad x_2 : A_2(x_1) \quad \ldots \quad x_n : A_n(x_1, x_2, \ldots, x_{n-1})$$

to which we have of course to add some *background* hypotheses, corresponding to the various presuppositions each sentence can have. Some of the variables may of course be left unspecified: it is here that text shows its essential function, to be a matrix for all the interpretations that can emerge from its reading. It is a small step to make to reach questions concerning dialogue.

In Ranta's view, speech acts in dialogue (assertions, questions...) assume a role taken by the hearer as well as by the speaker. For instance, for a propositional question, that Ranta represents by $A|B$, "the most straightforward semantic explanation is not a condition for making the question, but a rule stating how the hearer must react to it" (p. 138).

(|') You may ask the question $A|B$ if the hearer knows $A \vee B$ is true.

This is of course a good point, but how do you know that the hearer knows that $A \vee B$ is true? It is not by chance that communication succeeds. It seems that, in fact, you never knows such facts for sure, but you have indications from your hearer that you can ask him or her the question. This is one of the points in Ranta's theory that we think should be extended if we want to get a fruitful account of dialogue. Such an extension will be suggested in the next chapter, concerning ludics.

Other points are crucial in the observations made by Ranta, like the one according to which (p. 126)

> In type-theoretical grammar, we are of course not satisfied with formalizations correct up to logical equivalence only [...] As regards a text, we must pay attention to the fact that it consists of indicative sentences, each by itself formalized as an assertion, and not as a proposition.

This seems to put the emphasis (if we understand well the remark) on the *concreteness* of text as opposed to the abstraction of propositions. This opens the field to modern studies on *textuality*. An utterance is never the same when repeated several times. It depends on a precise location in time and space, and also on a precise address in the text, relative to other utterances. Type theory has difficulties rendering such locativity phenomena. We think that only a theory which directly takes addresses into account, as opposed to their content, may be an approach to them.

PART IV

Ludics

Chapter 12

Interaction and Dialogue

12.1 Dialogue and Games

12.1.1 *Introduction*

The previous chapter introduced new notions, in particular the idea of representing meaning in a more procedural way than its usual truth-conditional representation. Meanings are obtained as processes: namely as *proof processes* and as soon as they are obtained as such, they can be used and handled for building other processes, that is, other meanings. Nevertheless, we pointed out some difficulties with regard to the application of such a conception to ordinary language: we are simply left none the wiser since we have in general no precise idea concerning the proof of an ordinary sentence! As seen in the previous chapter, we can work around the issue by assuming for instance we are only dealing with artificial worlds inhabited by finite objects (that is, objects which all have a finite description) or we can dispense with explicit introduction and elimination rules. All these solutions are unsatisfactory when dealing with our daily conversations. Another issue consists in using approximations, thus assuming that we may always approximate potentially infinite objects by means of finite ones, or we may also take infinite objects as such. It is remarkable that, in our daily words, we always mention objects, states and events even if we have only a very partial knowledge about them. I may say *If the weather is bad tomorrow, I will take the tram* even if I have only a very imprecise idea about what it means to say that the weather is bad, even if I don't know which tram I will precisely take. It suffices to be able to answer a set of questions on what I mean by these words. Agreement between speakers comes from their decisions to share some contents of the words they use. The remarkable fact is that we have no idea, when beginning a discussion, of the degree of contents shared between two speakers. Here appears the question of misunderstanding and the fact that, as said by a famous French linguist,[1] "understanding is simply a particular case of misunderstanding". If we don't agree on some words or expressions, we have to *justify* ourselves, and we cannot predict until what point we shall have to argue!

We therefore suggest replacing finite processes with potential infinite ones, and therefore "proofs" "(potentially infinite) chains of justifications".

This moreover involves a *dialogical* conception according to which dual arguing processes are opposed to each other. My steps in arguing are precisely determined by your steps in asking for justifications. This opens the field to an interactive view of language. Of course, there is now a long tradition based on *game semantics* which has already developed similar views. It is therefore time to have a look also at this tradition.

12.1.2 *Game semantics*

12.1.2.1 *The notion of strategic meaning*

The most notorious school in game semantics applied to natural language is Hintikka's, enriched by many works of scholars like Kulas, Sandu, Pietarinen and others (Hintikka and Sandu, 1997; Hintikka and Kulas, 1983; Pietarinen and Sandu, 1999; Pietarinen, 2001; Sandu and Pietarinen, 2001). This school gives rise to important notions, like *strategic meaning*, that is, the aspect of meaning concerned with a set of decisions we have to take when understanding a discourse, in order to respect some coherence properties. According to Hintikka, for instance, the meaning of a proper name does not refer to a set of properties, neither does it refer to some original baptismal, but comes from strategic meaning. If I meet, say, in a novel, a proper name Lynne for the first time, Myself, as a player, chooses an individual as the value of the name. Since Myself tries to check a sentence S against another player (sometimes named "Nature" by Hintikka) by leading the semantic game $G(S)$ associated with S until an atomic subformula of S is observed to be true, it appears that in a winning strategy, Myself will have to choose, as the value of the name, the real bearer of this name. Hintikka proposes a similar theory for *anaphoric pronouns*.

This obviously rests on the notion we have of a game $G(S)$ associated with a sentence S.

[1] Antoine Culioli.

12.1.2.2 *Games associated with sentences*

Let us begin with a first-order language L (that is, a language containing only individual variables, constants, letters of predicate with various arities, endowed with the usual connectives and the two quantifiers $\forall$ and $\exists$), and let us suppose it is interpreted in a model M. Consider a sentence S of L. We now define a two-person semantic game $G(S\ ;\ M)$. The players are called Myself and Nature. Myself has the role of verifier and Nature has the one of falsifier, that is, Nature tries to put Myself in trouble. It is assumed there is a set of constants, interpreted in M, that each player may pick out when desired. The game starts from S and is governed by the following rules:

$\vee$: $G(S_1 \vee S_2)\colon M)$ begins with the verifier's choice of $i = 1$ or $i = 2$, and is continued as in $G(S_i\colon M)$.

$\wedge$: $G(S_1 \wedge S_2)\colon M)$ begins with the falsifier's choice of $i = 1$ or $i = 2$, and is continued as in $G(S_i\colon M)$.

$\exists$: $G((\exists x)S_0[x])\colon M)$ begins with the verifier's choice of an individual α from the domain of M, with name c_α (a constant), and is continued as in $G(S_0[c_\alpha]\colon M)$.

$\forall$: $G((\forall x)S_0[x])\colon M)$ begins with the falsifier's choice of an individual α from the domain of M, with name c_α (a constant), and is continued as in $G(S_0[c_\alpha]\colon M)$.

$\neg$: $G(\neg S\colon M)$ is like $G(S\colon M)$ except that the roles of the two players are inverted.

At: If S is an atomic formula or an identity, the player who is the verifier wins and that who is the falsifier loses, if S is true in M. The player who is the falsifier wins and that who is the verifier loses if S is false in M.

The originality of this view lies in the fact that *truth* and *falsity* receive a new characterization. Of course it does not depend on just one play of $G(S\ ;\ M)$. Like Hintikka and Sandu (1997) say:

> Truth in M is defined by reference to what the initial verifier *can* do in $G(S\colon M)$. It refers to the strategies, in the sense of game theory, that Myself and Nature have available in $G(S\ ;\ M)$. A strategy in this sense is a rule that tells a player what to do in every conceivable situation that might arise in a play of the given game. More exactly, a strategy for a player m in the game $G(S\ ;\ M)$ is a set F_m of functions f_Q corresponding to different logical constants Q

> which can prompt a move by player m in $G(S\ ;\ M)$. A winning strategy in a two-person zero-sum game is one which results in a win for that player no matter which strategy one's opponent uses.

This yields interesting logical results, like the equivalence of this notion of truth and the one defended by Tarski: assuming the *axiom of choice*, for any first-order sentence S and model M, S is true in M in the Tarskian sense if and only if there is a winning strategy for Myself in $G(S\ ;\ M)$.

12.1.2.3 *An extension: IF-logic*

Extensions may be provided, resting on this game semantics. For instance, it is usually observed that first-order language lacks the power of expressing some kind of independence between variables selected by quantifiers. As is well known, in a first-order sentence like

$$(\forall x)(\exists y)(\forall z)(\exists u)R(x,y,z,u) \tag{12.1}$$

y of course depends on x, and u on z, but u also depends on x, and sometimes, that is perhaps not desirable. Imagine a sentence like *For each village at the north of the river there is a man and for each village at the south of the river there is a woman such that the man and the woman will meet.*

It could be expressed by

$$\begin{aligned}&(\forall x \in North)(\exists y{:}\, man)(\forall z \in South)(\exists u{:}\, woman)\\ &inhabit(y,x) \wedge inhabit(u,z) \wedge willmeet(y,u)\end{aligned} \tag{12.2}$$

thus making the choice of the woman in z dependent on the village x, which may not be the case. This question was already addressed by Henkin, who used a special notation in order to express independence; he wrote

$$\begin{pmatrix}\forall x\ \exists y\\ \forall z\ \exists u\end{pmatrix} R(x,y,z,u) \tag{12.3}$$

to express that y is dependent on x, and u on z and not on x. Hintikka introduced the notation

$$(\forall x)(\forall z)(\exists y/\forall z)(\exists u/\forall x)R(x,y,z,u) \tag{12.4}$$

with the slash (/) expressing independence. That would be a mere notation if there was not some way to give sense to it. We must first notice that such a step makes us reach some degree of *second-order* in logic since a way of expressing the previous formula is also to use so-called *Skolem functions*. Skolem functions are precisely those functions which express the dependencies between variables when one is universally quantified and the other is existentially quantified. For instance, if we have $(\forall x)(\exists y)P(x, y)$ as a true sentence, we may assume that there is a function f such that the sentence $(\forall x)P(x, f(x))$ is true. A Skolemized formula is not necessarily equivalent to its original formula; nevertheless, every model of the Skolemized formula is a model of the initial one and every model of a formula may be extended in a model of the Skolemized one. Here, by using Skolem functions, the first formula is Skolemized into

$$(\exists f)(\exists g)(\forall x)(\forall z)R(x, f(x), z, g(x, z)) \tag{12.5}$$

and the second one into

$$(\exists f)(\exists g)(\forall x)(\forall z)R(x, f(x), z, g(z)) \tag{12.6}$$

The only difference is in the domain of g. Such formulae are of second-order since function-variables are used inside them (and not only individual variables like in first-order logic). It results from that that, if we wish to express this phenomenon of independence in the usual frameworks, we would have to adopt second-order logic. Hintikka shows that, by means of his game semantics, this is not necessary. He names *independence-friendly* logic (or *IF-logic*) the logic where we may use the kind of formulae he introduced, with an interpretation given by the game-theoretical semantic. When considering formula (12.6), we associate with it

> a game $G(S\ ;\ M)$ which will consist of two moves by Nature prompted by the two universal quantifiers and two moves by Myself prompted by the two existential quantifiers. The idea that the existential quantifier $(\exists y)$ depends only on the universal quantifier $(\forall x)$ (and not on $(\forall z)$) and the quantifier $(\exists u)$ depends only on $(\forall z)$ (and not on $(\forall x)$) can be captured in GTS by requiring that the game in question is one of *imperfect information.* In this particular case, this means that Myself, at his first move, does not know Nature's second choice, and at his second move, Myself does not have access to Nature's first choice (Hintikka and Sandu, 1997).

The example given above may seem a little unnatural, but such phenomena more frequently occur in other cases, using *modalities* and *negation* for instance. A particular case (Hintikka and Sandu, 1989) concerns the phenomenon known as *negation climbing* as observed in many languages like English, where a sentence like

Example 12.1. Bill does not believe that John is at home

in many uses does not mean the contradictory to

Example 12.2. Bill believes that John is at home

but rather means an equivalent of

Example 12.3. Bill believes that John is not at home

The point here concerns the respective scopes of negation and *believes.* A usual representation of Example 12.1 is

$$\neg \mathbf{B}_{Bill}(John\ is\ at\ home) \tag{12.7}$$

whereas we would like

$$\mathbf{B}_{Bill}(\neg John\ is\ at\ home) \tag{12.8}$$

but this is at odds with compositional views. Rather, Hintikka suggests it is expressed as

$$\neg(\mathbf{B}_{Bill}/\neg)(John\ is\ at\ home) \tag{12.9}$$

In other words, the *believe* operator would be independent of $\neg$. This seems to bring little with regard to the thesis of a simple inversion of scopes. However, it is possible to go further into the analyses of sentences, showing that this new way of seeing things is helpful. Let us look at for instance a sentence like

Example 12.4. Nobody does not believe that Homer existed

which has the same meaning as

Example 12.5. Nobody doubts that Homer existed

It may be translated into

$$\neg(\exists x)\neg \mathbf{B}_x((\exists y)(Homer\ =\ y)) \tag{12.10}$$

If negation climbing would amount to inverting the scopes, nothing then could prevent us from having

$$\neg(\exists x)\mathbf{B}_x(\neg(\exists y)(Homer\ =\ y)) \tag{12.11}$$

which would express

Example 12.6. Nobody believes that Homer did not exist

This would be different in meaning from (12.10). If the way to express negation climbing is the one proposed by Hintikka, when it applies to the sentence in question, it will simply give

$$\neg_1(\exists x)\neg_2(\mathbf{B}_x/\neg_1\ \neg_2)((\exists y)(Homer\ =\ y)) \tag{12.12}$$

In the game associated with this formula, the choice of an individual as a value of x is made by Nature, with then a permutation of the roles of the two players. If $\mathbf{B}_x$ is independent of $\neg_1$, then this exchange of roles may be done *after* the move linked to $\mathbf{B}_x$ so that the choice of a world associated with x's beliefs is made by Nature. Formula (12.12) is therefore equivalent, in terms of strategies, to

$$(\forall x)\mathbf{B}_x\neg_1\neg_2(\exists y)(Homer\ =\ y) \tag{12.13}$$

which is equivalent to (12.10).

Informational independence is also at work in the well-known distinction between *de dicto* and *de re* readings. The traditional view concerning this ambiguity is that, for a sentence like

Example 12.7. John believes Sarkozy is crazy

we have the two readings

$$\mathbf{B}_{John}(Sarkozy\ is\ crazy) - de\ dicto \tag{12.14}$$

and

$$(\exists x)(x\ =\ Sarkozy) \wedge \mathbf{B}_{John}(x\ is\ crazy) - de\ re \tag{12.15}$$

The first one means that there is, in the mental world of John, a belief that whatever individual can be named "Sarkozy", this person is crazy; it only concerns John's mental world. The second one means that there exists someone we know, whose name is Sarkozy, and John believes this person is crazy, even if he, himself, does not know that this person *is* Sarkozy (for instance he saw on TV a strange man who behaved absurdly but did not know he was Sarkozy!). This second meaning in some sense concerns the world, and not only John's beliefs. Hintikka questions why there is such a big difference in the syntax of the two formulae and where the existential quantifier comes from. And he claims that a better solution to exhibit this opposition of two readings would consist in expressing it in terms of IF-logic. The difference would be between the two following formulae:

$$\mathbf{B}_{John}(Sarkozy\ is\ crazy) - de\ dicto \tag{12.16}$$

and

$$\mathbf{B}_{John}(Sarkozy/\mathbf{B}_{John}\ is\ crazy) - de\ re \tag{12.17}$$

where the second formula simply means that the value associated with the proper name "Sarkozy" is chosen *independently* of John's beliefs, which is actually exactly what we mean by the *de re* reading of that sentence.

Other phenomena can be studied by means of IF-logic. As we expect, they are phenomena which involve going out of first-order. It is for instance well known that many generalized quantifiers cannot be expressed in first-order. It is the case of expressions like THERE ARE AS MANY As AS Bs (Boolos, 1981), which could be expressed in set theory by

$$Card\{x; A(x)\} = Card\{x; B(x)\}$$

In fact, the proper meaning of this sentence is that there exists a one-to-one mapping f from As to Bs, but how to express this knowledge? In naive terms, it is expressed by

$$\begin{aligned}(\exists f)(\exists f')(\forall x)(\forall y)(y = f(x) \Leftrightarrow (x = f'(y)))\\ \wedge (\forall x)(A(x) \Rightarrow B(f(x)) \wedge (\forall y)(B(y) \Rightarrow A(f'(y)))\end{aligned} \tag{12.18}$$

that is, a second-order sentence. The way to express it in IF-logic is then

$$\begin{aligned}(\forall x)(\exists y) \qquad (\forall z/x\ y)(\exists u/x\ y)((x = u) \Leftrightarrow (y = z)\\ \wedge (A(x) \Rightarrow B(y)) \wedge (B(z) \Rightarrow A(u))\end{aligned} \tag{12.19}$$

Game semantically, Nature chooses x, Myself chooses y, then, independently of these choices, Nature chooses z and Myself u so that if it happens that $u = x$ then $z = y$ and so that every time x is an A, then y is a B, and reciprocally, every time z is a B, u is an A.

We think that this approach brings a deep insight into the essentially *dialogical* character of language. If we adopt the view that *to assert* means to be engaged in the validity of what we say, I assert that there are as many As as Bs if and only if I commit myself to rightly answer to the other speaker, as long as she shows me a new element of the set B, by giving to her a new element of the set A. Such a game could be done even in a non-set-theoretical frame, if we wish, for instance, to express a vaguer quantifying. Suppose each player selects its members in a topology (that is, a set of open sets) and that there may exist some overlapping between open sets; then the AS MANY AS becomes approximate, like in some Amazonian languages.[2]

12.1.2.4 *Abstract and strategic meanings*

Let us conclude this section by emphasizing the distinction Hintikka draws between *abstract* and *strategic* meaning. According to Hintikka, there are two alternative views on the interpretation provided by games. In the first one, we interpret *truth* as the mere existence of a winning strategy (without necessarily knowing one); this is *abstract* meaning. In the second one, we are not content with the "Platonistic" existence of such a strategy: we wish to convey in the sentence indications of how to find it. "Such surplus meaning is called *strategic* meaning. It is not so much information about the world, as information about how a sentence can actually be verified" (Hintikka and Sandu, 1997, p. 398).

The best example of the role of strategic meaning resides in the binding problems with which we began this book. According to Hintikka and Kulas (1983), it is rather vain to look for "syntactic" considerations in the determination of co-reference links in the use of anaphoric pronouns, for instance. A sentence S which includes such a pronoun must be evaluated according to a game $G(S)$ and the interpreter has to make a choice during this evaluation for any pronoun that he or she meets. Of course, this choice is constrained by game rules according to which, for instance, some rules have a necessary priority with regard to others. This is not a question of *abstract* meaning in the sense that the hearer does not only need to know that there is a winning strategy for this sentence, he or she also needs to determine such a strategy: if he or she takes a wrong decision, he or she will simply lose the game. It is what happens in a sentence like

Example 12.8. Tom's friend Bill met Mary and understood that there was a barrier between him and her

where the verifier has to play in such a way that *her* refers to *Mary* and *him* to *Tom*, because if he or she makes another choice (like *him* referring to *Bill*), it is not that the sentence could not be abstractly true, but that the verifier could lose.

[2]See on this point the numerous studies by Pica *et al.* (2004) on the Mundurucus, a tribe living in Pará in Brazil.

12.1.3 *Internal games*

Another viewpoint on dialogism is provided by works by Lorenz, Lorenzen and Felscher which have been pursued by Rahman, Rückert, Keiff and others (Lorenzen, 1960; Lorenz, 1961; Felscher, 1986; Rahman, 2000; Rahman and Rückert, 2001; Keiff, 2007). Their initial objective was to give a new kind of foundations to logic, namely a dialogical one. The purpose of Lorenzen was, for instance, to show that intuitionistic logic was "more natural" when starting from this dialogical view. The game rules introduced in that trend are quite similar to Hintikka's, except that they never refer to any model M. The rule for atoms (our *At* rule in the previous presentation) is therefore necessarily different. Because there is no intervention of any kind of model, it is often said that there is *no outside* for this logic, hence the denomination of *internal* games, as opposed to *external* games in the case of GTS. If the *At* rule cannot be in the form of the rule given in GTS, what can it be? Here appears, from our viewpoint, the most interesting notion of this approach. In a game between two players, one Proponent (**P**) and one Opponent (**O**), *the proponent can assert an atomic proposition p if and only if the opponent already asserted it, in a previous move.* That is, in order to be in a winning position when asserting an atomic p, the proponent must have constrained in the previous moves his or her opponent to *concede* this p. We here reach a conception of logic where we get rid of the "semantic" notions like model, and where the semantics rests in fact on what the rules allow the players to do. A kind of *procedural* semantics, therefore.

How may logic be conceived as a game? Let us begin, as usual, by defining a language. It consists in a denumerable set of variables $x, y, z, ...$, the constant connectives $\wedge, \vee, \Rightarrow, \neg$ and the quantifiers $\forall$ and $\exists$. Let **P** and **O** be two constant symbols (respectively representing Proponent and Opponent). For every expression e, we have two distinct expressions: **P**: e and **O**: e which are said to be respectively signed by **P** and **O**. X and Y are two variables which take their values in **P, O**. We assume that in every situation, $X \neq Y$.

An argumentation form is a schematic presentation of how an assertion made by X can be attacked by Y and how, if possible, it can be defended by X. In order to have an argumentation form for each formula, it is enough to define argumentation forms for connectives and quantifiers.

We thus have for the connectives

$\wedge$	assertion	$X : w_1 \wedge w_2$
	attack	Y: $\wedge i$ (i = 1,2)
	defence	$X : w_i$

$\vee$	assertion	$X : w_1 \vee w_2$
	attack	$Y : \vee$
	defence	$X : w_i$ (i = 1,2)

$\Rightarrow$	assertion	$X : w_1 \Rightarrow w_2$
	attack	$Y : w_1$
	defence	$X : w_2$

$\neg$	assertion	$X : \neg w$
	attack	$Y : w$
	defence	no defence!

and for the quantifiers

$\forall$	assertion	$X : \forall x w$
	attack	$Y : t$ (Y chooses t)
	defence	$X : w(t)$

$\exists$	assertion	$X : \exists x w$
	attack	$Y : \exists$
	defence	$X : w(t)$ (X chooses t)

Technically, a dialogue is a sequence δ of signed expressions, alternatively supported by **P** and **O** and satisfying at each step some argumentation form. This needs some more precision. Let also η be a function on natural numbers. Let A and D be the two letters for Attack and Defence.

Definition 12.1. (Felscher, 1986) A dialogue is a couple (δ, η) such that

- δ is a sequence of signed expressions,
- η is a function on the ranks of the signed expressions in δ such that $\eta(n)$ is a couple $[m, Z]$ where $m \leq n$ and $Z = A$ or D, which satisfies the following three conditions (D00), (D01), (D02).

D00 $\delta(n)$ is **P**-signed if n is even and **O**-signed if n is odd; $\delta(n)$ is a compound formula;

D01 If $\eta(n) = [m, A]$, then $\delta(m)$ is a compound formula and $\delta(n)$ is an attack on $\delta(m)$ which corresponds to a regular argumentation form;

D02 If $\eta(p) = [n, D]$, then $\eta(n) = [m, A]$, and $\delta(p)$ is an answer to the attack $\delta(n)$ which corresponds to a regular argumentation form.

12.1.3.1 *Global rules*

Global rules are used according to the logic we aim at expressing. Let us envisage the following rules (some of which are not compatible with others):

D10 **P** may assert an atomic formula only if **O** asserted it previously;
D11 If at a position $p1$, several attacks are still not answered, it is only the last not already defended attack that can be answered at p;
D12 An attack can have at most one answer;
D12 (bis) **P** may repeat an attack or a defence if and only if **O** has introduced a new atomic formula (which can now be used by **P**);
D13 A **P**-signed formula can be attacked at most once;
D13 (bis) In any move, each player may attack a complex formula asserted by his partner or he may defend himself against any attack (including those which have already been defended).

Playing this kind of games obviously needs a winning rule:

D2 X wins if and only if it is Y's turn but Y cannot move.

Definition 12.2. A formula is said to be valid in a dialogical system (defined by the previous definition and some of the global rules) if and only if **P** has a winning strategy for it, that is, **P** can win whatever moves are done by **O**.

Theorem 12.1. *Classical theses coincide with formulae valid for the system which includes D10, D12 (bis), D13 (bis), D2.*

Theorem 12.2. *Intuitionistic theses coincide with formulae valid for the system which includes D10, D11, D12, D13, D2.*

12.1.3.2 *Examples*

It is fundamental to notice here that in order to defend a formula (either in classical or in intuitionistic logic) **P** not only has to use more or less automatically the argumentation forms, he has also to choose some strategy which compels **O** to give him or her as much information as possible. This is particularly clear with regard to the global rule D10 which allows him or her to assert an atomic formula, provided that his or her partner already asserted it. In this case he or she has therefore to oblige **O** to concede some fact. This is clearly shown in the following cases. (The two first examples come from Felscher (1986), the third one from Rahman and Rückert (2001).)

Example 1:

$0.\ P : (a \wedge b) \Rightarrow (a \wedge b)$			
$1.\ O : (a \wedge b)$	$[0, A]$		
$2.\ P : \wedge 1$	$[1, A]$		
$3.\ O : a$	$[2, D]$		
$4.\ P : \wedge 2$	$[1, A]$		
$5.\ O : b$	$[4, D]$		
$6.\ P : (a \wedge b)$	$[1, D]$		
$7.\ O : \wedge 1$	$[6, A]$	$7.\ O : \wedge 1$	$[6, A]$
$8.\ P : a$	$[7, D]$	$8.\ P : b$	$[7, D]$

Example 2:

$0.\ P : ((a \Rightarrow a) \Rightarrow b) \Rightarrow b$			
$1.\ O : (a \Rightarrow a) \Rightarrow b$	$[0, A]$		
$2.\ P : (a \Rightarrow a)$	$[1, A]$		
$3.\ O : b$	$[2, D]$	$3.\ O : a$	$[2, A]$
$4.\ P : b$	$[1, D]$	$4.\ P : a$	$[3, D]$
$5.\ O : a$	$[2, A]$	$5.\ O : b$	$[2, D]$
$6.\ P : a$	$[5, D]$	$6.\ P : b$	$[1, D]$

Example 3:

$0.\ P : \forall x(P(x) \vee \neg P(x))$	
$1.\ O : \tau$	$[0, A]$
$2.\ P : P(\tau) \vee \neg P(\tau)$	$[1, D]$
$3.\ O : \vee$	$[2, A]$
$4.\ P : \neg P(\tau)$	$[3, D]$
$5.\ O : P(\tau)$	$[4, A]$
$6.\ P : P(\tau)$	$[3, D]$

Notice that at step 4, **P** cannot choose $(P(\tau))$ because it is an atomic formula not already asserted by **O**. He therefore chooses $\neg P(\tau)$, which obliges **O** to assert $P(\tau)$ and then **P** can choose it, thus providing a complete argument for his or her thesis. Notice also that this is only possible in classical logic since in the intuitionistic one, **P** would only be allowed to answer **O**'s last attack, which is the attack performed at step 5.

12.1.4 *Comparison between the two frameworks*

Many similarities exist of course between Hintikka's GTS and Lorenzen's dialogical logic, but if GTS has been intensively used from a linguistic perspective, this is not the case for the second approach. We may in fact question the philosophical bases of the two frameworks. At first sight, GTS is quite hybrid, since it involves, at the same time, game rules and an exterior model $\mathcal{M}$. Truth is defined by means of strategies: as we saw above, there is an equivalence between "to be true" in the Tarskian sense, and to be such that there exists a strategy for the proponent in order to win. We may therefore classify Hintikka's viewpoint among the *verificationist* ones: how to reveal the truth of a sentence (otherwise existing independently of our efforts to make it apparent)? That seems to be the question asked by the theory. The rules for *quantifiers* are indicative of the way it works: in the $\forall$ rule, player 1 invites player 2 to select one object in $\mathcal{M}$, while in the $\exists$ rule, she selects it and asks player 2 to check whether it is convenient. Like Pietarinen and Tulenheimo (Tulenheimo, 2010) have emphasized, such rules govern a typical activity of *searching* and *finding*. A dialogical logician behaves differently: the rules she or he uses are really constitutive of the meaning of the logical expressions, which is not based on an exterior world. The distinction already present in GTS between two levels — the level of a play (where a player may win even if she has no winning strategy, but simply because her opponent made wrong moves), and the strategic level — is used to make a strong distinction between meaning (properly speaking), which only depends on the rule definitions concerning "particles" (that is, logical constants), and some "semantic attributes" (like truth, falsity or validity) which depend on the existing strategies which may be built starting from these rule definitions. *Meaning* is therefore not equated with *truth*, for instance. Moreover, as we may see in dialogical logic, structural rules may change (like it is the case in *substructural logics*, including linear logic): that results in working inside classical logic or inside intuitionistic logic, depending on the use of D11 or not (if it is the case that at a position p, when several attacks are still not answered, it is only the last not already defended attack that can be answered at p, then only intuitionistic theses are well defended, while classical ones may be defended if this restriction is removed). Of course, we may imagine several practices concerning the use of these rules, for instance it may happen that sometimes such restriction is applied and sometimes not, or it may happen that other restrictions are added (Rahman reconstructs free and partial logic by this means). We may conclude that there is no strictly defined notion of *truth* or *falsity*: these semantic attributes can be changed, opening the field to a so-called *pluralist* view in logic (see Keiff, 2007). At this point, we may wonder if the whole edifice of logic is not crumbling... If there are several notions of truth in the same framework, how to compare them? How to decide which is the correct one for the task at hand? Is it not time to finally get rid of *truth*?

Otherwise, a question which is often addressed concerns the *origin* of the rules. As pointed out by Tulenheimo, it seems that symmetry plays an important role. Of course, since the two players in a game are supposed to be interchangeable, the formulation of particle rules must not depend on the particular position each player has during the game, but Tulenheimo argues that the interchange of the players has consequences on the kind of actions performed on the subformulae. Let us take the case of the implication $A \Rightarrow B$. If **P** supports it, we may say that **P** *asserts* $A \Rightarrow B$ and that if **O** *grants* A, then, **P** may *assert* B, but if **O** supports it, then it is assumed that she *grants* $A \Rightarrow B$ and that if **P** *asserts* A, then **O** is required to *grant* B as well. Symmetry is respected but the price to pay is to accept an interpretation of dialogues where two successive kinds of attitudes exist, one as *positive* (to assert) and the other as *negative* (to grant). This is a point that we shall meet in Girard's ludics.

The last point we may consider is one of the kind of activities which are involved in these various kinds of games. Clearly, GTS moves in a play do not refer to speech acts. Rather, a formula is taken and then decomposed in order to find whether it may be proven to be true or not. The course of a play amounts to correlating elementary constituents (words) with the external reality. Quantifier rules are used to find extra-linguistic objects in accordance with the game rules. Only moves in dialogical logic may be seen as utterance acts. Hence the strong differences between the two approaches. This sheds light on why GTS is more used for a linguistic purpose than dialogical logic: the former theory is mostly concerned with a question of *reference*, the very same question which seems to be crucial for solving linguistic phenomena like proper names and anaphora, while the latter could be more efficient, if ever used, for dealing with *pragmatics* and material dialogues.

12.2 Ludics[3]

12.2.1 *Focusing proofs*

12.2.1.1 *Introduction*

Ludics (see Girard, 2001, 2006; Curien, 2003; Faggian, 2006) is a theory proposed by Girard as a new foundation for logic. It results from an analysis of the concept of *proof* in linear logic. As was established by Andréoli (1992), every proof in linear logic may be viewed as a process of polarized steps, alternately positive and negative. This process comes from the ability to concentrate several successive positive (resp. negative) steps — each of which are associated with a positive (resp. negative) rule — into only one "big" positive (resp. negative) step which consists in the application of a so-called "synthetic" operator. Of course an infinity of such operators exist; this leads to a more abstract way of seeing a logical system. Otherwise, because of this feature of any proof, and because we may associate a positive (resp. negative) step with the positive (resp. negative) action of one speaker, a proof may also be seen as a *strategy* in a game. We are thus able to reconcile the two aspects which have drawn our attention in these last sections: the *proof* aspect and the *game* aspect. In previous works (see Blass, 1992; Abramsky and Jagadeesan, 1994), a game semantics had been given for linear logic, thus making, as is traditionally the case in logic, a syntactic level (the rules of linear logic, as seen in Chapter 5) and a semantic one (an interpretation of connectives in terms of game connections) interact. Ludics rests on the willingness to collapse the opposition between the two levels: the semantics of rules is *in the rules themselves* since proofs (as syntactic objects) may now be viewed as strategies (the interpretive objects) at one unique level.

12.2.1.2 *Focalization and polarities*

Actually, the connectives of linear logic may be split into two groups which behave differently with regard to the active formula, at each step.

- The negative connectives:
  - the multiplicatives $\bot$, $\wp$,
  - the additives: $\top$, $\&$.

[3]This part of this chapter is co-authored by Myriam Quatrini.

- The positive connectives:
  - the multiplicatives: 1, $\otimes$,
  - the additives: 0, $\oplus$.

Negative connectives are deterministic and reversible. In effect, when the [$\wp$]- or the [&]-rule is applied, there is only one way to do it, and consequently, it is always possible to go back: no information is lost when the rule is applied. But when the [$\oplus$]- or the [$\otimes$]-rule is applied, there are many ways to continue the proof (always seeing the proof in a bottom-up direction). In case of $A \oplus B$, it is necessary to choose between continuing on A and continuing on B; in case of the [$\otimes$]-rule, splitting the context is not deterministic. This means that it will be difficult to reverse the process when a choice has been made: in the first case, B will be forgotten if A has been chosen. In the second case, if we reverse the rule, we obtain a false rule since any partition of the context does not fit the expected conclusion.

Hence the idea according to which an efficient proof-search must select at each step the main formula so that

- if the sequent contains negative formulae, then any of them may be chosen (at random) and the proof may be continued in this way until there is no more negative formula in the sequent to prove (this phase is deterministic);
- when all the negative formulae have been decomposed, a main formula must be selected in a non-deterministic manner, but as soon as it has been selected, it is necessary to focus on it in such a way that in the following, only subformulae of this formula are selected, as long as the subformula is positive.

It is then possible to present a proof in the form of an alternate sequence of steps: a sequence of negative steps followed by a sequence of positive ones, followed again by a sequence of negative steps and so on. Even better: we may consider a sequence of steps of the same sign as a single step of that sign, consisting in applying a rule associated with a synthetic connective.

With such connectives, and for a proof consisting in such a sequence of alternate steps, the focalization method preserves as an invariant the fact that a sequent contains at most one negative formula (all the negative connectives have been grouped together in the same formula). If the active formula is positive, then all the other formulae of the sequent are positive (if not, the negative one would have been chosen as the active formula) and all the premises in the application of the rule have exactly one negative formula, each of which arises from the decomposition of the active formula. If the active formula is negative, all the other formulae are positive and each premise is a sequent consisting only of positive formulae. At the beginning, we want to prove a sequent which contains a single formula, which conforms to the invariant.

To consider the polarized proofs is particularly interesting for the analysis of the objects which prevails in the development of ludics and which can be summarized in the following slogan: "keep of the proof object only what is concerned with cut-elimination". The relevant equivalences between proofs from the proof-search point of view are also relevant from the cut-elimination one. The quotient of the polarized proofs realized by using synthetic connectives is thus relevant for approximating good objects and it is exactly what ludics will do. Let us illustrate this matter by means of an example.

Let us consider the following equivalent LL-formulae: $(A \otimes C) \oplus (B \otimes C)$ and $(A \oplus B) \otimes C$, and let us suppose we are eliminating a cut on such a formula (by putting them in interaction with $(A^\perp \wp C^\perp) \& (B^\perp \wp C^\perp)$ or $(A^\perp \& B^\perp) \wp C^\perp$ for instance). The proof for $(A \otimes C) \oplus (B \otimes C)$ can be

$$\textit{either} \quad 1 = \dfrac{\dfrac{\vdash A \quad \vdash C}{\vdash A \otimes C}}{\vdash (A \otimes C) \oplus (B \otimes C)} \qquad \textit{or} \qquad 2 = \dfrac{\dfrac{\vdash B \quad \vdash C}{\vdash B \otimes C}}{\vdash (A \otimes C) \oplus (B \otimes C)}$$

The continuation of the cut-elimination procedure leads to

- eliminate one cut on $\vdash A$ and one cut on $\vdash C$ if the proof is 1,
- eliminate one cut on $\vdash B$ and one cut on $\vdash C$ if the proof is 2.

If we consider the other formula, namely $(A \oplus B) \otimes C$, again there are two possible proofs for this formula:

$$\textit{either} \quad 1' = \dfrac{\dfrac{\vdash A}{\vdash A \oplus B} \quad \vdash C}{\vdash (A \oplus B) \otimes C} \qquad \textit{or} \qquad 2' = \dfrac{\dfrac{\vdash B}{\vdash A \oplus B} \quad \vdash C}{\vdash (A \oplus B) \otimes C}$$

Once again there are two possibilities:

- if the proof is $1'$, continue on $\vdash A$ and $\vdash C$;
- if the proof is $2'$, continue on $\vdash B$ and $\vdash C$.

The proofs 1 and $1'$ on the one hand, 2 and $2'$ on the other hand are indistinguishable according to the cut-elimination procedure. Because ludics only preserves the information "to continue on A and C" or "to continue on B and C", it will make no difference between 1 and $1'$ on one side and 2 and $2'$ on the other.

The following example taken from Curien (2003) will make clear the role of synthetic connectives in displaying proofs in sequences of alternate steps.

Let us consider the formula $((N_1 \otimes N_2)\&Q)\wp R$ where Q, R are supposed to be positive, $P = N_1 \otimes N_2$ is positive and N_1, N_2 are negative. $(P\&Q)\wp R$ is negative. The rule for the synthetic connective (a synthesis of & and $\wp$) is

$$\frac{\vdash P, R, \Lambda \qquad \vdash Q, R, \Lambda}{\vdash (P\&Q)\wp R, \Lambda}$$

A focalized proof of $((N_1 \otimes N_2)\&Q)\wp R$ ends up like this:

$$\frac{\dfrac{\vdash N_1, R, \Lambda_1 \qquad \vdash N_2, \Lambda_2}{\vdash (N_1 \otimes N_2), R, \Lambda} \qquad \vdash Q, R, \Lambda}{\vdash ((N_1 \otimes N_2)\&Q)\wp R, \Lambda}$$

where we can see the alternation of steps: first a negative rule, then a positive one (here the tensor rule). The same proof may be presented in the following way:

$$\frac{\dfrac{N_1^\perp \vdash R, \Lambda_1 \qquad N_2^\perp \vdash \Lambda_2}{\vdash (N_1 \otimes N_2), R, \Lambda} \qquad \vdash Q, R, \Lambda}{((N_1 \otimes N_2)^\perp \oplus Q^\perp) \otimes R^\perp \vdash \Lambda}$$

In this presentation, there are only positive connectives. Moreover, there is at most one formula on the left of $\vdash$.

All these observations lead us to simplify the presentation of the calculus: we shall consider only sequents with only positive formulae and at most one formula on the left-hand side. We have therefore only positive rules, but for each synthetic connective, we have a left introduction rule, which is reversible, and a set of right introduction rules which are not reversible, for instance for the synthetic connective above, which may be written as $(P^\perp \oplus Q^\perp) \otimes R^\perp$:

$$\frac{\vdash P, R, \Lambda \qquad \vdash Q, R, \Lambda}{(P^\perp \oplus Q^\perp) \otimes R^\perp \vdash \Lambda}\{\{P, R\}, \{Q, R\}\} \qquad \frac{P \vdash \Gamma \qquad R \vdash \Delta}{\vdash (P^\perp \oplus Q^\perp) \otimes R^\perp, \Gamma, \Delta}\{P, R\}$$

$$\frac{Q \vdash \Gamma \qquad R \vdash \Delta}{\vdash (P^\perp \oplus Q^\perp) \otimes R^\perp, \Gamma, \Delta}\{Q, R\}$$

the rule at the top right comes from the following deduction:

$$\frac{\dfrac{\dfrac{P \vdash \Gamma}{\vdash P^\perp, \Gamma}}{\vdash P^\perp \oplus Q^\perp \Gamma} \qquad \dfrac{R \vdash \Delta}{\vdash R^\perp \Delta}}{\vdash (P^\perp \oplus Q^\perp) \otimes R^\perp, \Gamma, \Delta}$$

We may notice (cf. Curien, p. 17) that the connective isomorphic to it, $(P^\perp \otimes Q^\perp) \oplus (P^\perp \otimes R^\perp)$, has exactly the same rules.

12.2.2 *Paraproofs*

12.2.2.1 *A hypersequentialized version of the linear sequent calculus*

We summarize the rules which arise from the previous considerations.

12.2.2.2 *Formulae and sequents*

Synthetic connectives are still denoted by $\oplus$ and $\otimes$ but are of various arities. We will distinguish the case where both $\oplus$ and $\otimes$ are of arity 1 and denote it by $\downarrow$ (see below the interpretation of this unary connective).

- The only linear formulae which are considered are built from the set P of linear constants and propositional variables according to the following schema:

$$F = P|(F^\perp \otimes \cdots \otimes F^\perp) \oplus \cdots \oplus (F^\perp \otimes \cdots \otimes F^\perp)| \downarrow F^\perp$$

- The sequents are denoted $\Gamma \vdash \Delta$ where Δ is a multiset of formulae and Γ contains at most a formula (they are called forks, the left-hand side is called the handle).

Rules:

- There are some axioms (logical and non-logical axioms):

$$\overline{P \vdash P} \qquad \overline{\vdash 1} \qquad \overline{\vdash \downarrow \top, \Delta} \qquad \overline{\vdash \Delta}^{\dagger}$$

  where P is a propositional variable; 1 and $\top$ are the usual linear constants (respectively positive and negative).
- There is a cut rule:

$$\frac{\Gamma \vdash A, \Delta \qquad A \vdash \Delta'}{\Gamma \vdash \Delta, \Delta'} \, cut$$

- The "logical" rules are the following:
  Negative rule:

$$\frac{\vdash A_{11}, \ldots, A_{1n_1}, \Gamma \quad \ldots \quad \vdash A_{p1}, \ldots, A_{pn_p}, \Gamma}{(A_{11}^{\perp} \otimes \cdots \otimes A_{1n_1}^{\perp}) \oplus \cdots \oplus (A_{p1}^{\perp} \otimes \cdots \otimes A_{pn_p}^{\perp}) \vdash \Gamma}$$

  Positive rule:

$$\frac{A_{i1} \vdash \Gamma_1 \quad \ldots A_{in_i} \vdash \Gamma_p}{\vdash (A_{11}^{\perp} \otimes \cdots \otimes A_{1n_1}^{\perp}) \oplus \cdots \oplus (A_{p1}^{\perp} \otimes \cdots \otimes A_{pn_p}^{\perp}), \Gamma}$$

  where $\cup\Gamma_k \subset \Gamma$ and for any $k, l \in \{1, \ldots p\}$, $\Gamma_k \cap \Gamma_l = \emptyset$.

12.2.2.3 *Remarks on shifts*

Using the shift is a way to break a block of a given polarity. Separate steps may be enforced by using the shift operators $\downarrow$ and $\uparrow$ which change the negative (resp. positive) polarity into the positive (resp. negative) one. The rules introducing such shifted formulae are particular cases of the positive and the negative one:

$$\frac{A^{\perp} \vdash \Gamma}{\vdash \downarrow A, \Gamma} [+] \qquad\qquad \frac{\vdash A^{\perp}, \Gamma}{\downarrow A \vdash \Gamma} [-]$$

where A is a negative formula.

Example: In a block like $A \otimes B \otimes C$ in principle, A, B and C are negative, but if we don't want to deal with A, B, C simultaneously, we may change the polarity of $B \otimes C$ (which is positive) and make it negative by means of $\uparrow$. We write then $A \otimes \uparrow (B \otimes C)$.

Compare the two following partial proofs, where (1) does not use any shifts and (2) uses one:

$$\text{instead of (1):}\quad \frac{A^{\perp}\vdash \quad B^{\perp}\vdash \quad C^{\perp}\vdash}{\vdash A\otimes B\otimes C} \qquad \text{we get (2):}\quad \cfrac{A^{\perp}\vdash \quad \cfrac{\cfrac{B^{\perp}\vdash \quad C^{\perp}\vdash}{\vdash B\otimes C}}{\downarrow(B\otimes C)^{\perp}\vdash}}{\vdash A\otimes\uparrow(B\otimes C)}$$

We may use the notation $\oplus_y$ (and dually $\&_x$) instead of $F_{y_1}\oplus\cdots\oplus F_{y_n}$ or simply $\exists y$ (dually $\forall x$) when it is clear in the context that y belongs to a finite set.[4]

12.2.2.4 *Proofs and paraproofs*

A proof may be obtained in the usual way in the above hypersequentialized calculus: we start from a sequent and repeatedly apply the rules until we reach axiom cases, but this is not precisely what is aimed at in ludics. Rather, we focalize on the *proof-search* aspect. We imagine the search of a proof of a given sequent not as a direct process which would lead from a root to leaves, but as a way to construct an argumentation by answering or reacting to questions or tests which are displayed in another process. The final answer (to the question: is this sequent valid?) will come from the fact that all tests have been concluded successfully. This point is well explained in Girard (2006) and it has some relationship with constructive type theory and intuitionistic logic except that we speak of proving and testing instead of only proving.

> *A test for $A\wedge B$ is a pair (i,τ), where either $i=1$ and τ is a test for A, or $i=2$ and τ is a test for B.*
>
> *A test for $A\vee B$ is a pair (τ_1,τ_2) where τ_1 is a test for A and τ_2 is a test for B.*
>
> *A test for $A\Rightarrow B$ is a pair (θ,τ) where θ is a proof of A and τ a test for B.*

[4]Rigorously speaking, there are differences between $\&_x$, $\oplus_x$ on one hand and $\forall x$, $\exists y$ of the first-order on the other. This has been studied in Fleury and Quatrini (2004). First-order in ludics may be obtained, provided a condition of *uniformity* on proofs is added. Namely, all the proofs of the formulae $F(x)$ must be identical to get $\forall xF(x)$ or $\exists xF(x)$.

By using temporarily the notation A^t for "a test for A", we may summarize these observations by

$$(A \wedge B)^t \ = \ A^t \vee B^t \quad (A \vee B)^t \ = \ A^t \wedge B^t \quad (A \Rightarrow B)^t \ = \ A \wedge B^t$$

That suggests the identification between A^t and $\neg A$, based on the principle

$$\text{a test for } A = \text{a proof of } \neg A$$

Nevertheless if we look carefully at these statements, we of course notice that various instances of "and" are used. When dealing with implication, for instance, the proof of A and the test for B must be performed together, while when dealing with disjunction, the two conjuncts (τ_1, τ_2) are only potential, only one of the two is activated: there is a choice, depending on whether it is A which is tested or B. This leads to a sequence of improved statements, where we recognize the usual connectives of linear logic:

$$(A\&B)^t \ = \ A^t \oplus B^t \quad (A \oplus B)^t \ = \ A^t\&B^t \quad (A \multimap B)^t \ = \ A \otimes B^t$$

and to the principle according to which

$$\text{a test for } A = \text{a proof of } A^{\perp}$$

Of course, this introduces a strong change with regard to our habits: if there is a proof of A, there is no proof of $\neg A$! Therefore, we are led to the question of the status of these strange things which are supposed to be "proofs" of $\neg A$ (or of $A^{\perp}$) when we want to prove A. Intuitively, both are attempts to prove something... except that, on good occasions, one proof is succeeding and the other not. Let us stay neutral for a while: both are said to be *paraproofs* (or, in French, *épreuves*, according to a suggestion by Livet). What we gain with that is the idea that paraproofs are interacting in such a way that, in good cases, one is succeeding and the other not, but this compels us to give *failure* a status. This is done by a new particular positive rule which may apply to any positive sequent. It is called a *daimon* and is denoted by †. Of course, from the logical viewpoint, it is not a correct rule; it may be written as

$$\frac{}{\vdash \Gamma}\ \dagger$$

where Γ is any sequence of formulae. It is a *paralogism*, something which can be interpreted as an *abandon* in the proof-search. Of course, when there are two proof processes activated such that one is trying to prove $\vdash A$ and the other $\vdash A^{\perp}$, if one of the two ends up with an application of †, then the other wins.

It may be shown that the hypersequentialized calculus (from now on HS) with the † rule is strongly normalizable (Girard, 2006, p. 298).

12.2.3 *An easy application to natural language*

Let us take as an illustration the following sentence (see Lecomte and Quatrini, 2010):

Example 12.9. Every linguist speaks an African language

The Montagovian framework leads us to consider that it has two logical representations:

$$S_1 = \forall x(L(x) \Rightarrow \exists y(A(y) \wedge P(x,y))) \tag{12.20}$$

$$S_2 = \exists y(A(y) \wedge \forall x(L(x) \Rightarrow P(x,y))) \tag{12.21}$$

where $L(x)$ means "x is a linguist", $A(y)$ means "y is an African language" and $P(x,y)$ means "x speaks y".

Let us assume for a while that we may change classical connectives into linear ones.[5]

In HS, the beginning of a proof for S_1 is provided by

$$\dfrac{\dfrac{\begin{matrix}\mathcal{D}_{d'} \\ \vdots\end{matrix} \quad \dfrac{\downarrow A^{\perp}(e_d) \vdash \qquad \downarrow P^{\perp}(d,e_d) \vdash}{\vdash \downarrow L^{\perp}(d), \exists y(\uparrow A(y) \otimes \uparrow P(d,y))} \quad \begin{matrix}\mathcal{D}_{d''} \\ \vdots\end{matrix}}{(\forall x(\uparrow L(x) \multimap \exists y(\uparrow A(y) \otimes \uparrow P(x,y))))^{\perp} \vdash}}{\vdash S_1}$$

and it may be opposed to a "proof" of $\vdash S_1{}^{\perp}$ or $S_1 \vdash$, which has the following aspect:

$$\dfrac{\dfrac{\downarrow L^{\perp}(f) \vdash \qquad \dfrac{\begin{matrix}\vdots \\ \vdash \downarrow A(a')^{\perp}, P(f,a')^{\perp}\end{matrix} \quad \begin{matrix}\vdots \\ \vdash \downarrow A(a)^{\perp}, P(f,a)^{\perp}\end{matrix} \quad \ldots}{(\forall y \uparrow A(y) \multimap \downarrow P^{\perp}(f,y))^{\perp} \vdash}}{\vdash \exists x(\uparrow L(x) \otimes \forall y(\uparrow A(y) \multimap \downarrow P^{\perp}(x,y))}}{S_1 \vdash}$$

[5] We know that classical logic may be embedded into linear logic, by using the exponentials; this is not attempted here since the proper ludical interpretation of exponentials would lead us too far. In any case, linear logic connectives are more fine-grained.

These two (para)proofs may be put "in interaction" (see below). At the first level (starting from the bottom), the right formula behaves as a *test* for the left one. Since they are dual, we could put a cut between them; when the cut is eliminated, we are sent to a second level of the interaction, where roles are exchanged.

First step:

P

$$\dfrac{\begin{matrix}\mathcal{D}_{d'} \\ \vdots\end{matrix} \quad \begin{matrix}\mathcal{D}_{d} \\ \vdots \\ \vdash\downarrow L^{\perp}(d), \exists y(\uparrow A(y)\otimes\uparrow P(d,y)))\end{matrix} \quad \begin{matrix}\mathcal{D}_{d''} \\ \vdots\end{matrix}}{(\forall x(\uparrow L(x)\multimap\exists y(\uparrow A(y)\otimes\uparrow P(x,y))))^{\perp}\vdash}$$

P is ready to give justifications for every individual: $d, d', \ldots$

O

$$\dfrac{\downarrow L^{\perp}(f)\vdash \quad (\forall y\uparrow A(y)\multimap\downarrow P^{\perp}(f,y))^{\perp}\vdash}{\vdash\exists x(\uparrow L(x)\otimes\forall y(\uparrow A(y)\multimap\downarrow P^{\perp}(x,y))}$$

O proposes an individual f (claiming that f is a linguist and doesn't know any African language)

Second step:

P

$$\dfrac{\downarrow A(e_f)\vdash \quad \downarrow P(f,e_f)\vdash}{\vdash\exists y(\uparrow A_y\otimes\uparrow P(f,y)))}$$

P shows a language e_f (saying that e_f is an African language and that f speaks e_f)

O

$$\dfrac{\begin{matrix}\vdots \\ \vdash\downarrow A(a')^{\perp}, P(f,a')^{\perp}\end{matrix} \quad \begin{matrix}\vdots \\ \vdash\downarrow A(a)^{\perp}, P(f,a)^{\perp}\end{matrix} \ \ldots}{(\forall y\uparrow A_y\multimap\downarrow P^{\perp}(f,y))^{\perp}\vdash}$$

O is ready to receive a new claim from P, for some language a or $a'\ldots$

Third step:
Continuing to go towards the top of the paraproofs, it may happen in the interaction that a proof (here, the right one) ends up with positive daimons, thus stopping the research. The process of finding a proof terminates.

P

$$\dfrac{\downarrow A(e_f)\vdash \quad \downarrow P(f,e_f)\vdash}{\vdash\exists y(\uparrow A_y\otimes\uparrow P(f,y)))}$$

P shows an African language

O

$$\dfrac{\dfrac{}{\vdash\downarrow A(a')^{\perp}, P(f,a')^{\perp}}\dagger \quad \dfrac{}{\vdash\downarrow A(a)^{\perp}, P(f,a)^{\perp}}\dagger \ \ldots}{(\forall y\uparrow A(y)\multimap\downarrow P^{\perp}(f,y))^{\perp}\vdash}$$

O is ready to withdraw

Nevertheless, the interaction could continue. We may notice that the fact that "f is a linguist" was not questioned; it is something that may be tested. Because it is an atomic fact, the claim may be checked by means of a data base; in that case, the branch does not stop on a daimon, but on a particular positive rule, of null arity:

P

$$\dfrac{\mathcal{D}_{d'} \quad \dfrac{\dfrac{\dfrac{\begin{matrix}\mathcal{D}_f\\ \vdots\end{matrix}}{\vdash \exists y(\uparrow A(y) \otimes \uparrow P(f,y)))}}{L(f) \vdash \exists y(\uparrow A(y) \otimes \uparrow P(f,y)))}}{\vdash \downarrow L^{\perp}(f), \exists y(\uparrow A(y) \otimes \uparrow P(f,y)))} \quad \mathcal{D}_d}{(\forall x(\uparrow L(x) \multimap \exists y(\uparrow A(y) \otimes \uparrow P(x,y))))^{\perp} \vdash}$$

P checks
that f is a linguist

O

$$\dfrac{\dfrac{\dfrac{}{\vdash L(f)}\emptyset}{\downarrow L^{\perp}(f) \vdash} \quad (\forall y \uparrow A(y) \multimap \downarrow P^{\perp}(f,y))^{\perp} \vdash}{\vdash \exists x(\uparrow L(x) \otimes \forall y(\uparrow A_y \multimap \downarrow P^{\perp}(x,y))}$$

O may ensure that f
is a linguist (by means of a datum)

The fact that e_f is an African language, or that f really speaks e_f may also be challenged and the relative branches may also stop on "facts" attested by some data base, which yields the following proof $\mathcal{D}'$:

$$\dfrac{\begin{matrix}\mathcal{D}_{d'}\\ \vdots\end{matrix} \quad \dfrac{\dfrac{\dfrac{\dfrac{}{\vdash A(e_f)}\emptyset}{\downarrow A^{\perp}(e_f) \vdash} \quad \dfrac{\dfrac{}{\vdash P(f,e_f)}\emptyset}{\downarrow P^{\perp}(f,e_f) \vdash}}{\cdots}}{\vdash \downarrow L^{\perp}(f), \exists y(\uparrow A(y) \otimes \uparrow P(f,y))} \quad \begin{matrix}\mathcal{D}_{d''}\\ \vdots\end{matrix}}{(\forall x(\uparrow L(x) \multimap \exists y(\uparrow A(y) \otimes \uparrow P(x,y))))^{\perp} \vdash}$$

We shall see in the following that this proof interacts with the paraproof $\mathcal{E}$ in Fig. 12.1. Of course, it may be objected that nothing very new appears in this presentation since, like in the usual frameworks, we are starting

$$\dfrac{\dfrac{\dfrac{}{\vdash L(f)}\emptyset}{\downarrow L^{\perp}(f) \vdash} \quad \dfrac{\cdots \quad \dfrac{\dfrac{\dfrac{\dfrac{\dfrac{\dfrac{}{\vdash}\dagger}{P(f,e) \vdash}(P,\{\emptyset\})}{\vdash \downarrow P^{\perp}(f,e)}}{A(e) \vdash \downarrow P^{\perp}(f,e)}(A,\{\emptyset\})}{\vdash \uparrow A^{\perp}(e), \uparrow P^{\perp}(f,e) \ldots}}{(\forall y \downarrow A(y) \multimap \uparrow P^{\perp}(f,y))^{\perp} \vdash}}{\vdash \exists x(\uparrow L(x) \otimes (\forall y(\uparrow A(y) \multimap \downarrow P^{\perp}(x,y))}$$

Fig. 12.1 The paraproof $\mathcal{E}$

from formulae which are supposed to be correct translations of the sentence at hand. The question "where does such formulae come from?" may still be posed. At the present moment, nevertheless, we may observe that we need neither truth values nor models in order to interpret our formulae: the notion of (para)proof is enough for that. We may say that the interpretation of the formula S_1 associated with the sentence is provided by all the potential interactive dialogues it contains and which may arise in various situations. In some sense, if we see these interaction steps as inference steps, the *meaning* of the formula is in the kind of inference it allows when confronted with a testing procedure. For instance, when I assert that *Every linguist speaks an African language*, I am committed to justify that some particular language L is an African one, and to justify that somebody I chose, f, really speaks that language. That also assumes that I am able to recognize a language, to recognize a linguist and to recognize what it means to speak such and such language. It may be presumed that on this basis, I can answer any question like *What African language does Chomsky speak?* (for instance). This view is more in accordance with an *inferentialist* conception of semantics (*à la* Brandom, 2000) than with a *referentialist* one. This is in fact the track that we will pursue in this chapter.

However, we may not only content ourselves with this view; we would like the process to be inverted, that is, from the potential dialogues arising from a sentence to infer a formula which encompasses all these potentialities. That is in fact what we shall be able to do in the next sections.

12.2.4 *Designs*

12.2.4.1 *Introduction*

The jump from the above (hypersequentialized) calculus to ludics properly speaking rests on a decisive step which gets rid of formulae as bases of a logical calculus. Formulae are replaced by locations (*loci*) and paraproofs by so-called *designs* (that is, at first sight, trees of addresses) on which a procedure of interaction may be defined, which can be geometrically characterized as *orthogonality*. Since in the *formulae as types* principle, formulae may be understood as types, designs, the nodes of which are not labelled by formulae, can be compared with *untyped λ-terms*. Ludics is therefore to the hypersequentialized calculus what the untyped λ-calculus is to the typed one. How to recover types (or formulae)? If a type is supposed to embody the behaviour of an object with regard to others in a same space, we shall therefore take the set of interactions of a given object to be its type, or, if we wish, the formula it inhabits. Behaviours are here seen in reference with a precise procedure: interaction. Hence the fundamental result that we need (and that is demonstrated in Girard, 2001): the fact that in ludics, objects may be discriminated by means of the interactions they have with the other objects (*separation theorem*).

12.2.4.2 *Designs as proof skeletons*

Let us take again the example of the following rules for synthetic connectives:

$$\frac{\vdash P, R, \Lambda \qquad \vdash Q, R, \Lambda}{(P^\perp \oplus Q^\perp) \otimes R^\perp \vdash \Lambda}\{\{P, R\}, \{Q, R\}\} \qquad \frac{P \vdash \Gamma \qquad R \vdash \Delta}{\vdash (P^\perp \oplus Q^\perp) \otimes R^\perp, \Gamma, \Delta}\{P, R\}$$

$$\frac{Q \vdash \Gamma \qquad R \vdash \Delta}{\vdash (P^\perp \oplus Q^\perp) \otimes R^\perp, \Gamma, \Delta}\{Q, R\}$$

Let us assume that the atomic formulae are stored at various addresses, these ones being words on $\mathbb{N}$ (the set of integers), and that $P^\perp, Q^\perp, R^\perp$ are located at the respective addresses 1, 2, 3. Let ξ be a variable of address. $\xi\star I$ denotes the set $\{\xi i; i \in I\}$ (ξi is often denoted $\xi \star i$). Then, if the formula $(P^\perp \oplus Q^\perp) \otimes R^\perp$ is supposed to be stored at the address ξ, the atomic formulae P, Q, R which occur in the premises are supposed to be stored at the respective addresses $\xi 1$, $\xi 2$ and $\xi 3$. 1, 2 and 3 thus occur as relative addresses. Sequents are normalized, in the sense that they are written down in the form of what Girard calls a *fork*, with at most one address on the left-hand side. The two types of sequents are therefore $\xi \vdash \Lambda$ and $\vdash \Lambda$. The proofs of sequents of the first (resp. second) type are said to have a negative (resp. positive) base.

Thus the first rule at the top left is a negative base rule (the positive connective is introduced on the left), whereas the two others are positive rules (the positive connective is introduced on the right). A negative rule is therefore a rule with several sequents as premises, each one containing several addresses on their right (and nothing on their left); these sequents are of the form $\vdash \xi\star J, \Lambda_J$. A positive rule is a rule with several premises, each one having one address on its left. Those premises are therefore of the $\xi \star i \vdash \Lambda_i$ form.

The three rules above become

$$\frac{\vdash \xi_1, \xi_3, \Lambda \quad \vdash \xi_2, \xi_3, \Lambda}{\xi \vdash \Lambda}(-, \xi\{\{1,3\},\{2,3\}\}) \qquad \frac{\xi_1 \vdash \Gamma \quad \xi_3 \vdash \Delta}{\vdash \xi, \Gamma, \Delta}(+, \xi, \{1,3\})$$

$$\frac{\xi_2 \vdash \Gamma \quad \xi_3 \vdash \Delta}{\vdash \xi, \Gamma, \Delta}(+, \xi, \{2,3\})$$

and the general schemata of the rules are the following:
Positive rule:

$$\frac{\ldots \quad \xi \star i \vdash \Lambda_i \quad \ldots}{\vdash \xi, \Lambda}(+, \xi, I)$$

Negative rule:

$$\frac{\ldots \quad \vdash \xi \star J, \Lambda_J \quad \ldots}{\xi \vdash \Lambda}(-, \xi, \mathcal{N})$$

where I and J are finite subsets of $\mathbb{N}$, $i \in I$, with the Λ_i pairwise disjoints, $\mathcal{N}$ is a set (possibly infinite) of finite subsets of $\mathbb{N}$, each J of the negative rule being an element of this set, all the Λ_J are included in Λ, and moreover each base sequent is well formed in the sense that all addresses are pairwise disjoint.

To that, we add the daimon:

$$\frac{}{\vdash \Lambda}\dagger$$

The (para)proofs which are drawn in this system are called *designs*. A design is therefore an alternate sequence of negative and positive rules, with occasionally a daimon to close it, but a design may be infinite (as we shall see later on, a dispute may be endless!). In fact, there is an infinity in both dimensions: in breadth and in depth. There may be infinity in breadth also because the negative rule admits cases where there would be an infinite amount of J (infinite branching).

The general form of a design is thus the following one:

$$\frac{\ldots \qquad \dfrac{\dfrac{\ldots \ \dfrac{\ldots}{\vdash \xi_{k_{11}} \star J_2, \Lambda_{J_m}} \ \ldots}{\xi_{k_{11}} \vdash \Lambda_{i_1}}(-, \xi_{k_{11}}, \mathcal{N}_2) \quad \dfrac{\ldots}{\xi_{k_{12}} \vdash \Lambda_{i_2}} \ \ldots \ \dfrac{\ldots}{\xi_{k_{1j}} \vdash \Lambda_{i_j}}}{\vdash \Lambda_{I_k}, \xi_{k_1}, \xi_{k_2}, \ldots, \xi_{k_{n_k}}}(+, \xi_{k_1}, J_1) \qquad \ldots}{\xi \vdash \Lambda}$$

At the first step, it is assumed that $\mathcal{N} = \{I_1, \ldots, I_{N_1}\}$. Only the positive sequent associated with k, that is relative to the choice of I_k in $\mathcal{N}$, has been represented as a premise. It contains the elements $k_1, k_2, \ldots, k_{n_k}$. At the second step, we have a positive rule for each premise of the previous step, chosen for instance in taking as the focus the first ξ_{k_l}, that is ξ_{k_1}, and that gives for each a sequence of negative premises indexed by a set J_1, which, itself, contains the elements $1, 2, \ldots, j$. The Λ_{i_p} of the numerator absorb the ξ_{k_l} which have not yet been used as foci (but they will do!). Then, the design is continued by making the steps alternate.

The positive rule may be interpreted as

- the player selects a locus ξ to pursue the interaction, and
- from this locus, he asks a finite set of questions i;

and the negative rule as

- after having asked a question, the player considers a directory of possible answers $\mathcal{N}$ from his/her partner.

There is no axiom rule in this system. It is replaced by *Fax*, which allows us to make a correspondence between "identical" formulae occurring at different addresses (therefore not identical properly speaking, but simply "equiform"). *Fax* is the design which displays the proof of

$$A \vdash A$$

by performing all the possible decompositions of the formula A. It has therefore the recursive definition

$$Fax_{\xi,\xi'} = \cfrac{\cfrac{\cfrac{\ldots \quad Fax_{\xi_{i1},\xi'_{i1}} \quad \ldots}{\ldots \xi' \star i \vdash \xi \star i \ldots}\ (+,\xi',J_1)}{\ldots \quad \vdash \xi \star J_1, \xi' \quad \ldots}\ (-,\xi,\mathcal{P}_f(\mathbb{N}))}{\xi \vdash \xi'}$$

12.2.4.3 *Designs and chronicles*

Thinking of a **design** as a tree, the nodes of which are **forks** $\Gamma \vdash \Delta$, the root of which is its basis, built only by means of the three rules above (positive rule, negative rule and daïmon), a **design** can be viewed as a set of chronicles, where a chronicle is an alternate sequence of moves (or *actions*). More precisely:

Definition 12.3. A chronicle of basis $\Gamma \vdash \Delta$ is a non-empty sequence of alternate actions $k_0, \ldots, k_n$, where $k_i = (\epsilon, \xi_i, I_i)$, with ϵ a polarity, ξ_i an address and I_i a set of integers, such that

- if the basis is negative (resp. positive), k_0 has the polarity $-$ (resp. $+$),
- only the last action, k_n may be the daimon $= (+, \dagger)$,
- if we call the focus the locus of the active formula, a negative action k_i has, as its focus, either the unique element of Γ (in this case, the first action), or an element of $\xi_{i-1} \star I_{i-1}$,
- the focus of a positive action k_i is either an element ξ_i of Δ or an element of $\xi_q \star I_q$, where $(-, \xi_q, I_q)$ is a previous negative action,
- the foci are pairwise distinct.

We may make more precise the meaning of design:

Definition 12.4. An (e)-design of basis $\Gamma \vdash \Delta$ is a set $\mathcal{D}$ of basic chronicles $\Gamma \vdash \Delta$, such that

- $\mathcal{D}$ is a forest,
- the chronicles of $\mathcal{D}$ are pairwise coherent (if two chronicles differ, it is at first on a negative action, and if these negative actions have not the same focus, in the continuation, the two chronicles have never the same foci),
- if a chronicle ends, its last action is positive,
- if the basis is positive, then $\mathcal{D}$ is non-empty.

We have now objects which can be associated with "players", assuming that player 1 (resp. 2) is characterized by a set of chronicles which tell him/her what to do when facing an action of player 2 (resp. 1).

The following example will be named "the real estate example" from now on.

Example: A person X wishes to know whether another one, Y, who is supposed to sell some houses and flats, will sell a goods denoted by 1. X knows that Y possesses three goods: 1, 2 and 3. X wishes to know

(1) whether Y will sell 1;
(2) in case Y would sell 1, at what price?

The strategy of X consists in beginning the exchange by asking the question: $0 =$ *What goods do you wish to sell?* This is represented by a positive move, starting from an empty address, here denoted by $\langle\ \rangle$.

$$\frac{0 \vdash}{\vdash \langle\rangle}$$

Starting from there, X must foresee all the possible answers by Y (therefore the set of the non-empty subsets of $\{1, 2, 3\}$):

$$\frac{\vdash 01 \quad \vdash 01, 02 \quad \vdash 01, 03 \quad \vdash 01, 02, 03 \quad \vdash 02 \quad \vdash 02, 03 \quad \vdash 03}{0 \vdash}$$

X then foresees that for some of the answers by Y, (s)he will not pursue the dialogue, because they don't interest him or her, hence,

$$\frac{\vdash 01 \quad \vdash 01, 02 \quad \vdash 01, 03 \quad \vdash 01, 02, 03 \quad \frac{}{\vdash 02}\dagger \quad \frac{}{\vdash 02, 03}\dagger \quad \frac{}{\vdash 03}\dagger}{0 \vdash}$$

In the cases where 1 is for sale, X must plan to ask the price. For instance, if 1 is the only good for sale, (s)he plans to ask: 1 = *At what price*? Hence,

$$\frac{011 \vdash}{\vdash 01}$$

On the left-hand side of the sequent, there is a representation of the history of the game: X asked the question 0 and obtained the answer 1; now (s)he asks the question 1. X must be ready to receive any price i:

$$\frac{\dfrac{... \vdash 011i ...}{011 \vdash}}{\vdash 01}$$

and when (s)he gets the value of i, X stops the dialogue:

$$\frac{\dfrac{... \quad \frac{}{\vdash 011i}\dagger \quad ...}{011 \vdash}}{\vdash 01}$$

Because every branch is treated in this way, X's strategy (his or her design) may be represented by,

$$\frac{\dfrac{\dfrac{... \frac{}{\vdash 011i}\dagger ...}{011 \vdash}}{\vdash 01} \quad \dfrac{\dfrac{... \frac{}{\vdash 011i, 02}\dagger ...}{011 \vdash 02}}{\vdash 01, 02} \quad \dfrac{\dfrac{... \frac{}{\vdash 011i, 03}\dagger ...}{011 \vdash 03}}{\vdash 01, 03} \quad \dfrac{\dfrac{... \frac{}{\vdash 011i, 02, 03}\dagger ...}{011 \vdash 02, 03}}{\vdash 01, 02, 03} \quad \frac{}{\vdash 02}\dagger \quad \frac{}{\vdash 02, 03}\dagger \quad \frac{}{\vdash 03}\dagger}{\dfrac{0 \vdash}{\vdash \langle\rangle}}$$

Of course, on his or her side, Y has also his or her own strategy, which consists first in being ready to receive any question:

$$\frac{\vdash 0}{\langle\rangle \vdash}$$

It is the dual move with regard to X's move:

$$\frac{0 \vdash}{\vdash \langle\rangle}$$

Then (s)he answers the question by choosing a subset of $\{1, 2, 3\}$: that amounts to applying a positive rule.

$$\cfrac{\cfrac{0n \vdash \ldots \quad 0m \vdash}{\vdash 0}}{\langle\rangle \vdash}$$

In the next move, Y must be ready to receive the question 1 and (s)he is free to give the integer i, which corresponds to some price, if desired.

A design of Y may therefore be (among others)

$$\cfrac{\cfrac{\cfrac{\cfrac{011i \vdash}{\vdash 011}}{01 \vdash} \qquad 03 \vdash}{\vdash 0}}{\langle\rangle \vdash}$$

A game (a dialogue) may then be represented by a chronicle, in the sense given above. For instance,

$$(+, \langle\rangle, \{0\}), (-, 0, \{1, 3\}), (+, 01, \{1\}), (-, 011, \{i\})\dagger$$

is a chronicle. It consists for X in

(1) first, positive, move: X chooses a question (= a set I, here $\{0\}$),
(2) second, negative, move: Y chooses starting from the address 0 the subset $\{1, 3\}$,
(3) third, positive, move: X chooses $\{1\}$, extending the address proposed by Y (here 01),
(4) fourth, negative, move: Y chooses a price i, extending the address 011,
(5) the daimon stops the dialogue: X got all the information (s)he wanted.

Another chronicle is

$$(+, \langle\rangle, \{0\}), (-, 0, \{2, 3\}), \dagger$$

It is worth noticing that X must also foresee a design from Y which would consist in a refusal to sell anything. This design would be represented by

$$\cfrac{\cfrac{}{\vdash 0}\dagger}{\langle\rangle \vdash}$$

However, there is no one-to-one correspondence between i-designs and e-designs, even if every i-design uniquely defines an e-design. There are cases where an e-design allows several solutions for a corresponding i-design. Let us take, like Girard (2006, Chapter 13), the following i-design:

$$\dfrac{\dfrac{\dfrac{\dfrac{}{\sigma 2 \vdash \xi 30}\,(\sigma 2, \emptyset)}{\vdash \xi 30, \sigma}\,(\sigma, \{2\}) \qquad \dfrac{}{\vdash \xi 30, \xi 35, \sigma}\,\dagger}{\xi 3 \vdash \sigma}\,(\xi 3, \{\{0\}, \{0, 5\}\}) \qquad \dfrac{\dfrac{}{\vdash \tau}\,\dagger}{\xi 7 \vdash \tau}\,(\xi 7, \{\emptyset\})}{\vdash \xi, \sigma, \tau}\,(\xi, \{3, 7\})$$

Because τ is never focalized, its distribution does not matter: it could have been "given" to 3 or even destroyed; the result in terms of chronicles (and therefore e-designs) would have been the same, something that we can represent also by a tree, which collects *actions*, and summarizes the set of chronicles:

$$\dfrac{\dfrac{(\sigma, \{2\})}{(\xi 3, \{0\})} \quad \dfrac{\dagger}{(\xi 3, \{0, 5\})} \quad \dfrac{\dagger}{(\xi 7, \emptyset)}}{(\xi, \{3, 7\})}$$

We now face a situation where we may change at wish our viewpoint on designs. Do we want a "syntactic" object, like a proof? We have i-designs; do we want "semantic" objects, like strategies in a game? We have e-designs.

12.2.4.4 *Defining interaction*

As it can be noticed, there is no explicit formulation of the cut rule in ludics. Nevertheless, as was seen, two designs may be confronted with each other via the fact that they contain the same loci but with opposite polarities. It is thus possible to characterize interaction in the following way:

Interaction is the coincidence of two dual loci in two different designs.

Like normalization in sequent calculus amounts to eliminating occurrences of the cut rule, normalization here also will coincide with resolving a locus with its dual. Then, *orthogonality* is defined via the notion of normalization, thus revealing the importance of geometrical notions. Let us recall that in a vector space, two subspaces are orthogonal if and only if their intersection is simply the null subspace $\{0\}$, and of course $\{0\}$ is self-orthogonal. In ludics, this role must be played by a particular design: it is obviously the daimon.

Two paraproofs are orthogonals if their interaction leads to it, or in other words, if there is a consensus: the game stops because the two players agree to end it. It is also said that the normalization of the interaction between the two objects is convergent. This notion is made more precise in what follows.

Namely, a closed net consists of a cut between the two following designs:

$$\dfrac{\dfrac{\begin{matrix}\mathcal{D}\\ \vdots\end{matrix}}{\vdash \xi}\,\kappa \qquad \dfrac{\begin{matrix}\mathcal{E}\\ \vdots\end{matrix}}{\xi \vdash}\,(\xi, \mathcal{N})}{}$$

The normalization of such a closed net is such that

- if κ is the daimon, then the normalized form is

$$\dfrac{}{\vdash}\,\dagger$$

  (this normalized net is called *Dai*),
- if $\kappa = (\xi, I)$, then if $I \notin \mathcal{N}$, normalization fails,
- if $\kappa = (\xi, I)$ and $I \in \mathcal{N}$, then we consider, for all $i \in I$, the design $\mathcal{D}_i$, sub-design of $\mathcal{D}$ of basis $\xi \star i \vdash$, and the sub-design $\mathcal{E}'$ of $\mathcal{E}$, of basis $\vdash \xi \star I$, and we replace $\mathcal{D}$ and $\mathcal{E}$ with, respectively, the sequences of $\mathcal{D}_i$ and $\mathcal{E}'$. In other words, the initial net is replaced by

$$\dfrac{\dfrac{\begin{matrix}\mathcal{D}_{i_1}\\ \vdots\end{matrix}}{\xi \star i_1 \vdash} \quad \ldots \quad \dfrac{\begin{matrix}\mathcal{E}'\\ \vdots\end{matrix}}{\vdash \xi \star i_1, \ldots, \xi \star i_n} \quad \dfrac{\begin{matrix}\mathcal{D}_{i_n}\\ \vdots\end{matrix}}{\xi \star i_n \vdash}}{}$$

  with a cut between each $\xi \star i_j \vdash$ and the corresponding "formula" $\xi \star i_j$ in the design $\mathcal{E}'$.

This definition formalizes the intuition we mentioned above, of an interaction which ends with a consensus (or, to the contrary, which diverges, like it is the case when $I \notin \mathcal{N}$).

Moreover, orthogonality may be now well defined: A design $\mathcal{D}$ is orthogonal to a design $\mathcal{E}$ ($\mathcal{D} \perp \mathcal{E}$) if and only if the elimination of the cut between the two leads to the net *Dai*. Let us look how to prove orthogonality of two designs, by taking again the real estate example. Let us consider the two following designs, linked by a cut at their basis:

$$\dfrac{\dfrac{\dfrac{\dfrac{\dfrac{\dfrac{}{\ldots \vdash 011i, 02\ldots}\,\dagger}{011 \vdash 02}\,(011, \{i; i \in \mathbb{N}\})}{\vdash 01, 02}}{0 \vdash}}{\vdash \langle\rangle} \qquad \dfrac{\dfrac{\dfrac{\dfrac{011i \vdash}{\vdash 011}}{01 \vdash \quad 02 \vdash}}{\vdash 0}}{\langle\rangle \vdash}}{}$$

It is possible, in several stages, to make the cut go up. For instance, a first stage leads to

$$\dfrac{\dfrac{\dfrac{\dfrac{\dfrac{}{\ldots \vdash 011i, 02\ldots}\,\dagger}{011 \vdash 02}\,(011, \{i; i \in \mathbb{N}\})}{\vdash 01, 02}}{0 \vdash} \qquad \dfrac{\dfrac{\dfrac{011i \vdash}{\vdash 011}}{01 \vdash \quad 02 \vdash}}{\vdash 0}}{}$$

a second stage to

$$\dfrac{\dfrac{\dfrac{011i \vdash}{\vdash 011}}{01 \vdash} \qquad \dfrac{\dfrac{\dfrac{}{\ldots \vdash 011i, 02\ldots}\,\dagger}{011 \vdash 02}\,(011, \{i; i \in \mathbb{N}\})}{\vdash 01, 02} \qquad 02 \vdash}{}$$

where X's design has been split into two parts, for readability, and where there is a cut between $01 \vdash$ and 01 which is in $\vdash 01, 02$ on one side, and another one between 02 which is in $\vdash 01, 02$ and $02 \vdash$ on the other side.

A third stage leads to

$$\dfrac{\dfrac{011i \vdash}{\vdash 011} \qquad \dfrac{\dfrac{}{\ldots \vdash 011i, 02\ldots}\,\dagger}{011 \vdash 02}\,(011, \{i; i \in \mathbb{N}\}) \qquad 02 \vdash}{}$$

then we get

$$\dfrac{011i \vdash \quad \dfrac{}{\vdash 011i, 02}\,\dagger \quad 02 \vdash}{}$$

with a cut between $011i \vdash$ and the $011i$ which occurs in the sequence of sequents $\{\vdash 011i, 02; i \in \mathbb{N}\}$ and between 02 and $02 \vdash$. Because each cut possesses a member obtained by the daimon, the interaction converges towards Dai.

12.2.4.5 *Separation*

How to use a design $\mathcal{D}$ of base $\vdash \sigma$? As Girard says: "simply by cutting it with a *counter-design* $\mathcal{E}$ of base $\sigma \vdash$ and normalizing the net they constitute". What is very important here is that designs are determined by their orthogonal, that means *their use*. Hence the importance of the *separation theorem*:

> If $\mathcal{D} \neq \mathcal{D}'$ then there exists a counter-design $\mathcal{E}$ which is orthogonal to one of $\mathcal{D}, \mathcal{D}'$ but not the other.

Otherwise, in order to play at his best, the player needs to organize his/her designs according to their possible uses. We therefore define a partial order on designs, made possible by the separation theorem which guarantees that the relation is indeed an order.

> $\mathcal{D} \preceq \mathcal{D}'$ if and only if every design orthogonal to $\mathcal{D}$ is also orthogonal to $\mathcal{D}'$ ($\mathcal{D}^{\perp} \subset \mathcal{D}'^{\perp}$).

In *From Foundations to Ludics* Girard (2003) makes this definition more intuitive: $\mathcal{D} \preceq \mathcal{D}'$ means that $\mathcal{D}'$ is "more convergent" than $\mathcal{D}$. We may think of $\mathcal{D}'$ as obtained from $\mathcal{D}$ by means of two operations:

- Enlargement: We add more premises for positive rules; in other words we replace $\mathcal{N}$ with $\mathcal{N}' \supset \mathcal{N}$ (in other words, fewer branches of the dialogue will lead to a dissensus).
- Shortening: Some positive rules (ξ, I) are replaced by simple occurrences of the daimon (on some branches, the dialogue ends earlier).

This definition gives sense to the notions of a smallest and a biggest design. The smallest one is not a design properly speaking; it leads the most often to dissensus, it has no branches and it is the empty set of chronicles. Girard calls it *Fid*. The biggest one always converges and at once; it is the daimon.

We may distinguish between a positive daimon and a negative one. Positive daimon:

$$\frac{}{\vdash \Lambda}\dagger$$

Negative daimon:

$$\dfrac{\ldots \quad \dfrac{}{\vdash \xi \star I, \Lambda}\,\dagger \quad \ldots}{\xi \vdash \Lambda}\;(\xi, \wp_f(\mathbb{N}))$$

If we take $\mathcal{N} = \emptyset$ in the negative rule, then obviously, for any move made by the other player, there will be a dissensus (the proponent has not enough information to answer the attacks of the opponent). This case is represented by the *skunk*:

$$\dfrac{}{\xi \vdash}\;(\xi, \emptyset)$$

Looking at the real estate example, we see that the design of X ($\mathcal{D}_X$) is orthogonal to several possible designs of Y, and its orthogonal therefore contains all these designs. Let us now imagine another design, $\mathcal{D}'_X$, which is orthogonal to still more designs; in other words, where X would have planned still more moves by his or her partner. This design would contain more chronicles ending with the daimon, and chronicles that would develop starting from other negative actions and also leading to the daimon.

Now, in the general case, when we consider the orthogonal of the set $\mathcal{D}^{\perp}$, then, obviously, it contains $\mathcal{D}$, but it also contains all the designs $\mathcal{D}'$ such that $\mathcal{D} \preceq \mathcal{D}'$. In other words, it is *stable by super-design.*

12.3 Behaviours

When we have a non-empty set of designs, we are interested in all the ways of extending it by means of other ones which would behave similarly with regard to normalization. The completed set is called a *behaviour*: we may actually think of it as the extension of a real behaviour (in the sense of a set of coherent primitive actions, which intends to reach some goal).

Definition 12.5. A *behaviour* is a set of designs (of the same basis) $\mathbf{C}$ which is equal to its bi-orthogonal ($\mathbf{C} = \mathbf{C}^{\perp\perp}$).

12.3.1 *Some particular behaviours*

We may now define the behaviour induced by a design $\mathcal{D}$, which is the bi-orthogonal of the set $\{\mathcal{D}\}$. In the example we have been dealing with until now, it contains all the designs of $\mathcal{D}_X$, and particularly the daimon of basis $\vdash \langle\rangle$, a design which did not occur in our example as a design of X, but that could be now chosen: it would correspond to an immediate issue from the dialogue, even before having asked the initial question.

Another example is given by the following design:

$$\frac{}{\vdash \xi}\,(\xi, \emptyset)$$

It is very particular. It is a positive design, therefore held by the proponent. If (s)he aims at a consensus, the opponent has not many choices since there is no address for answering (the proponent gave none). The opponent can therefore only oppose the daimon, but in doing so, (s)he admits that (s)he has lost the game! With such a design, the proponent always wins! Girard therefore calls it the "Nuclear Bomb", because it is the absolute weapon! Its only orthogonal design is therefore

$$\frac{\dfrac{}{\vdash}\,\dagger}{\xi \vdash}$$

but this last design is also orthogonal to

$$\frac{}{\vdash \xi}\,\dagger$$

Therefore we have a behaviour which contains *two* designs: the Bomb and the daimon. Let us denote by **1** this behaviour: it will be the neutral element of the tensor. But instead of the Bomb, we may take the skunk and look for the negative behaviour which contains it. It will be denoted by $\top$. What is the orthogonal of $\top$? Since the other player does not give me any set of addresses to play, I can only play the daimon; in other words I lose. We get

$$\top^{\perp} = Dai$$

and

$$\top = Dai^{\perp}$$

that is, all the negative designs of the same basis.

$$\top = \left\{ \frac{\vdots}{\langle\,\rangle \vdash}\,(\langle\,\rangle, \mathcal{N})\ ;\ \mathcal{N} \subset \wp_f(\mathbb{N}) \right\}$$

Finally, if we content ourselves simply with the design

$$\frac{}{\langle\,\rangle \vdash}\,(\langle\,\rangle, \emptyset)$$

we shall obtain the same behaviour since it has also, as its orthogonal, $\mathcal{D}ai$. That means that all the branches of $\top$ which are associated with a non-empty directory are useless from the viewpoint of building a behaviour! We shall say that the *incarnation* of $\top$ is

$$|\top| = \left\{\frac{}{\langle\ \rangle \vdash}(\langle\ \rangle, \emptyset)\right\}$$

Given a design $\mathcal{D}$ belonging to a behaviour **C**, there exists in **C** the smallest element included in $\mathcal{D}$ which is, in some way, its best witness; we call it the *incarnation of $\mathcal{D}$ with regard to* **C**. It is the smallest part of a design which guarantees its belongness to a behaviour. This notion proves useful when looking for a new interpretation of connectives in ludics terms.

12.3.2 *What logic for ludics?*

12.3.2.1 *"Truth" and "falsehood"*

Having a notion of *win* (a design is said to be *winning* if it does not play the daimon and it is parsimonious and uniform) allows us to get back a notion of "truth": a behaviour could be said to be "true" when it contains a winning design, "false" when its orthogonal contains one. With this characterization, two designs which incarnate two orthogonal behaviours cannot be both true. Or else, if they are both winning, they are not orthogonal, which means that *their "dispute" goes on indefinitely*. But we can have behaviours which are neither true nor false, which is the case if the two players give up.

Probably this attribution of "truth" to a behaviour is not particularly adequate. Perhaps it would be better to speak of a "fruitful" behaviour. But this is only to show that even if we adopt an anti-realist conception,[6] it is still possible to introduce a duality which can be compared to "true vs false".

12.3.2.2 *Formulae and behaviours*

We may notice here that behaviours are sets of designs exactly like in intuitionistic logic, formulae are interpreted as sets of proofs. This

[6] An adoption compatible with but not required by ludics.

leads to the following correspondence:

formula	behaviour
proof	design

When we speak of the intersection of two behaviours, it is as if we spoke of the intersection of two formulae. In intuitionistic logic, this has an obvious meaning: an element of this set proves A and proves B and therefore it proves $A \wedge B$, but in ludics, things go differently. The notion of formula as a "spiritualist" notion evaporates and is replaced by that of locus. There are therefore two ways of speaking of intersection and union; one takes into account this concrete aspect of the formulae (their location) and the other one takes into account the possibility we have to "relocate" them (particularly by means of *Fax*).

12.3.2.3 *A localist viewpoint against a spiritualist one*

As explained by Girard (2006), it is clear that in set theory, the union is *localist*: it crucially depends on the elements of the sets. If we take two isomorphisms ϕ and ψ, we shall not have in general $X \cup Y = \phi(X) \cup \psi(Y)$. For instance, it is not enough to know the cardinality of X and that of Y to know that of $X \cup Y$. But it is possible to make the disjoint union of two sets X and Y. For that we shall have to take two one-to-one mappings ϕ and ψ such that $\phi(X)$ and $\psi(Y)$ are disjoint and we shall have of course $Card(X+Y) = Card(X) + Card(Y)$. The disjoint sum is therefore *spiritualist*, which means it does not depend on concrete location.

We have the same thing for the product. The classical Cartesian product is spiritualist, because we have $Card(X \times Y) = Card(X) \times Card(Y)$, but what is, in this case, the analogue of the localist union? Girard gives as an example the product of two power sets $\wp(X)$ et $\wp(Y)$, denoted by $\wp(X) \boxtimes \wp(Y)$ and defined as $\wp(X \cup Y)$, in other words the set $\{x \cup y;\ x \in X \wedge y \in Y\}$.

In ludics, it is possible to perform some operations like in the case of the disjoint union. If a locus σ is given, we may define two applications ϕ and ψ which will be seen as *relocations*:

$$\phi(\sigma \star i \star \tau) = \sigma \star 2i \star \tau \qquad \psi(\sigma \star i \star \tau) = \sigma \star (2i+1) \star \tau$$

The images by ϕ and ψ are still designs. If $\mathbf{G}$ and $\mathbf{H}$ are behaviours of the same basis σ, then $\phi(\mathbf{G})$ and $\psi(\mathbf{H})$ are disjoint. Or, more exactly, when they are positive behaviours, their intersection is the smallest behaviour on a positive basis, that is, $\mathbf{0} = \{Dai\}$, and when they are negative behaviours, every design $\mathcal{D}$ which belongs to both is such that the intersection between its incarnation with regard to one and its incarnation with regard to the other is empty.

12.3.3 *Sums and products*

If A is an arbitrary set of designs, then $A^{\perp}$ is a behaviour (since $A^{\perp\perp\perp} = A^{\perp}$). Reciprocally, every behaviour on a basis $\langle\rangle \vdash$ or on a basis $\vdash \langle\rangle$ is the orthogonal of a set of designs. For instance, if $A = \emptyset$, $A^{\perp}$ is the set of designs which converge with no other one, that is, which converge only with *Dai*. It is $\top$, the behaviour that the skunk is the incarnation of.

Otherwise, since behaviours are sets, nothing prevents us from making unions, intersections and products of them. It is possible to show

The intersection of two behaviours is a behaviour.

We may also define a behaviour from a union. Let us write

$$\mathbf{G} \sqcup \mathbf{H} = (\mathbf{G} \cup \mathbf{H})^{\perp\perp}$$

which will allow us to retrieve the connectives of linear logic.

12.3.3.1 *Additives*

The case of additives corresponds to the case where behaviours are disjoint.

Definition 12.6. A *directory* is a subset of $\wp_f(\mathbb{N})$ (= a set of "ramifications"). If $\mathbf{G}$ is a positive behaviour on $\vdash \langle\rangle$, then the directory of $\mathbf{G}$, denoted by $Dir(\mathbf{G})$, is the set of the sets of indexes I such that $(+, \langle\rangle, I)$, that is, the first action of a design belonging to $\mathbf{G}$.

If $\mathbf{G}$ is a negative behaviour on $\xi \vdash$, then the directory of $\mathbf{G}$ is the set of sets of indexes I for which, for any design incarnated in $\mathbf{G}$, there is an action $(-, \xi, I)$ as the initial action.

This definition entails (Curien, 2003, p. 43) that the directory of $\mathbf{G}$ and that of $\mathbf{G}^{\perp}$ are identical.

Definition 12.7. Two behaviours $\mathbf{G}$ and $\mathbf{G}'$ on the same basis are said to be disjoint if their directories are disjoint.

If two negative behaviours **G** and **H** are disjoint and if $\mathcal{D}_1$ and $\mathcal{D}_2$ are incarnations of these behaviours, then their union is well defined, and it is a design of $\mathbf{G} \cap \mathbf{H}$. Its first action has, as its set of ramifications, the union of that of $\mathcal{D}_1$ and of that of $\mathcal{D}_2$; it is therefore a design which is inside **G** and inside **H**, but it is at the same time a pair of designs, thus leading to

$$|\mathbf{G}| \times |\mathbf{H}| = |\mathbf{G} \cap \mathbf{H}|$$

where $|\mathbf{G}|$ denotes the set of incarnations with regard to **G**. In other words, every incarnated design of $\mathbf{G} \cap \mathbf{H}$ is obtained by taking a design incarnated in **G** and a design incarnated in **H**, making the union of the chronicles. It is what Girard calls the mystery of incarnation (the fact that intersection and Cartesian product coincide).

It is then possible to define the additive connectives.

Definition 12.8. If **G** and **H** are two disjoint negative behaviours, then, we put

$$\mathbf{G} \& \mathbf{H} = \mathbf{G} \cap \mathbf{H}$$

If they are positive, we put

$$\mathbf{G} \oplus \mathbf{H} = \mathbf{G} \cup \mathbf{H}$$

$\oplus$ is therefore a spiritualist connective.

12.3.3.2 *Multiplicatives*

Definition 12.9. Let $\mathcal{U}$ and $\mathcal{B}$ be two positive designs; we define the tensor product $\mathcal{U} \odot \mathcal{B}$ by

- if one of the two is the daimon, then $\mathcal{U} \odot \mathcal{B} = Dai$,
- if not, let $(+, \langle\rangle, I)$ and $(+, \langle\rangle, J)$ be the first actions of respectively $\mathcal{B}$ and $\mathcal{U}$, if $I \cap J \neq \emptyset$, then $\mathcal{U} \odot \mathcal{B} = Dai$. If not, we replace in each chronicle of $\mathcal{B}$ and $\mathcal{U}$ the first action with $(+, \langle\rangle, I \cup J)$, which gives respectively $\mathcal{B}'$ and $\mathcal{U}'$, then $\mathcal{U} \odot \mathcal{B} = \mathcal{U}' \cup \mathcal{B}'$.

It is then possible to define the tensor product $\otimes$ of two behaviours by means of relocations, but notice that ludics also provides us with new connectives, with regard to linear logic. These relocations (for instance the previous ϕ and ψ which send biases following the locus of the basis either to odd integers or to even ones) allow the first actions respective to the two designs which incarnate the two behaviours to have disjoint ramifications. We may define the product $\odot$ of two behaviours by $\mathbf{F} \odot \mathbf{G} = \{\mathcal{A} \odot \mathcal{B}; \mathcal{A} \in \mathbf{F}, \mathcal{B} \in \mathbf{G}\}$, and then the tensor properly speaking by $\mathbf{F} \otimes \mathbf{G} = (\mathbf{F} \odot \mathbf{G})^{\perp\perp}$.

Under full completeness (that is, under the assumption that the bi-orthogonal does not contain more than the initial behaviour), every design $\mathcal{D}$ in $\mathbf{F} \otimes \mathbf{G}$ may be decomposed according to $\mathcal{D} = \mathcal{D}_1 \odot \mathcal{D}_2$ with $\mathcal{D}_1 \in \mathbf{F}$ and $\mathcal{D}_2 \in \mathbf{G}$.

Nevertheless, let us go back to our new connective by means of a striking example where we see that the locative viewpoint enables one to take into account some "non-logical phenomena", like the possibility to obtain *true* $\odot$ *true* = *false*. Let $\mathbf{G} = \{\mathcal{D}^{\perp\perp}\}$, where $\mathcal{D}$ is the design:

$$\cfrac{\cfrac{}{0 \vdash}\,(0, \emptyset)}{\vdash \langle\rangle}\,(\langle\rangle, \{0\})$$

In other words, I ask the question 0, then I don't open any possible answer for my opponent, therefore, of course, I win! Let us calculate $\mathbf{G} \odot \mathbf{G}$. Since $I \cap J \neq \emptyset$, the definition gives us $\mathbf{G} \odot \mathbf{G} = Dai$; it is the behaviour which contains the only design:

$$\cfrac{}{\vdash \langle\rangle}\,\dagger$$

It is the behaviour which immediately loses. Therefore we may very well have

$$true \odot true = false$$

We can have the following interpretation. Let us imagine $\mathbf{G}$ to be the behaviour generated by a design $\mathcal{D}$ which consists in the possibility to assert 0 from a situation located in $\langle\rangle$. Let $\mathbf{H}$ be the behaviour generated by the following $\mathcal{E}$:

$$\cfrac{\ldots \quad \cfrac{}{\vdash n}\,\dagger \quad \ldots}{\langle\rangle \vdash}\,(-, \langle\rangle, \{\{n\}; n \in \mathbb{N}\})$$

which can be interpreted as "in the situation $\langle\ \rangle$ as soon as you give me an argument, I shall give up".

Of course, $\mathcal{D}$ and $\mathcal{E}$ are orthogonal, therefore $\mathcal{D} \in \mathbf{H}^{\perp}$, something that we translate by saying that $\mathbf{G}$ wins against $\mathbf{H}$. Moreover, $\mathbf{G}$ is *true* since it has a winning design. But $\mathbf{G}$ does not contain only $\mathcal{D}$, it also contains de-located designs of $\mathcal{D}$. For instance, we have seen that the argument given in $\mathcal{D}$ was 0, we can now imagine a variant with the argument 1. Let $\mathcal{D}'$ be this variant; it is a design of basis $\vdash \langle\rangle$ and its only chronicle is $(\langle\rangle, \{1\})$.

Let us compare $\mathcal{D} \odot \mathcal{D}$ and $\mathcal{D} \odot \mathcal{D}'$. $\mathcal{D} \odot \mathcal{D}'$ is no longer orthogonal to $\mathcal{E}$, therefore it no longer belongs to $\mathbf{H}^{\perp}$. It is now orthogonal to what we get with another design $\mathcal{E}'$ of the kind

$$\frac{\dots \frac{}{\vdash n, n'}\dagger \quad \dots}{\langle\rangle \vdash} \left(\{\{n, n'\}; n, n' \in \mathbb{N}, n \neq n'\}\right)$$

which is now interpreted as "give me two arguments, and I shall give up". Because $\mathcal{D} \odot \mathcal{D}'$ is orthogonal to $\mathcal{E}'$, $\mathcal{D} \odot \mathcal{D}'$ belongs to the orthogonal of the behaviour $\mathbf{H}'$ generated by $\mathcal{E}'$. But let us remember that $\mathcal{D} \odot \mathcal{D} = Dai$, and $\mathbf{G} \odot \mathbf{G} = \{Dai\}$. It follows that $\mathcal{D} \odot \mathcal{D}$ cannot be winning. If I have only one argument to give, I cannot give two. Hence the fact that if $\mathbf{G}$ is winning against *give me an argument*, in order for $\mathbf{G} \odot \mathbf{G}$ to be true, it should converge with a $\mathbf{H}'$ which would be interpreted as *give me two arguments*, which is not the case.

The tensor product thus expresses, like we expected, a resource sensitivity. From the viewpoint of a "perfect dialogue", each question calls for one answer and an answer does not satisfy two questions at the same time, or there must be two instances (situated in two different loci) of the answer, each answering a separate question.

12.3.3.3 *Back to the "logical form" example*

We said above that Example 12.9 (*Every linguist speaks an African language*) could be translated into logical formulae pertaining to hypersequentialized logic, then giving rise to potential dialogues, but we asked to what extent these formulae could be justified. Let us admit from now on that the *meaning* of a sentence (or the *proposition* that it expresses) is a behaviour, and that an utterance of it is some design $\mathcal{D}$ belonging to it (or the incarnation of such $\mathcal{D}$ with regard to the whole behaviour). That means that there exists a family of designs $\mathbb{E}$ such that this meaning coincides with the orthogonal $\mathbb{E}^{\perp}$ of this family. The designs $\mathcal{E}_i$ belonging to this family may be interpreted as tests or questions which positively interact (that is, converge) with $\mathcal{D}$ (or its incarnation). Let $\mathbb{H}$ be such a behaviour. Finding a formula (or a type) corresponding with this meaning simply amounts to trying *to decompose* $\mathbb{H}$ *into more elementary behaviours*, relying on the operations on behaviours we introduced in the previous section. Let us only sketch the approach:

(1) When Example 12.9 is uttered in one sense, it may be objected by the "second player" that either

- *There is a linguist who speaks no African language* or
- *There is no African language spoken by every linguist.*

(2) In the first case, the interaction has the following form:

$$\dfrac{\dfrac{\ldots\vdash x.1.n\ldots\vdash x.1.5\ldots\vdash x.1.m\ldots}{x.1\vdash}}{\vdash x}(1) \quad\Bigg|\quad \dfrac{\dfrac{x.1.5\vdash}{\vdash x.1}(c) \quad \vdash x.2}{x\vdash}$$

and it converges.

(3) In the second case, it has the form

$$\dfrac{\dfrac{\ldots\vdash x.1.n\ldots\vdash x.1.5\ldots\vdash x.1.m\ldots}{x.1\vdash}}{\vdash x}(1) \quad\Bigg|\quad \dfrac{\vdash x.1 \quad \dfrac{x.2.0\vdash}{\vdash x.2}(d)}{x\vdash}$$

and it diverges.

We draw as a conclusion that (1) is ambiguous and can be translated into a formula, the form of which is $\mathbb{S}_1 \oplus \mathbb{S}_2$. Pursuing this exploration, we reach the level of "atomic" behaviours $\mathbb{L}(x)$ (for "x is a linguist"), $\mathbb{A}(y)$ (for "y is an African language") and $\mathbb{P}(x, y)$ (for "x speaks y"), and we obtain the decomposition

$$\mathbb{S} = (\&_x(\downarrow\mathbb{L}(x) \multimap \oplus_y(\downarrow\mathbb{A}(y)\otimes \downarrow\mathbb{P}(x,y)))) \oplus \downarrow (\oplus_y(\downarrow\mathbb{A}(y) \otimes \&_x(\downarrow\mathbb{L}(x) \multimap\downarrow \mathbb{P}(x,y))))$$

However, we speak of "atomic" behaviours by *abus de langage* since we have no atom, properly speaking! These behaviours are more elementary but they still can be complex behaviours: all depend on the set of justifications that they convey. As seen above, various justifications may be asked at this level. Sometimes, so-called "facts" in a data base may suffice, but this is not the absolute law. On other occasions, disputations may be held, leading to unforeseen results (for instance, infinity).

Another point deserves attention with regard to normalization. There are many cases where the "disputation" does not terminate on a *Dai*, but on some *reduced net*. The two designs are not in this case really orthogonal but neither do they diverge. We are strongly interested in these cases since it seems that normalization provides side effects we are concerned with. The situation looks like the case of *speech acts* (Fleury and Tronçon, 2010) where words are uttered, resulting in side changes of the world seen as the set of pre- and post-conditions of the utterance.

The question *Which is the African language that Chomsky speaks?* can be associated with the following design (Lecomte and Quatrini, 2009):

$$
\dfrac{
  \dfrac{\dfrac{}{\vdash L(ch)}\,\emptyset}{\downarrow L^{\perp}(ch) \vdash}
  \qquad
  \dfrac{
    \dots \quad
    \dfrac{\dfrac{\dfrac{\dfrac{\dfrac{\begin{matrix}\mathcal{F}ax \\ A(e) \vdash A(e)\end{matrix}}{\vdash A(e), \downarrow A^{\perp}(e)}}{P(ch,e) \vdash A(e), \downarrow A^{\perp}(e)}}{\vdash A(e), \uparrow A^{\perp}(e), \downarrow P^{\perp}(ch,e)}}{\downarrow A(e)^{\perp} \vdash \downarrow A^{\perp}(e), \downarrow P^{\perp}(ch,e)}}{\vdash \uparrow A^{\perp}(e), \downarrow P^{\perp}(ch,e)), A}
    \quad \dots
  }{(\forall y(\uparrow A(y) \multimap \uparrow P^{\perp}(ch,y))^{\perp} \vdash A}
}{
  \dfrac{\vdash \exists x(\uparrow L(x) \otimes (\forall y(\uparrow A(y) \multimap \downarrow P^{\perp}(x,y)), A}{S \vdash A}
}
$$

in such a way that, by interaction with $\mathcal{D}'$, here reproduced,

$$
\dfrac{
  \dfrac{
    \begin{matrix}\mathcal{D}_{d'} \\ \vdots\end{matrix}
    \quad
    \dfrac{\dfrac{\dfrac{\dfrac{}{\vdash A(e_f)}\,\emptyset}{\downarrow A^{\perp}(e_f) \vdash} \quad \dfrac{\dfrac{}{\vdash P(f,e_f)}\,\emptyset}{\downarrow P^{\perp}(f,e_f) \vdash}}{\dots}}{\vdash \downarrow L^{\perp}(f), \exists y(\uparrow A(y) \otimes \uparrow P(f,y))}
    \quad
    \begin{matrix}\mathcal{D}_{d''} \\ \vdots\end{matrix}
  }{(\forall x(\uparrow L(x) \multimap \exists y(\uparrow A(y) \otimes \uparrow P(x,y))))^{\perp} \vdash}
}{\vdash S}
$$

it gives

$$
\dfrac{\dfrac{\dfrac{}{\vdash A(e)}\,\emptyset}{\downarrow A(e)^{\perp} \vdash}}{\vdash A}
$$

which is precisely a design associated with the African language which is spoken by Chomsky! The key point here resides in the use of a particular locus for A, which is the locus for the answer to the question, and in the use of the $\mathcal{F}ax$ procedure at the top of the questioning paraproof.

12.3.4 *Discursive relations*

After exploring the properties of ludics, the idea appears to apply them to discourse and dialogue, where the notions of *game* and *strategy* are central. Most previous approaches which attempted to take these aspects into account were based on *a priori* game rules. Games in their extensive forms are defined by means of pre-defined rules and a *trade-off* function, and they are planned to have a winner. The logic of conversation is not so rigid: players are not submitted to some external jurisdiction and they can play with the rules they assume, and can even change the rules during their interaction. Girard suggests a notion of game which is quite different. Let us suppose two players called Myself and You, where the former takes the positive viewpoint. If Myself has the design $\mathcal{D}$ and You the design $\mathcal{E}$, a *play* is the sequence of conversions used in the normalization of $[[\mathcal{D}|\mathcal{E}]]$ (that Girard calls a *dispute*). Designs are here strategies, and to lose is simply playing the daimon. Of course, players have not only one design each at their disposal; they have several. If Myself has the set of designs $\mathbf{G}$, then, in order to have convergent disputes, it is assumed that You has the set of designs $\mathbf{G}^{\perp}$. For each design $\mathcal{D}$ played by Myself, You has the choice of an answer among the set of her designs, which are all orthogonal to $\mathcal{D}$. Of course, if one of these designs ends with the daimon, that means she loses. Hence the proposition held by Girard:

> Myself and You behave relatively to two sets $\mathbf{G}$ and $\mathbf{G}^{\perp}$. The polarity condition must be seen as a constraint: the designs of $\mathbf{G}$ must be polar with regard to those of $\mathbf{G}^{\perp}$, and hence, $\mathbf{G}^{\perp}$ is "the game rule for $\mathbf{G}$", while $\mathbf{G}$ is "the game rule for $\mathbf{G}^{\perp}$".

For instance, Myself may have the Nuclear Weapon $(+, \xi, \emptyset)$ which obliges You to play her negative daimon,

$$\mathcal{D}ai^{-} = \cfrac{\cfrac{}{\vdash}\,\dagger}{\xi \vdash}$$

However, You can prevent the use of this Bomb by including in her set

$$\cfrac{\ldots \quad \cfrac{}{\vdash \xi \star I}\,\dagger \quad \ldots}{\xi \vdash}\ (\xi, \wp_f(\mathbb{N}) - \{\emptyset\})$$

since now the two sets are no longer orthogonal. If Myself plays the Bomb, the result will not be a win but a dissensus. Nevertheless, Myself wins, since all the sub-designs associated with a non-empty ramification end with the daimon.

This is a mere illustration of the fact that by envisaging situations of opposition (polarity) between two speakers, we are able to take into account the actions they perform against each other.

We claim that so-called "rules of discourse" may be explained by such interaction situations. For that, we argue that even monologic discourse is governed by two enunciative instances: the speaker herself and the *image* she has of a (potential) interlocutor. Only in concrete dialogues, this image is realized by a real other speaker but even then, images of the speakers enter into the field, that is, when there is a dialogue between A and B, we may argue that A has a dialogue with the image she has of B (let us give it the name $i(B)$) and B has a dialogue with the image she has of A or $i(A)$. The easy (but generally incorrect) situation occurs when $A = i(A)$ and $B = i(B)$. We argue in this section that discursive relations (Asher and Lascarides, 2003) (like *topicalization*, *elaboration*, *narration*, as well as *assertion*, *interrogation* and (*de*)*negation*) may be explained by the relationships these enunciative instances have with one another. These relationships are well represented by situations of interaction between designs.

12.3.4.1 *Topicalization*

Let us take topicalization as a first example, the relation according to which a speaker succeeds in anchoring one's discourse in a discursive topic. For instance, the first action in a discourse, for a speaker A, is a positive action, which starts from a positive fork:

$$\vdash \xi_1, \xi_2, \dots, \xi_n$$

The action can be $(+, \xi_1, \{0, 1, \dots, k\})$ if we wish to express the fact that the speaker A selects the focus ξ_1 and then, from this focus, develops some of its several aspects (for instance, asks questions on different aspects). We may see $\xi_1, \xi_2, \dots, \xi_n$ as potential *themes* which can be developed in the continuation of the discourse. When one of them is chosen, the remaining ones appear as the *context* in which it is developed. This situation occurs when the speaker plans to speak about her holidays, telling where she goes, with whom and by what means. Her design is

$$\dfrac{\dfrac{\dots}{\xi_{10} \vdash \Lambda_0} \; \dots \; \dfrac{\dots}{\xi_{1k} \vdash \Lambda_k}}{\vdash \xi_1, \xi_2, \dots, \xi_n}$$

where the Λ_i are pairwise disjoint and contained in $\{\xi_2, \dots, \xi_n\}$.

In fact, the starting point of the discourse may always be a single positive $\vdash \tau$ such that τ is the locus of the whole discourse which will take place. It only suffices to imagine *downward* actions: $\xi_1, \xi_2, \dots, \xi_n$ are renamed

$\tau_{01}, \tau_{02}, \ldots, \tau_{0n}$ and they appear as a premise for a negative rule, the basis of which is $\tau_0 \vdash$, which itself comes from a fictitious positive action $(+, \tau, \{0\})$, so that we get

$$\dfrac{\dfrac{\dfrac{\dfrac{\ldots}{\tau_{010} \vdash \Lambda_0} \quad \ldots \quad \dfrac{\ldots}{\tau_{01k} \vdash \Lambda_k}}{\vdash \tau_{01}, \tau_{02}, \ldots, \tau_{0n}}}{\tau_0 \vdash}}{\vdash \tau}$$

These actions may be seen as a reaction against the complementary actions of another speaker (or of an image of him) which consists in asking what the topic of the discourse is, and whose first actions are therefore

$$\dfrac{\dfrac{\ldots}{\tau \vdash}}{\vdash \langle\, \rangle}$$

12.3.4.2 *Elaboration and narration*

There are at least two ways a discourse may unfold. Asher and Lascarides (2003) define elaboration as a discursive relation which holds between two discursive structures α and β when β "describes in more detail some part of α" (p. 162). This relation is connected with *explanation*. They provide two examples:

a. *Max fell. John pushed him.*

b. *Alexis did really well in school this year. She got As in every subject.*

Intuitively, they say, "the proposition that John pushed him explains why the proposition that Max fell is true. And the proposition that Alexis got As in every subject elaborates the proposition that she did well in school this year". In ludics terms, we say that elaboration is a process, which can be represented by the following kind of design:

$$\dfrac{\dfrac{\dfrac{\dfrac{\dfrac{\ldots}{\xi_{0i0} \vdash \xi_{01}, \ldots, \xi_{0n}, \Delta}}{\vdash \xi_{01}, \ldots, \xi_{0i}, \ldots, \xi_{0n}, \Delta}\, a_2}{\xi_0 \vdash \Delta}}{\vdash \xi, \Delta}\, a_1}{\mathsf{TOP} : \textit{holidays}}$$

The steps a_1 and a_2 may be materialized by concrete utterances:

a_1. *I go to the mountain.*

a_2. *I like skiing.*

The design may be commented on in the following way:

(1) starting from a topic "holidays",
(2) the speaker A focalizes on one locus belonging to that topic: where she goes for holidays,
(3) planning various sub-themes (or subparts) of this focus,
(4) and finally choosing one: *skiing*,
(5) and still planning various ways of addressing this new focus.

What is relevant in this process is that each focus is selected among the loci describing the subparts of the previous one. This fact is marked in the indices which keep track of the history of the process: the sequence of integers which appears on the focus put on the negative side at some positive step is kept as a prefix for the next focus.

However, this description would be insufficient without its orthogonal since it would remain a mere description. If the process is so, it is because it has an orthogonal one which looks like

$$\dfrac{\dfrac{\dfrac{\dots}{\xi_{01} \vdash \Delta_1}, \dots \quad \dfrac{\dfrac{\dots}{\vdash \xi_{0i0}, \dots, \xi_{0im}, \Delta_i}}{\xi_{0i} \vdash \Delta_i} \quad \dots \dfrac{\dots}{\xi_{0n} \vdash \Delta_n}}{\vdash \xi_0, \Delta}}{\xi \vdash \Delta}$$

that is,

(1) the focus ξ_0 is expected,
(2) several questions are asked, according to the subparts of that focus,
(3) among which the question expressed on the focus of *skiing*,
(4) this question itself expects several possible answers.

The premises $\xi_{01}, \dots, \xi_{0n}$, different from ξ_{0i}, represent potential foci which may be developed in the future. The net made of these two designs normalizes: the process of normalization expresses at the same time the choices made by the first speaker (positive actions a_1 and a_2) and the limits into which she has to make her choices (the supposed positive actions of the second, maybe virtual, speaker).

Another discursive relation is *narration*, where "the temporal order of the events matches their textual order". Asher and Lascarides give, for instance,

Max fell. John helped him up.

The foci of a narration generally have a short prefix in common, but this prefix is not regularly growing like it is the case for elaboration: after a short common prefix, the remaining parts may be totally pairwise disjoint. That results in selecting at each positive step a focus among not yet explored loci, and giving up the previous ones. The following positive design is an example:

$$\dfrac{\dfrac{\dfrac{\dfrac{\cdots}{\xi_{20} \vdash \xi_3, ..., \xi_n}\,(\xi_{20}, \{\emptyset\})}{\vdash \xi_2, ..., \xi_n}\,a_2}{\xi_{10} \vdash \xi_2, ..., \xi_n}\,(\xi_{10}, \{\emptyset\})}{\vdash \xi_1, \xi_2, ..., \xi_n}\,a_1$$

The negative rule labelled by $\{\emptyset\}$ expresses the particular case where the only ramification in the directory is empty: the focus is then ignored and it remains only the context, for a continuation of the discourse. This design therefore positively interacts (converges) with

$$\dfrac{\dfrac{}{\vdash \xi_{10}}\,\emptyset}{\xi_1 \vdash} \qquad \cdots \qquad \dfrac{\dfrac{}{\vdash \xi_{i0}}\,\emptyset}{\xi_i \vdash} \qquad \cdots$$

The role of the virtual second speaker is exactly (by using the empty positive rule) to enforce the first one to quit her previous focus and to choose one in the list of non-explored ones. This list is of course planned by the virtual speaker, as if the first one had always in mind her complete narration even before beginning to speak.

12.3.4.3 *Assertion, denegation and interrogation*

The previous remarks concerning the necessity of a two-faced process for a representation of discursive relations opens the field to a deeper reflection on elementary speech acts. Discourse is above all action and commitment: action of Myself on Yourself and reciprocally (Beyssade and Marandin, 2006). Like it is said by Walton (2000), "asserting" is "willing to defend the proposition that makes up the content of the assertion, if challenged to do so". This results in the fact that when Myself asserts P, I must have in mind all justifications for predictable objections. That is, "I" have a design like the following one:

$$\cfrac{\cfrac{\mathcal{D}_1}{\vdash \xi_0.I_1} \quad \dots \quad \cfrac{\mathcal{D}_n}{\vdash \xi_0.I_n}}{\cfrac{\xi_0 \vdash}{\vdash \xi}} \mathcal{N}$$

where $\mathcal{N}$ is a set of predictable thematic variations on the theme "I" introduce, and where every $\mathcal{D}_i$ is a design which never ends with a †.

Denegation is slightly different. We can refer to works by Ducrot (1984) in the 1980s, according to whom discursive negation (that we shall name denegation in order to avoid confusion with standard logical negation) is necessarily *polyphonic*. We may for instance have the following utterance in a dialogue:

Example 12.10. Mary is not nice, on the contrary, she is execrable

where the second part of the utterance may be understood only if we think that it denegates not the first part of the sentence but a dual utterance, the one which the present one is confronted with and which could be *Mary is nice*. We are therefore led to conclude that a (de)negation like *Mary is not nice* is always opposed to a virtual positive statement like *Mary is nice*. A possible modelling of such a case consists in having a positive action by Myself which compels the other speaker to accept my denegation by playing her †. If not, she enters into a diverging process.[7] The only way for Myself to force the other speaker to play her † is to use the $\emptyset$ positive rule. On the other hand, the denegation assumes that the virtual other speaker produced a statement which is now denied by Myself. This statement is in fact a paradoxal assertion since the set $\mathcal{N}$ is reduced to $\{\emptyset\}$! (The virtual speaker has no plan to sustain the claim she makes.) Denegation therefore supposes we make a step downward, to the fictitious claim

$$\cfrac{}{\vdash \xi, \Lambda}\emptyset \qquad vs \qquad \cfrac{\cfrac{}{\vdash \Gamma}\dagger}{\xi \vdash \Gamma}$$

$$\downarrow$$

[7]That could be repaired in a dynamic way.

$$\dfrac{\dfrac{}{\vdash \xi_0, \Lambda}\ \emptyset\ a_2}{\xi \vdash \Lambda}\ \{\{0\}\} \qquad vs \qquad \dfrac{\dfrac{\dfrac{}{\vdash \Gamma}\ \dagger}{\xi_0 \vdash \Gamma}\ \{\emptyset\}}{\vdash \xi, \Gamma}\ \{0\}\ (a_1)$$

where

$$a_1 : Mary\ is\ nice$$
$$a_2 : Mary\ is\ not\ nice$$

Interrogation is still another game. If other speech acts can always be represented as anchored at a single locus (modulo some "shift" which makes us go downward, searching the topic or the basis for a denegation), we assume questions always start from two loci, among which one is called the locus of the answer. The design of a question has therefore a basis $\vdash \xi, \sigma$ with σ devoted to an answer, and ends with a $\mathcal{F}ax_\sigma$, so that, in interaction with a dual design $\mathcal{E}$, the answer to the question is moved to σ.

Let us now take as examples two elementary dialogues consisting of sequences of *Questions-Answers*, where one is well formed and the other ill formed. The first one is

Example 12.11.

- You: Have you a car?
- I: Yes.
- You: Of what make?

$$\dfrac{\dfrac{\dfrac{}{\vdash \sigma}\ \dagger \quad \dfrac{\dfrac{\dfrac{Fax_{\xi 010,\sigma}}{\xi 010 \vdash \sigma}}{\vdash \xi 01, \sigma}\ You_3}{}}{\xi 0 \vdash \sigma}\ \{\emptyset, \{1\}\}}{\vdash \xi, \sigma}\ You_1 \qquad \text{vs} \qquad \dfrac{\dfrac{\dfrac{\vdots}{\vdash \xi 010}}{\dfrac{\xi 01 \vdash}{\vdash \xi 0}}}{\xi \vdash}$$

The answer "yes" is represented by $\{1\}$, creating hence a locus from which the speaker may continue the interaction on the topic of the car and for example may ask what is its make. The answer "no" is represented by $\emptyset$ (there is no more to say about this car).

The second dialogue is

Example 12.12.

- Have you a car?
- No, I have no car.
- *Of what make?

and it may be represented either by the following dispute, which fails since You did not plan a negative answer:

$$\dfrac{\dfrac{\dfrac{\dfrac{Fax_{\xi010,\sigma}}{\xi010 \vdash \xi01, \sigma}}{\vdash \xi01, \sigma}\,{\scriptstyle You_3}}{\xi0 \vdash \sigma}\,{\scriptstyle \{\{1\}\}}}{\vdash \xi, \sigma}\,{\scriptstyle You_1} \qquad \text{vs} \qquad \dfrac{\dfrac{}{\vdash \xi0}\,{\scriptstyle \emptyset}}{\xi \vdash}$$

or by this one, where the interaction also fails since You can only play her left branch, but in this case, she may not focus on ξ_{01}:

$$\dfrac{\dfrac{\dfrac{\xi010 \vdash}{\vdash \sigma} \quad \dfrac{\dfrac{Fax_{\xi010,\sigma}}{\xi010 \vdash \sigma}}{\vdash \xi01, \sigma}\,{\scriptstyle You_3}}{\xi0 \vdash \sigma}\,{\scriptstyle \{\emptyset,\{1\}\}}}{\vdash \xi, \sigma}\,{\scriptstyle You_1} \qquad \text{vs} \qquad \dfrac{\dfrac{}{\vdash \xi0}\,{\scriptstyle \emptyset}}{\xi \vdash}$$

12.3.4.4 *Conclusion*

Ludics provides a frame in which we can explore the speech acts realized in discourse as really *two-faced.* This is mainly because ludics, as a locative framework, makes it possible for two parallel processes to interact, thus generalizing the well-known dynamics of proofs (that we have already in Gentzen's sequent calculus, by means of the procedure of cut-elimination) to the dynamics of paraproofs. In such a framework, there is no *truth* properly speaking but only ways for a proof-candidate to pass *tests* which are themselves other proof-candidates. In a concrete dialogue situation, our utterance is a proof-candidate: it has necessarily to cope with counter proof-candidates, which are either the reactions of the other speaker or some kind of virtual reaction that we have in mind. This way, our interventions are doubly driven: once by our positive acts, and second by the positive actions of the other speaker or of such a virtual partner and by the way we record these reactions, that is, by negative acts. Of course, whereas in a dialogue each participant has to take into consideration the expectations and reactions of the other, in monologues, utterances are co-determined by the speaker herself and by her virtual interlocutor. It is this interaction which drives the speech until a tacit agreement occurs either coming directly from the speaker or indirectly via the image she has of her speech.

Chapter 13

The Future in Conclusion

This book started from a standard presentation of works in formal semantics, and finally arrived at a novel view which, from our viewpoint, may considerably enlarge our vision of meaning. The approach concentrates on the procedural aspect of meaning, something which has been only studied on the surface in the past. The main tool to address this aspect seems to be proof theory. In Ranta's type-theoretical grammar, meanings of lexical items, like verbs, are not seen as truth conditions (conditions in which sentences using these items are "true") but as functions which may be applied only if proof elements are provided. For instance, the meaning of *to repair* depends on a variable z which takes as its value a proof, the proof that the object was broken. This draws a canvas where meanings are dependent on themselves (and not on an exterior reality, even if, at the end, of course, a connection is made with such a reality, but this connection is holistic and perhaps overpasses our means of knowledge). This view seems to be well in accordance with strong remarks made by Chomsky, according to whom semantics is not concerned with questions of truth or falsity, but with possibilities or impossibilities to combine words and expressions and with the kind of interpretation these combinations allow us to make.[1] In some sense, it may be argued that it is still syntax! Yes, of course, but

[1] In Chomsky (2007), the linguist underlines that "even the simplest words and concepts of human language and thought lack the relation to mind-independent entities that has been reported for animal communication". He pursues that by saying: "what we understand to be a house, a river, a person, a tree, water, and so on, consistently turns out to be a creation of what 17th century investigators called the 'cognoscitive powers', which provide us with rich means to refer to the outside world from certain perspectives. The objects of thought they construct are individuated by mental operations that cannot be reduced to a 'peculiar nature belonging' to the thing we are talking about".

why not. Such views are in harmony with the evolution of logic we tried to describe in this book, which leads to giving up the opposition between syntax and model, suggesting that the right opposition is between proofs and counter-proofs. The model-theoretic account of language has always been taken in default, at least since its formulation by Tarski (1931) who said that the concept of truth was only definable in formal languages. But we know that even in this case, the model-theoretic view is problematic since it may be only founded on a *fuite en avant*, always going from a meta-level to a meta-meta-level.

This proof-theoretic approach is, from our viewpoint, the main contribution that logic may provide to semantics, but that does not exclude other contributions. We have studied in depth how investigations in logic have shed new light on syntax and on questions of the interface between syntax and semantics. Residuation may be, besides recursivity, one of the main mathematical structures at work in language. It is the reason why we emphasized Moortgat's works, trying to put a link between them and the more orthodox formalization of syntax (with regard to Chomsky's minimalist theory) proposed by Stabler. In this conception, linear logic, due to Girard, has been shown to be an efficient frame to take the syntactic process into account, to the extent that this process is basically one of resource consumption. As we have seen, exponentials give a sufficient flexibility to systems based on linear logic, in order to represent processes of binding (by means of contraction, for instance). Co-reference of nouns and pronouns is thus explained by hypothetical reasoning: pronouns and traces are hypotheses which are discharged at the same time. Pollard's convergent grammars replace this mechanism by a more *ad hoc* one, relying on Cooper storage, but they still remain based on (various flavours of) linear implication.

We don't think there is a strong necessity to argue in favour of logic in language studies. Chomsky himself recognizes that the definition of the two fundamental operations in syntax, that he now names "external" and "internal" Merge,[2] is just a "matter of logic".

These views on syntax and the syntax-semantics interface were not directly related to the "proofs as meanings" viewpoint. And this viewpoint has no particular link with those views, since we saw in particular that Ranta's work, on its syntactic aspect, remains dependent on a very traditional conception of grammar, based on rules. The link between new

[2]Which respectively correspond to what was named before Merge and Move.

conceptions on syntax and the proof-theoretical approach must therefore be a point to further explore in the future. But we may suggest that there are already existing views on syntax which converge with this viewpoint. Kempson (Kempson *et al.*, 2001, 2003, 2010), when she addresses the "split utterance challenge", makes two strong claims:

- one is that "the formal models of language and language processing should be evaluated by how good an explanation they provide of conversational dialogue, this being THE central use of language";
- another is that "concepts of sentence-meaning and logical form for a sentence sui generis have to be abandoned".

Concerning Kempson's second claim, we must understand it as a much more radical step than the one we took in this book (where we gave some place to the so-called "logical forms", even in the chapter devoted to ludics), but a step which seems to us in harmony with the theses here defended. It is hard to think about language in "representational terms", as if in the cognitive processes we were always inclined to contemplate our thoughts before expressing them! Rather, we are ready to jump from our own perspective to our hearer's one, without taking time to precisely evaluate our utterances. Kempson *et al.* in many papers give examples of utterances which are split in dialogue, and nevertheless perfectly acceptable for the grammarian, like

Example 13.1.

A: What shall I give...
B: Eliot? A toy.

or

Example 13.2.

A: Where have you got to with...
B: Your book? Up to Chapter 4.

She claims that, first of all, of course, to take into account such examples necessitates conceiving the processing of utterances as incremental, since, at almost any point, the discourse may be taken charge of by the other speaker, thus revealing that (s)he has a perfect knowledge[3] of the partial information provided by the first words.

[3]In the sense of "know how", and not of "know that".

Work remains to be done in order to establish a link with our ludical view. We can tentatively suggest that two speakers in interaction co-build their dialogue, the first one beginning a "design" and the other one immediately making an attempt at a counter-design, such that, at any time, the second design may be inverted and become the continuation of the first one. But this continuation leaves some elements constant, for instance, in Example 13.2, "your" does not transform into "my".

Many elements are absent from this book, which deserve to be at least mentioned. We of course leave them for another volume. Among those things a presentation of the so-called *coherence semantics* of linear logic.

In a few words, coherent spaces are analogous to topological sets. Whereas the latter are endowed with a family of open sets, the former are endowed with a particular relation, reflexive and symmetric, but not necessarily transitive, denoted by "$\frown\!\!\!\!\smile$", which is the coherence relation. An alternative definition amounts to providing the original set $\mathcal{X}$ with a set of sets X such that

- $a \in X$ and $b \subset a$ imply $b \in X$,
- if A is a subset of X formed by elements which are pairwise disjoint, then $\bigcup A \in X$ (in particular, $\emptyset \in X$).

A coherence relation may be in effect defined on $\mathcal{X}$ by putting

$$x \frown\!\!\!\!\smile y \quad [mod\ X]\ iff\ \{x, y\} \in X \tag{13.1}$$

The relation "$\frown$" is defined by

$$x \frown y \quad [mod\ X]\ iff\ x \frown\!\!\!\!\smile y \quad [mod\ X]\ and\ x \neq y \tag{13.2}$$

The relation "$\asymp$" is defined by

$$x \asymp y \quad [mod\ X]\ iff\ \neg(x \frown y \quad [mod\ X]) \tag{13.3}$$

Linear connectives may be defined in coherent space terms:

- Negation: $x \frown\!\!\!\!\smile y\ [mod\ X^{\perp}]$ iff $x \asymp y\ [mod\ X]$;
- Tensor product: $(x, y) \frown\!\!\!\!\smile (x', y')\ [mod\ X \otimes Y]$ iff $x \frown\!\!\!\!\smile x'\ [mod\ X]$ and $y \frown\!\!\!\!\smile y'\ [mod\ Y]$;
- Multiplicative disjunction: $(x, y) \frown\!\!\!\!\smile (x', y')\ [mod\ X \wp Y]$ iff $x \frown x'\ [mod\ X]$ or $y \frown y'\ [mod\ Y]$;

- Linear implication: $(x, y) \frown (x', y')$ $[mod\ X \multimap Y]$ iff $(x \frown x'\ [mod\ X] \Rightarrow y \frown y'\ [mod\ Y])$ and $(x \smile x'\ [mod\ X] \Rightarrow y \smile y'\ [mod\ Y])$.

We do not enter into more detail here (in particular, definitions can also be provided for additives, constant elements and exponentials). Moreover, proofs are interpreted in coherent spaces, which is the main result we are expecting since linear logic is essentially based on having a semantics of proofs. What seems to us particularly important is the help that such spaces could provide to us for dealing with lexical semantics. As we know, predicate logic is often not very satisfactory for representing lexical knowledge. Must we interpret an utterance like *Paris is fun* as expressing the fact that the object "Paris" belongs to the set of "fun" objects? No. When asserting such a claim, the speaker only expresses that there is some compatibility (or coherence) relation between "Paris" and "fun", something that we could translate as "the coherent space P associated with Paris and the coherent space F associated with 'fun' have a non-empty intersection". But nothing says that Paris is fun for everybody ... in particular, it is not fun for workers taking the RATP line number 13 every morning!

We did not mention either new, interesting approaches of dialogue, contained in Groenendijk's *inquisitive semantics* (Groenendijk and Roelofsen, 2010), where interpretation is given in terms which could be close to this coherent semantics. This is also left to another volume.

At the end of the chapter "Proofs as Meanings", we suggested that a promising view was given on texts. We emphasized the point that text necessitates locativity, in the sense that for interpreting a particular utterance, not only the words count but also their proper addresses, inside the text itself, and perhaps also inside the collection of texts. We have seen in the last chapter how locativity may be taken into account in logic. Such views may help us not only to understand the progression of a text, but also the play on words that there can be in a textual, or discursive, practice. Fallacies, for instance, have always been a critical point for any theory of argumentation since at least Aristotle. As pointed out by Hamblin (2004), most fallacies are due to the fact that a stability of meaning is presupposed when the same utterance is repeated. But we know that it is not always the case. For instance, in *All metals are elements, brass is a metal, therefore brass is an element*, the false inference is made because of a slight change of meaning between the first and the second occurrence of the word "metal". Summing up all the fallacies under the term of "equivocation", he claims that (p. 286)

> By locating most of the properties of the locutions in propositional letters such as 'A', 'B', 'S' and 'T', [we smuggle] in the fiction that the question of meaning can be isolated from that of dialectical properties [...]. The question of whether the meaning of a given symbol changes is to be answered *a posteriori* and the question should not be begged by writing in an assumption of constancy.

This, from our viewpoint, puts the emphasis on what we have named *locativity*, that is, the principle according to which addresses are prior to contents. As we investigated in ludics, the question of identity is not answered *a priori*: there is no "identity axiom" in ludics since there is no formula and therefore no atomic formula! Instead of it, there is $\mathcal{F}ax$, which is not an axiom but a procedure, a procedure by which we attempt to identify contents of addresses by means of their design structure.

Another point deserves attention, which is consistent with this locativity principle, it is the fundamental geometrical character of such a conception. If only addresses stay under consideration, we can only detect geometrical configurations: that was already what we noticed about proof nets.

Finally, much could be said about the connection this conception of meaning has with *inferentialism*, a doctrine mainly developed by the philosopher Robert Brandom. According to Brandom (2000), the meaning of an assertion basically resides in the inferences we may draw from it. An assertion must therefore include, as soon as it is uttered, the set of possible continuations it may have, but also the potential reactions from the interlocutor. When we speak, we are committed to justifications of what we assert, which precisely amounts to saying, in another language, that we are always calling for proofs.

Bibliography

Abramsky, S. and Jagadeesan, R. (1994). Games and full completeness for multiplicative linear logic, *Journal of Symbolic Logic* **59**, pp. 543–574.

Abrusci, M. (1991). Phase semantics and sequent calculus for pure noncommutative classical linear propositional logic, *Journal of Symbolic Logic* **56**, pp. 1403–1451.

Abrusci, M. and Ruet, P. (2000). Non-commutative logic I: The multiplicative fragment, *Annals of Pure and Applied Logic* **101**, pp. 29–64.

ACL (2001). *Proceedings of the 39th Meeting of the ACL* (ACL 2001, Toulouse).

Ades, A. and Steedman, M. (1982). On the order of words, *Linguistics and Philosophy* **4**, pp. 517–558.

Ajdukiewicz, K. (1935). Die syntaktische konnexität, *Studia Philosophica* **1**, pp. 1–27, English translation in S. McCall (ed.), *Polish Logic, 1920–1939* (Oxford University Press, Oxford) 1967, pp. 207–231.

Amblard, M. (2007). *Calculs de représentations sémantiques et syntaxe générative: Les grammaires minimalistes catégorielles*, PhD thesis, Université Bordeaux 1.

Amblard, M., Lecomte, A. and Retoré, C. (in press). Categorial minimalist grammars: From generative syntax to logical forms, *Jim Lambek Festschrift* **5**.

Andréoli, J.-M. (1992). Logic programming with focusing proofs in linear logic, *Journal of Logic and Computation* **2**, 3, pp. 297–347.

Anoun, H. (2007). *Approche logique des grammaires pour les langues naturelles*, PhD thesis, Université Bordeaux 1.

Anoun, H. and Lecomte, A. (2006). Linear grammars with labels, in G. S. P. Monachesi, G. Penn and S. Wintner (eds.), *Proceedings of the Conference on Formal Grammar 2006* (CSLI Publications, Stanford, CA).

Asher, N. and Lascarides, A. (2003). *Logics of Conversation* (Cambridge University Press, Cambridge).

Baldridge, J. and Kruijff, G.-J. (2003). Multi-modal combinatory categorial grammar, in EACL (ed.), *Proceedings of the 10th Meeting of the EACL* (Budapest), pp. 211–218.

Bar-Hillel, Y. (1964). *Language and Information* (Addison-Wesley, Reading, MA).

Bar-Hillel, Y., Gaifman, H. and Shamir, E. (1960). On categorial and phrase structure grammars, *Bulletin of the Research Council of Israel* **9**, pp. 1–16.

Barker, C. (2002). Continuations and the nature of quantification, *Natural Language Semantics* **10**, pp. 211–242.

Belnap, N. and Anderson, A. (1975). *Entailment*, vol. 1 (Princeton University Press, Princeton).

Bernardi, R. and Moortgat, M. (2010). Continuation semantics for the Lambek–Grishin calculus, *Information and Computation* **208**, pp. 397–416.

Beyssade, C. and Marandin, J.-M. (2006). The speech act assignment problem revisited: Disentangling speaker's commitment from speaker's call on addressee, in *Proceedings of the Colloque de Syntaxe et Sémantique de Paris* (CNRS).

Blass, A. (1992). A game semantics for linear logic, *Annals of Pure and Applied Logic* **56**, pp. 183–220.

Boolos, G. (1981). For every A there is a B, *Linguistic Inquiry* **12**, pp. 465–467.

Brandom, R. (2000). *Articulating Reasons: An Introduction to Inferentialism* (Harvard University Press, Cambridge, MA).

Bransen, J. (2010). The Lambek–Grishin calculus is NP-complete, *CoRR*, abs/1005.4697.

Brody, M. (1995). *Lexico-Logical Form: A Radically Minimalist Theory* (MIT Press, Cambridge, MA).

Carpenter, B. (1996). The Turing-completeness of multimodal categorial grammars, in *JFAK: Essays Dedicated to Johan van Benthem on the Occasion of his 50th Birthday* (Institute for Logic, Language, and Computation, University of Amsterdam). Available on CD-ROM at http://turing.wins.uva.nl.

Chomsky, N. (1981). *Lectures on Government and Binding* (Foris Pub., Dordrecht).

Chomsky, N. (1996). *The Minimalist Program* (MIT Press, Cambridge, MA).

Chomsky, N. (2000). Minimalist inquiries: The framework, in R. Martin, D. Michaels and J. Uriagereka (eds.), *Step by Step: Essays on Minimalist Syntax in Honor of Howard Lasnik* (MIT Press, Cambridge, MA), pp. 89–155.

Chomsky, N. (2001). Derivation by phases, in M. Kenstowicz (ed.), *Ken Hale: A Life in Language* (MIT Press, Cambridge, MA), pp. 1–52.

Chomsky, N. (2007). Of minds and languages, *Biolinguistics* **1**, pp. 9–27.

Church, A. (1941). *The Calculi of Lambda-Conversion* (Princeton University Press, Princeton).

Culicover, P. and Jackendoff, R. (2005). *Simpler Syntax* (Oxford University Press, Oxford).

Curien, P. (2003). Introduction to linear logic and ludics, part ii, Tech. rep., CNRS & Université Paris VII, http://www.pps.jussieu.fr/curien/LL-ludintroII.pdf.

Curien, P. and Herbelin, H. (2000). Duality of computation, *Proceedings of the Fifth AGM SIGPLAN* (SIGPLAN Notices, Montreal, Canada), pp. 233–243.

Curry, H. (1961). Some logical aspects of grammatical structure, in R. Jakobson (ed.), *Structure of Language and its Mathematical Aspects* (AMS, Providence), pp. 56–68.

Curry, H. and Feys, R. (1958). *Combinatory Logic*, vol. 1 (North-Holland, Amsterdam).

Danos, V., Joinet, J.-B. and Schellinx, H. (2003). Computational isomorphisms in classical logic, *Theoretical Computer Science* **294**, pp. 353–378.

Danos, V. and Régnier, L. (1995). Proof-nets and the Hilbert space, in [Girard *et al.* (1995)], pp. 307–328.

de Groote, P. (1999). The non-associative Lambek calculus with product in polynomial time, *Lecture Notes in Computer Science* **1617**, pp. 128–139.

de Groote, P. (2001a). Towards abstract categorial grammars, in [ACL (2001)], pp. 148–155.

de Groote, P. (2001b). Type raising, continuations, and classical logic, in R. van Rooy and M. Stokhof (eds.), *Proceedings of the 13th Amsterdam Colloquium*, pp. 97–101.

de Groote, P. (2006). Towards a Montagovian account of dynamics, in *Proceedings of the 16th Conference on Semantics and Linguistic Theory* (CLC, Tokyo), pp. 1–16.

Desclés, J.-P. (1990). *Langages Applicatifs, Langues Naturelles et Cognition* (Hermes, Paris).

Dobrovie-Saurin, C. and Beyssade, C. (2004). *Définir les Indéfinis* (Editions du CNRS, Paris).

Dǒsen, K. (1992). A brief survey of frames for the Lambek calculus, *Zeitschrift fur Mathematische logik und Grundlagen der Mathematik* **38**, pp. 179–187.

Ducrot, O. (1984). *Le Dire et le Dit* (Editions de Minuit, Paris).

Faggian, C. (2006). Ludics and interactive observability: The geometry of tests, *Theoretical Computer Science* **350**, pp. 213–233.

Felscher, W. (1986). Dialogues as a foundation for intuitionistic logic, in D. Gabbay and F. Guenthner (eds.), *Handbook of Philosophical Logic*, (Kluwer, Dordrecht), pp. 341–372.

Fleury, M.-R. and Quatrini, M. (2004). First order in ludics, *Mathematical Structures in Computer Science* **14**, pp. 189–213.

Fleury, M.-R. and Tronçon, S. (2010). Speech acts in ludics, in [Lecomte and Tronçon (2010)], pp. 1–25.

Frey, W. and Gärtner, H.-M. (2002). On the treatment of scrambling and adjunction in minimalist grammars, in G. Penn (ed.), *Proceedings of the Conference on Formal Grammar* (CSLI Publications, Stanford), pp. 41–52.

Gamut, L.-T.-F. (1991). *Logic, Language and Meaning*, vol. I and II (University of Chicago Press, Chicago).

Gärtner, H.-M. and Michaelis, J. (2008). A note on countercyclicity and minimalist grammars, in G. Penn (ed.), *Proceedings of the Conference on Formal Grammar* (CSLI Publications, Stanford), pp. 95–109.

Girard, J.-Y. (1987). Linear logic, *Theoretical Computer Science* **50**, 1, pp. 1–102.

Girard, J.-Y. (1991). A new constructive logic: Classical logic, *Mathematical Structures in Computer Science* **1**, 3, pp. 255–296.

Girard, J.-Y. (1995). Linear logic: Its syntax and semantics, in [Girard *et al.* (1995)], pp. 1–42.

Girard, J.-Y. (2001). Locus solum, *Mathematical Structures in Computer Science* **11**, 3, pp. 301–506.

Girard, J.-Y. (2003). From foundations to ludics, *Bulletin of Symbolic Logic* **9**, 2, pp. 131–168.

Girard, J.-Y. (2006). *Le Point Aveugle* (Hermann, Paris).

Girard, J.-Y., Lafont, Y. and Régnier, L. (1995). *Advances in Linear Logic* (Cambridge University Press, Cambridge).

Girard, J.-Y., Lafont, Y. and Taylor, P. (1989). *Proofs and Types* (Cambridge University Press, Cambridge).

Griffin, T. (1990). A formulae-as-type notion of control, in *Proceedings of the 17th Symposium on Principles of Programming Languages* (ACM, San Francisco), pp. 47–58.

Grishin, V. (1983). On a generalization of the Ajdukiewicz–Lambek system, in A.I. Mikhailov (ed.), *Studies in Nonclassical Logics and Formal Systems* (Nauka, Moscow), pp. 315–334, English translation in V. M. Abrusci and C. Casadio (eds.), *Proceedings of the 5th Rome Workshop* (Bulzoni Editore, Rome, 2002).

Groenendijk, J. and Roelofsen, F. (2010). Compliance, in [Lecomte and Tronçon (2010)], pp. 172–185.

Hamblin, C. L. (2004). *Fallacies* (Vale Press, Newport News), a reedition of a book dating from 1970 (Methuen, London).

Hauser, M.-D., Chomsky, N. and Fitch, T. (2002). The faculty of language: What is it, who has it, and how did it evolve? *Science* **298**, pp. 1569–1579.

Heim, I. and Kratzer, A. (1998). *Semantics in Generative Grammar* (Blackwell, Malden, Mass).

Hendricks, H. (1993). *Studied flexibility*, PhD thesis, ILLC, Amsterdam.

Hepple, M. (1990). *The grammar and processing of order and dependency*, PhD thesis, University of Edinburgh.

Herbelin, H. (2005). *Au coeur de la dualité*, PhD thesis, Université Paris 11.

Hilbert, D. (1922). Neubegründung der Mathematik: Erste Mitteilung, *Abhandlungen aus dem Seminar der Hamburgischen Universität* **1**, pp. 157–177, English translation in Mancosu, 1998, pp. 198–214, and Ewald, 1996, pp. 1115–1134.

Hintikka, J. and Kulas, J. (1983). *The Game of Language: Studies in Game Theoretical Semantics and Its Applications* (D. Reidel, Dordrecht).

Hintikka, J. and Sandu, G. (1989). Informational independence as a semantical phenomenon, in J.-E. Fenstad (ed.), *Logic, Methodology and Philosophy of Science VII* (Elsevier, Amsterdam), pp. 571–589.

Hintikka, J. and Sandu, G. (1997). Game-theoretical semantics, in [van Benthem and ter Meulen (1997)], chap. 6, pp. 361–410.

Howard, W. (1980). The formulae-as-types notion of construction, in J. Seldin and J. Hindley (eds.), *To H. B. Curry: Essays on Combinatory Logic, Lambda Calculus and Formalism* (Academic Press, London).

Husserl, E. (1958). *Recherches Logiques* (P.U.F., Paris), translated from German by Hubert Elie.

Jacobson, P. (1996). The syntax/semantics interface in categorial grammar, in S. Lappin (ed.), *The Handbook of Contemporary Semantic Theory* (Blackwell, Malden), pp. 89–116.

Jacobson, P. (2000). Towards a variable-free semantics, *Linguistics and Philosophy* **22**, 2, pp. 117–185.

Janssen, T. (1997). Compositionality, in [van Benthem and ter Meulen (1997)], chap. 7, pp. 417–475.

Kayne, R. (1994). *The Antisymmetry of Syntax* (MIT Press, Cambridge, MA).

Kayne, R. (2002). Pronouns and their antecedents, in S. D. Epstein and T. D. Seely (eds.), *Derivation and Explanation in the Minimalist Program* (Blackwell, Oxford), pp. 133–166.

Keiff, L. (2007). *Le pluralisme dialogique*, PhD thesis, Université Lille 3.

Kempson, R., Gregoromichelaki, E., Meyer-Viol, W., Purver, M., White, G. and Cann, R. (2010). Natural language syntax as procedures for interpretation, in [Lecomte and Tronçon (2010)], pp. 121–141.

Kempson, R., Meyer-Viol, W. and Gabbay, D. (2001). *Dynamic Syntax: The Flow of Language Understanding* (Blackwell, Oxford).

Kempson, R., Meyer-Viol, W. and Otsuka, M. (2003). Growth of logical form: The dynamics of syntax, in J. Peregrin (ed.), *Logic, Methodology and Philosophy of Science VII* (Elsevier, Amsterdam), pp. 121–147.

Kobele, G. (2009). Syntax and semantics in minimalist grammars, Tech. rep., ESSLLI 2009, Bordeaux.

Kratzer, A. (1994). External arguments, in E. Benedicto and J. Runner (eds.), *Functional Projections* (University of Massachussets, Occasional Papers, Amherst).

Kurtonina, N. and Moortgat, M. (1997). Structural control, in P. Blackburn and M. de Rijke (eds.), *Specifying Syntactic Structures* (CSLI Publications, Stanford), pp. 75–113.

Lambek, J. (1958). The mathematics of sentence structure, *American Mathematical Monthly* **65**, pp. 154–170.

Lambek, J. (1961). On the calculus of syntactic types, in R. Jacobson (ed.), *Structure of Language and its Mathematical Aspects* (AMS, Providence).

Lambek, J. (1988). Categorial and categorical grammars, in E. Bach, R. Oehrle and D. Wheeler (eds.), *Categorial Grammars and Natural Language Structures* (D. Reidel, Dordrecht), pp. 297–317.

Lambek, J. (1993). From categorial to bilinear logic, in [Schroeder-Heister and Dosen (1993)], pp. 207–237.

Lambek, J. and Scott, P. (1986). *Introduction to Higher-Order Categorical Logic* (Cambridge University Press, Cambridge).

Lasnik, H. and Uriagereka, J. (2005). *A Course in Minimalist Syntax* (Blackwell, Malden).

Lecomte, A. (2005). Categorial grammar for minimalism, in P. S. C. Casadio and R. Seely (eds.), *Language and Grammar: Studies in Mathematical Linguistics and Natural Language* (CSLI Publications, Stanford), pp. 163–188.

Lecomte, A. (2008). Semantics in minimalist-categorial grammar, in P. de Groote, L. Kallmeyer, G. Penn and G. Satta (eds.), *Proceedings of the 13th Conference on Formal Grammar* (CSLI Publications, Stanford).

Lecomte, A. and Quatrini, M. (2009). Ludics and its applications to natural language semantics, in H. Ono, M. Kanazawa and R. de Queiroz (eds.), *Logic, Language, Information and Computation* (Springer-Folli, Berlin), lNAI 5514.

Lecomte, A. and Quatrini, M. (2010). Pour une étude du langage via l'interaction: Dialogues et sémantique en ludique, *Mathématiques et Sciences Humaines — Mathematics and Social Sciences* **189**, pp. 37–67.

Lecomte, A. and Retoré, C. (1999). Towards a minimal logic for minimalism, in G. van Kruijf (ed.), *Formal Grammar'99* (ESSLLI'99, Utrecht).

Lecomte, A. and Retoré, C. (2001). Extending Lambek grammars: A logical account of minimalist grammars, in [ACL (2001)], pp. 354–362.

Lecomte, A. and Tronçon, S. (eds.) (2010). *Games, Dialogues and Interaction: Proceedings of the Prelude Workshop*, LNAI–Folli (Springer).

Lesniewski, S. (1929). Grundzüge eines neuen Systems der Grundlagen der Mathematik, in *Fundamenta Mathematicae* **24**, pp. 1–81.

Lorenz, K. (1961). *Arithmetik und Logik als Spiele*, PhD thesis, Universität Kiel.

Lorenzen, P. (1960). Logik und Agon, *Atti del Congresso Internazionale di Filosofia* **4**, pp. 187–194.

Martin-Löf, P. (1984). *Intuitionistic Type Theory* (Bibliopolis, Naples).

May, R. (1977). *The grammar of quantification*, PhD thesis, MIT.

Michaelis, J. (2001). Derivational minimalism is mildly context-sensitive, in M. Moortgat (ed.), *Logical Aspects of Computational Linguistics, LNCS/LNAI*, vol. 2014 (Springer, Berlin), pp. 179–198.

Montague, R. (1974a). English as a formal language, in [Thomason (1974)].

Montague, R. (1974b). The proper treatment of quantification in ordinary English, in [Thomason (1974)].

Moortgat, M. (1988). *Categorial Investigations, Logical and Linguistic Aspects of the Lambek Calculus* (Foris, Dordrecht).

Moortgat, M. (1996a). In situ binding: A modal analysis, in P. Dekker and M. Stokhof (eds.), *Proceedings of the Tenth Amsterdam Colloquium* (ILLC, Amsterdam).

Moortgat, M. (1996b). Multimodal linguistic inference, *JoLLI* **5**, pp. 349–385.

Moortgat, M. (1997). Categorial type logics, in [van Benthem and ter Meulen (1997)], chap. 2, pp. 93–178.

Moortgat, M. (2007). Symmetries in natural language syntax and semantics: The Lambek–Grishin calculus, in D. Leivant and R. de Queiroz (eds.), *Proceedings of the 14th International Conference on Logic, Language, Information and Computation, Lecture Notes in Computer Science*, vol. 4576 (Springer, Berlin), pp. 264–284.

Moortgat, M. (2009). Symmetric categorial grammar: *Journal of Philosophical Logic* **38**, 6, pp. 681–710.

Morrill, G. (1994). *Type Logical Grammar: Categorial Logic of Signs* (Kluwer, Dordrecht).

Muskens, R. (2003). Languages, lambdas and logic, in G.-J. Kruijff and R. Oehrle (eds.), *Resource-Sensitivity in Binding and Anaphora* (Kluwer, Amsterdam).

Oehrle, R. T. (1994). Term-labelled categorial type systems, *Linguistics and Philosophy* **17**, pp. 633–678.

Oehrle, R. T. (1995). Some 3-dimensional systems of labelled deduction, *Bulletin of the IGPL* **3**, 2, pp. 429–448.

Parigot, M. (1992). $\lambda\mu$-calculus: An algorithmic interpretation of classical natural deduction, in A. Voronkov (ed.), *Proceedings of the International Conference on Logic Programming and Automated Reasoning* (Springer, Berlin).

Partee, B. (2004). *Compositionality in Formal Semantics: Selected Papers by Barbara Partee* (Wiley-Blackwell, Oxford).

Pentus, M. (1993). Lambek grammars are context free, in *Proceedings of the Eighth Annual IEEE Symposium on Logic in Computer Science* (Montreal).

Pentus, M. (2003). Lambek calculus is NP-complete, CUNY Ph.D. Program in Computer Science Technical Report TR–2003005, CUNY Graduate Center, New York, http://www.cs.gc.cuny.edu/tr/techreport.php?id=79.

Pica, P., Lemer, C., Izard, V. and Dehaene, S. (2004). Exact and approximate arithmetic in an Amazonian indigene group, *Science* **306**, 5695, pp. 499–503.

Pietarinen, A. (2001). Peirce's game-theoretic ideas in logic, *Semiotica* **144**, pp. 33–47.

Pietarinen, A. and Sandu, G. (1999). Games in philosophical logic, *Nordic Journal of Philosophical Logic* **4**, pp. 143–173.

Pinker, S. (1994). *The Language Instinct* (Harper Perennial, New York).

Pinker, S. (2007). *The Stuff of Thought, Language as a Window into Human Nature* (Viking, New York).

Plotkin, G. (1975). Call-by-name, call-by-value and the λ-calculus, *Theoretical Computer Science* **1**, pp. 125–159.

Pollard, C. (2007). Convergent grammars, Tech. rep., Ohio State University.

Pollard, C. and Hana, J. (2003). Ambiguity, neutrality, and coordination in higher-order grammar, in *Proceedings of Formal Grammar 2003* (Vienna).

Pollard, C. and Sag, I. (1994). *Head-driven Phrase Structure Grammar* (University of Chicago Press, Chicago).

Rahman, S. (2000). On Frege's nightmare: A combination of intuitionistic, free and paraconsistent logics, in H. Wansing (ed.), *Essays on Non-Classical Logics* (King's College University Press, London).

Rahman, S. and Rückert, H. (2001). Dialogical connexive logic, *Synthese* **127**, pp. 105–139.

Ranta, A. (1994). *Type-Theoretical Grammar* (Oxford University Press, Oxford).

Ranta, A. (2004). Computational semantics in type theory, *Mathematics and Social Sciences* **165**, pp. 31–57.

Récanati, F. (2008). *Philosophie du Langage (et de l'Esprit)* (Gallimard, Paris).

Retoré, C. (1993). *Calcul des séquents ordonnés*, PhD thesis, Universite Paris-7.

Retoré, C. (1996). Calcul de Lambek et logique linéaire, in A. Lecomte (ed.), *Grammaire et Théorie de la Preuve*, vol. 37 (Revue de l'ATALA, Paris), pp. 39–70.

Roorda, D. (1991). *Resource logics, proof-theoretical investigations*, PhD thesis, University of Amsterdam.

Sandu, G. and Pietarinen, A. (2001). Informationally independent connectives, in I. van Loon, G. Mints and R. Mustens (eds.), *Logic, Language and Computation*, vol. 9 (CSLI, Stanford).

Schönfinkel, M. (1924). Building blocks of logic, in J. van Heijenoort (ed.), *From Frege to Gödel* (Harvard University Press, Cambridge, MA), pp. 355–366.

Schrödinger, E. (1958). *Mind and Matter* (Cambridge University Press, Cambridge).

Schroeder-Heister, P. and Dosen, K. (1993). *Substructural Logics* (Oxford University Press, Oxford).

Shan, C.-C. (2004). Delimited continuations in natural language: Quantification and polarity sensitivity, in H. Thielieke (ed.), *Proceedings of the Fourth ACM SIGPLAN Continuations Workshop*, pp. 1–11.

Shan, C.-C. (2008). Inverse scope as metalinguistic quotation in operational semantics, in K. Satoh, A. Inokuchi, K. Nagao and T. Kawamura (eds.), *New Frontiers in Artificial Intelligence, JSAI 2007 Conference and Workshops, LNCS*, vol. 4914 (Springer, Berlin), pp. 123–134.

Stabler, E. (1997). Derivational minimalism, in C. Retoré (ed.), *Logical Aspects of Computational Linguistics, LNCS/LNAI*, vol. 1328 (Springer, Berlin), pp. 68–95.

Stabler, E. (2003). Tupled pregroup grammars, Tech. rep., University of California, Los Angeles.

Steedman, M. (1996). *Surface Structure and Interpretation* (MIT Press, Cambridge, MA).

Steedman, M. (2000). *The Syntactic Process* (MIT Press, Cambridge, MA).

Steedman, M. and Baldridge, J. (2007). Combinatory categorial grammar, in R. Borsley and K. Borjars (eds.), *Non-transformational Syntax* (Blackwell, Oxford).

Strachey, C. and Wadsworth, C.-P. (1974). Continuations: A mathematical semantics for handling full jumps, Tech. rep., Oxford University Computing Laboratory, Oxford.

Szabolcsi, A. (1987). On combinatory categorial grammar, in *Proceedings of the Symposium on Logic and Language* (Akademiai Kiado, Debrecen), pp. 151–162.

Tarski, A. (1931). Le concept de vérité dans les langages formalisés, in G.-G. Granger (ed.), *Logique, Sémantique, Métamathématique*, vol. 1 (Armand Colin, Paris), pp. 157–274.

Thomason, R. (ed.) (1974). *Formal Philosophy: Selected Papers of Richard Montague* (Yale University Press, New Haven).

Tulenheimo, T. (2010). On some logic games in their philosophical context, in [Lecomte and Tronçon (2010)], pp. 94–120.

van Benthem, J. (1986). *Essays in Logical Semantics* (Reidel, Dordrecht).

van Benthem, J. and ter Meulen, A. (eds.) (1997). *Handbook of Logic and Language* (Elsevier, Amsterdam).

van der Sandt, R. (1992). Presupposition projection as anaphora resolution, *Journal of Semantics* **9**, 4, pp. 333–377.

Vauzeilles, J. (1993). Cut elimination for the unified logic, *Annals of Pure and Applied Logic* **62**, pp. 1–16.

Vermaat, W. (1999). Controlling movement: Minimalism in a deductive perspective, Tech. rep., OTS, Utrecht.

Wadler, P. (2003). Call-by-value is dual to call-by-name, in *Proceedings of the International Conference on Functional Programming, ICFP'03* (Uppsala).

Walton, D. (2000). The place of dialogue theory in logic, computer science and communication studies, *Synthese* **123**, pp. 327–346.

Yetter, D. N. (1990). Quantales and non-commutative linear logic, *Journal of Symbolic Logic* **55**, pp. 41–64.

General Index

Author Index